SPRINGER SERIES ON ENVIRONMENTAL MANAGEMENT

DAVID E. ALEXANDER
Series Editor

Springer Science+Business Media, LLC

Springer Series on Environmental Management
David E. Alexander, Series Editor

Alan Miller
Department of Psychology
University of New Brunswick
Fredericton, New Brunswick, Canada

Environmental Problem Solving

Psychosocial Barriers to Adaptive Change

Springer

Alan Miller
Department of Psychology
University of New Brunswick
Fredericton, NB E3B 6E4
Canada
amiller @ unb.ca

Series Editor:
David E. Alexander
Department of Geology and Geography
University of Massachusetts
Amherst, MA 01003
USA

Cover art: The arctic drama of two musk-ox bulls fighting for control of the herd captures the flavor of many environmental problem situations in which protagonists confront one another in the battle to decide whose ideology will control events—while the great mass of people look on in bemusement. Art by Elsie Klengenberg, *Out of My Way,* 1997, stencil print.

Library of Congress Cataloging-in-Publication Data
Miller, Alan, 1938–
    Environmental problem solving : psychosocial barriers to adaptive
change / Alan Miller.
      p.   cm. — (Springer series on environmental management)
    Includes bibliographical references and index.
    ISBN 978-0-387-40297-0   ISB N 978-1-4612-1440-3 (eBook)
    DOI 10.1007/978-1-4612-1440-3
    1. Environmental management.   2. Environmental sciences—Social
aspects.   3. Environmental sciences—Psychological aspects.
    I. Title.   II. Series
    GE300.M55   1999   98-11430
    363.7'05—DC21

Printed on acid-free paper.

Production coordinated by Impressions Book and Journal Services, Inc., and managed by Bill Imbornoni; manufacturing supervised by Nancy Wu.
Typeset by Impressions Book and Journal Services, Inc., Madison, WI.

9 8 7 6 5 4 3 2 1

ISBN 978-0-387-40297-0

Series Preface

This series is concerned with humanity's stewardship of the environment, our use of natural resources, and the ways in which we can mitigate environmental hazards and reduce risks. Thus it is concerned with applied ecology in the widest sense of the term, in theory and in practice, and above all in the marriage of sound principles with pragmatic innovation. It focuses on the definition and monitoring of environmental problems and the search for solutions to them at scales that vary from the global to the local according to the scope of analysis. No particular academic discipline dominates the series, for environmental problems are interdisciplinary almost by definition. Hence a wide variety of specialties are represented, from oceanography to economics, sociology to silviculture, toxicology to policy studies.

In the modern world, increasing rates of resource use, population growth, and armed conflict have tended to magnify and complicate environmental problems that were aready difficult to solve a century ago. Moreover, attempts to modify nature for the benefit of humankind have often had unintended consequences, especially in the disruption of natural equilibria. Yet, at the same time, human ingenuity has been brought to bear in developing a new range of sophisticated and powerful techniques for solving environmental problems, for example, pollution monitoring, restoration ecology, landscape planning, risk management, and impact assessment. Books in this series will shed light on the problems of the modern environment and contribute to the further development of the solutions. They will contribute to the immense effort by ecologists of all persuasions to nurture an environment that is both stable and productive.

David E. Alexander
Amherst, Massachusetts

Acknowledgments

I would like to thank the Social Sciences and Humanities Research Council of Canada for three grants in support of the research that provided the basis for this book, as well as the University of New Brunswick for a sabbatical year free from the travails of academic life, during which time it was possible to finish writing it.

David Alexander, the series editor, and Janet Slobodien of Springer-Verlag have made the production of this book relatively painless. I thank them for their encouragement and understanding at crucial moments.

Vickie MacLeod was kind enough to reproduce the figures used here, while Bill Liebhardt's comments on Chapter 2 were much appreciated. I benefited greatly from conversations with, and the work of, Paul Rusnock and Hugh Williams, both of whom made my exploration of forest policy an enjoyable and productive experience.

Finally, I would like to thank Janet Stoppard for her emotional support and intellectual honesty, without which academic life and book writing would have become impossible.

Credits

Extracts from Miller, A. (1984) Professional dissent and environmental management, *The Environmentalist*, volume 4, pp. 143–152, Reprinted with permission Chapman & Hall, Hants, England.

Extracts from Miller, A. & Rusnock, P. (1993) The rise and fall of the silvicultural hypothesis in spruce budworm management. *Forest Ecology and Management*, volume 671, pp. 171–189, reprinted with kind permission from Elsevier Science—NL, Sara Burgerhartstraat 25, 1055 KV Amsterdam, The Netherlands.

Extracts from Miller, A. (1993) The role of citizen scientist in nature resource decision-making, *The Environmentalist*, volume 13(1), pp. 47–59 reprinted with permission of Chapman & Hall, Hants, England.

Extracts from Miller, A. & Rusnock, P. (1993) The ironical role of science in policymaking: The case of the spruce budworm. *International Journal of Environmental Studies*, volume 43, pp. 239–251, reprinted with permission of Gordon and Breach Publishers, Switzerland.

Contents

Lists of Illustrations and Tables

1

Introduction

We call ourselves *Homo sapiens,* literally "Wise Man," a splendid example of self-delusion, given the way we treat our own species and the biosphere on which we depend for our survival.[1] It might be better to refer to ourselves as *Homo pragmaticus,* an interfering, myopic species wonderfully adept at solving problems in the short term while creating more problems for ourselves in the long term. Although this pragmatic ability has enabled us to survive and flourish as a species, history is littered with the wreckage of societies that have been unable to cope with the changes they themselves have wrought on their environment.[1] For this reason, it is sobering to find that recent warnings about the deteriorating state of the global environment have become more insistent and ominous. It is now widely, if not unanimously, accepted that the situation is extremely grave.[2] Evidently, the burgeoning human population is depleting natural resources, generating pollution, initiating climate change, and reducing biodiversity at unsustainable rates. Coupled with the growing disparity between rich and poor, these changes face us with the prospect of unprecedented ecological decline and sociopolitical turmoil,[3] the interlocking nature of these environmental and social problems making the global ecosocial crisis enormously complex. The question addressed here is whether we have the collective capacity to extricate ourselves from such a predicament.

Environmental Problems

Until recently, an environmental problem might have been defined as some change in the biosphere that threatens human well being or, in extreme cases, our very survival. However, because this is anthropocentric in its preoccupation with human welfare, radical environmentalists have argued that we need to extend our concern to include the well-being of other species. In this way of thinking, an environmental problem is the result of human actions that threaten the health of the biosphere including all the life-forms therein. This moral concern for other species, however, raises practical problems. It is easy enough to become concerned about the fate of charismatic species, such as some of the larger carnivores, but broadening this concern to include such lowly creatures as slime molds may be more difficult to achieve.[4] It is likely, therefore, that there will be continuing disagreement over exactly what an environmental problem is.

For most of human history, environmental problems had natural causes, but in more recent times, as human populations and technologies have proliferated, we have added human-caused (anthropogenic) problems to the array of threats that confront us. Typically, what we call "environmental problems" are the biological expressions of ill-considered human actions.[5] Because the root causes of many of our serious environmental problems are human in origin, it follows that "environmental problem solving," particularly as it is practiced in "environmental management," should be concerned with managing human behavior as well as the biological impacts of that behavior on the natural world. Unfortunately, conventional environmental management is often limited to ameliorating the biophysical symptoms, rather than the underlying psychosocial causes, of the problem. As a result, environmental problems persist and grow worse. The evident failure of modern societies to cope with environmental deterioration suggests that either there is something intractable about environmental problems or the way in which we tackle them is inadequate.

Environmental problems are certainly intractable, in large part because of their complexity and unpredictability. As Dryzek[6] explains, environmental problems are *complex* in the sense that they involve a great number and variety of elements. The multitude of interactions between these parts is difficult to both understand and predict, especially because they exhibit *organized* complexity. In situations where the elements of a system interact at random, it is possible to use statistical procedures to predict the consequences with some accuracy. However, both human and natural ecosystems exhibit a more organized form of complexity in which interactions are teleological or purposeful in nature. Unfortunately, the behavior of such systems cannot be so readily predicted by statistical procedures. Because environmental problems are also *nonreducible,* their resolution cannot be guaranteed through piecemeal attention to their component parts. This is because problem components are interconnected in such complex ways that problems have emergent properties that cannot be understood solely in terms of their constituent parts. As a result, environmental problems exhibit great *variability,* in both time and space.

Events in one part of an ecosystem can have distant consequences both temporally and spatially. In addition, the dynamic nature of ecosystems means that the same kind of problem can exhibit itself in different ways in different places because of the unique features of local ecosystems. The combination of complexity, nonreducibility, and dynamism makes environmental systems opaque and resistant to commonsense understanding.[6,7] It is difficult, therefore, either to capture the present state of a problem or to predict its future course, another way of saying that environmental problems show great *uncertainty*. Finally, it is human systems that produce and attempt to resolve environmental problems. A large number of people may have been involved in creating the problem and have a stake in its resolution. Thus, there is a *collective* quality to environmental problems; they are not amenable to solution by individuals.[6]

Because of these structural properties, environmental problems are embedded in a great deal of informational noise. They do not appear on the scene clearly defined in some easily recognizable form but are socially constructed by factions striving to impose their version of the truth on others. While technically trained professionals are inclined to emphasize the importance of the biophysical symptoms of environmental problems,[8] the more humanistically inclined insist that it is crucial to attend to the psychosocial roots of the matter at hand. Subsequent struggles over which viewpoint gains ascendancy is a very serious matter indeed, because the problem definition finally adopted in a situation determines who acquires control over events.

Given the structural characteristics of environmental problems (which create uncertainty) and their socially constructed nature (which results in conflict), it is clear that they belong to a class of difficult problems that often resist well-informed attempts at solution. Indeed, they are often referred to, in a spirit of exasperation, as transcientific messes with wicked properties, the implication being that they are resistant to the methods of traditional analytic-reductive science.[9,10] Many complex problems are manageable because they are reasonably easy to define; a lot is known about them, and methods are readily available for their solution. Solutions to these "tame" problems can usually be found through competent application of conventional wisdom, aided by some creative thinking and the occasional technical fix. "Wicked" problems are another matter entirely. They cannot be defined precisely because there is seldom enough known about them, and it may take years, or decades, before a problem is widely recognized. Even then, it may not be clear what form a solution might take, because it is difficult to reach any agreement on what should be done about the poorly understood "problem." Thus, wicked problems are annoyingly intractable. Needless to say, most environmental (and social) problems are wicked in nature, yet they are often mistaken for, and treated as, tame problems. For example, in chapter 4 I discuss a 40-year aerial-spraying campaign against a destructive forest insect pest. This massive and highly contentious program is an effort to impose a technological fix on a wicked problem. The net result has been an endless controversy that threatens to explode once again as the insect population increases to yet another epidemic. It might be useful before proceeding, therefore, to clarify the difference between tame and wicked problems.

The destruction of the *Challenger* space shuttle is a good example of a complex but relatively tame problem. On January 28, 1986, shortly after takeoff, *Challenger* exploded, killing the seven-member crew. Subsequent investigation determined that the immediate cause of the accident was a failure of the O-ring seals in a booster rocket. The potential problems with these seals in cold weather had been recognized before the launch by engineers working for the manufacturer of the booster rockets. On that fateful day, they had argued against launching the shuttle because of the dangerously low temperatures at the launch site, only to have their position overridden by a management decision.[11] It later became apparent that the technical problem with the O-ring seals was embedded in a much broader set of human shortcomings that were found by the Presidential Commission inves-

tigating the accident to include, among other things, flawed decision-making procedures.[12] The *Challenger* accident, therefore, is a prototypical example of a complex sociotechnical problem.

While environmental problems (such as deforestation, desertification, pollution, global warming, and so on) exhibit the same sociotechnical complexity, they are made even more intractable by numerous additional complicating factors. Chief among these is the difficulty in achieving consensus on what the "problem" actually is and what a politically acceptable "solution" might look like. In contrast, the *Challenger* explosion provided a single catastrophic event with an easily identifiable technical cause, about which there could be unanimous agreement and which could be brought to a speedy resolution through appropriate engineering changes. The human dimensions of the story were another matter, however. They were more contentious, more open to different interpretations, making it more difficult to effect change.[11,13] Nevertheless, it was possible for the shuttle program to proceed with appropriate technical adjustments, leaving the various psychosocial dilemmas inherent in the system to be dealt with in the fullness of time. In other words, what made the *Challenger* situation a relatively "tame" problem was that essential aspects of it could be "solved" by technical means.

Environmental problems seldom have these defining moments when the "problem" is clear to all. More frequently, the problem unfolds slowly amid the mixed signals provided by ambiguous, often contentious, data.[14] Global warming is a case in point. As with many important issues, there is an intense dispute in the scientific community over the nature of, and potential threat posed by, atmospheric changes. Although the basic mechanisms of warming are known and the various greenhouse gases have been identified, the prognosis for actual change in global temperatures remains shrouded in the mysteries of atmospheric modeling.[4] Even if consensus were to be reached on a plausible model of global dynamics, the prospects for some concerted action to remedy the situation seems dim. In the

Challenger disaster, the engineering problems faced by the manufacturers of the booster rockets were soluble by a relatively small group of technical and managerial experts. Environmental problems, however, require for their solution not only technical ingenuity but also sustained cooperation by a much wider audience. This kind of long-term mobilization of society around a single, unequivocal goal is difficult to achieve, to say the least. Although it is sometimes managed in wartime through the arousal of nationalistic passions, the same emotional response to something as nebulous as global warming, or the extinction of obscure tropical species, or the loss of topsoil from the Iowa cornfields is less evident.

Wicked problems resist solution because of the uncertainty and conflict surrounding them, one consequence being that they tend to be displaced rather than solved.[4,6] That is, "cleaning up" a toxic dump by shifting its contents elsewhere merely displaces the problem spatially. Storing nuclear waste in depositories displaces the problem in time to future generations, conveniently postponing the day of reckoning. Scrubbers used to reduce smokestack emissions may simply change one form of pollution, gaseous particulates, into another, sulfurous sludge.[4,6] Much of this "environmental problem solving," therefore, is really symptom amelioration. While this is important, a more effective and lasting form of problem solving must necessarily reach beyond environmental symptoms and grapple with the root causes of such problems in human behavior.

Conventional Problem Solving

Problem solving is simply an attempt to think one's way out of a difficult situation. While individuals contribute to the process, environmental problem solving is a collective exercise involving varying degrees of conflict and cooperation between groups of people thrown together, often temporarily, into "problem-solving systems." Typically, the broad policy statements and regulations that provide a framework for problem-solving efforts are the domain of political elites (elected politicians and upper-level bureaucrats) who hold ultimate decision-making authority. They have the onerous task of juggling the conflicting demands of, and advice from, the corporate world, the bureaucratic and scientific establishments, public interest groups, and the silent (voting) majority. The resulting cacophony tends to make legislators cautious. In an effort to avoid catastrophic decisions, they usually prefer minimal changes in policy. The terms used to describe this are "disjointed incrementalism" and "groping along," that is, small, ad hoc changes in policy that do not require any major shifts in direction, thereby avoiding the risk of some major blunder.[15,16] Thus, problem solving in the real world is dominated by economic and political considerations, with little attempt to engage in the comprehensive, but idealized, problem-solving process so beloved by the technical professional.

Implementation of policy devolves to agency and corporate managers who must translate policy into action. In natural resource management, the technical training of those involved predisposes them to the use of technical modes of reasoning and data-intensive approaches to problem solving.[17] In other words, there is an inclination to treat environmental problems as tame puzzles that can be resolved by technical fixes. Although technical data collection and information processing occupy a prominent place in the affections of environmental professionals, the psychosocial context of their endeavors continually intrudes into this rational world. Schon[18,19] has written at length about the dilemma that this poses. Professionals can stay on the "high ground" of technical purity, avoiding the lowland swamp with its messy psychosocial quicksand. In doing so, however, they risk becoming preoccupied with technical trivia at the expense of practical solutions. On the other hand, descending into the psychosocial swamp is not an option that is attractive to the majority of technical professionals.

Hovering around the edges of policy formulation and implementation is a variety of marginalized interest groups, devoid of any real political power and, to varying degrees, furious that they have relatively little impact on the problem-solving process. As these groups have become better organized and more knowledgeable, they have brought to the problem-solving process an alternative vision of the future and radically different notions of how problem solving should be conducted. At the same time, they have also generated a great deal of conflict. Whether this has benefitted the problem-solving process is a matter discussed at length subsequently.

Conventional problem solving, therefore, involves a mixture of groups using different forms of reasoning and subscribing to different ideologies and goals. The resulting conflict doesn't appear to please anyone, nor does it seem to result in effective problem solving. For instance, there is a persistent dispute in all western societies over the adequacy of traditional forms of decision making and problem solving, especially in the face of global change. The present arrangements in which political-economic elites, supported to varying degrees by scientific and managerial experts, control decision making is under considerable attack. Orthodox decision making, as it muddles through one crisis after another, is seen as being increasingly inadequate in dealing with the complexity and conflict facing us. Waddington[20] (p. xi) catches this sense of malaise:

I doubt if there has ever been a period in history when a greater proportion of people have found themselves frankly puzzled by the way the world reacts to their best efforts to change it, if possible for the better. We knock down some dilapidated slums and put up reasonably smart new buildings in their place, only to find a few years later that the inhabitants of the area are just as badly off and living in as great squalor as before. We lend considerable sums of money to a tropical country and show it how to organize public health, and even provide it with medical staff for some years, and the result is that the level of nutrition falls alarmingly and the babies are dying of starvation, instead of the infectious diseases that killed them before. If things go unexpectedly wrong once or twice, that is . . . only to be expected; but recently they seem to have been going wrong so often and in so many different contexts, that many people are beginning to feel that

they must be thinking in some wrong way about how the world works. I believe that this suspicion is probably correct. The ways of looking at things that we have in the past accepted as common sense really do not work under all circumstances, and it is very likely that we have reached a period in history when they do not match the type of processes which are going on in the world at large.

Problem-Solving Reform

Numerous reforms have been proposed in response to the inadequacies of conventional problem-solving systems. Unfortunately, many such reforms are limited by their ideological assumptions and further cramped by reliance on the conventional wisdom of single academic disciplines. For example, many scientific and technical professionals are inclined to believe that decision making would be improved if decision makers were to make better use of the available scientific data. In other words, their solution to our problems is technocratic and elitist. They emphasize their own disciplinary contribution while discounting the need for sociopolitical reform of decision-making structures or attention to the psychological factors that distort and warp information use. However, opinions differ widely on the prospective role of science in problem solving and have done so for a very long time. Speaking in 1946, Morgenthau[21] (p. 1) observed:

Two moods determine the attitude of our civilization to the social world: confidence in the power of reason, as represented by modern science, to solve the social problems of our age and despair at the ever renewed failure of scientific reason to solve them. That mood of despair is not new to our civilization, nor is it peculiar to it. . . . What is new in the present situation is not the existence of these anxieties in popular feeling but their strength and confusion. . . .

This continuing dilemma over how best to respond to complex ecosocial problems has not been resolved in the intervening half century but has intensified as problems appear to worsen. The same chasm between technical optimists and technoskeptics that was evident in the 1940s persisted through the 1970s[22] and remains a source of conflict in the present day.[23,24] One sees, for example, technical optimists engaged in massive international data-collection exercises in the belief that effective problem solutions are impossible without such data.[25,26] In this way of thinking, environmental problems are primarily perturbations in the biophysical environment that are amenable to technical manipulation based on an appropriate understanding of ecological and biophysical processes.[27] The growing conviction that it may not be possible to achieve a comprehensive understanding of complex biophysical systems seems not to have dampened the enthusiasm of technical optimists. Instead, they now grapple with chaos theory in an attempt to develop the levels of understanding and control they believe are within their grasp.[28,29]

A more technoskeptical view, however, is offered by Ludwig et al.[30] who, in light of the historical record, question the usefulness of yet more technical data. Their main point is that despite the availability of abundant scientific information, natural resources have been overexploited to the point of destruction "at many times and in many places, under a variety of political, social, and economic systems" (p. 36). Using the history of agricultural irrigation as an example, they note that the problem of salination has been known since the ancient Sumerians, and methods for avoiding this problem have been known since the late 1800s. Unfortunately, ". . .3000 years of experience and a good scientific understanding of the phenomena, their causes, and the appropriate prophylactic measures are not sufficient to prevent the misuse and consequent destruction of resources"[30] (p. 36). They conclude that natural resource degradation (and other resource problems) is not an environmental but a human problem that manifests itself in a biological form. At the root of "environmental problems," therefore, is human frailty, particularly greed and shortsightedness. In addition, they suggest that scientists have played a significant role in misleading policymakers by perpetuating the "illusion of sus-

tainable development through scientific and technological progress" (p. 36). The implication of this viewpoint is that environmental management is really a matter of managing people rather than the biophysical environment.[31]

Thus, the debate presents us with widely disparate views about how to proceed. While technical optimists place their faith in the salutary effect of scientific and technical information judiciously used, technoskeptics insist that it is human irrationality that is the major stumbling block to the resolution of environmental problems. This difference of opinion is an example of the much older philosophical debate about the role of rationality and irrationality in human affairs. There has always been a lively disagreement about the extent to which reason can, and should, prevail over human passions. Recent environmental debates are simply expressions of this ancient dilemma. Thus, while the neo-Malthusians and catastrophists believe that our more irrational behaviors could well lead us to disaster[32] believers in the power of human reason see a cornucopian future of peace and prosperity.[24,33] Fortunately for all of us, the predictions of the catastrophists in this debate never quite come to pass. The worldwide famines and plagues so confidently expected have not occurred (yet), nor has the golden future of the optimists unfolded. All this leads me to the conclusion that the way ahead, if there is a way ahead, must lie somewhere in the middle ground of this ancient debate. The extent to which it is possible to achieve some feasible compromise is the substance of later discussion.

Adaptive problem solving requires an enormous capacity to manage both ecological complexity and psychosocial conflict.[34,35] At the same time, solutions to the problems that confront us are unlikely to be achieved merely by technical tinkering within the prevailing socioeconomic system.[36] Something more drastic seems to be called for,[4,37] the implication being that we need to develop a much better understanding of the psychosocial as well as the technical aspects of environmental problems.

Recent discussions of sustainable development have taken a step in this direction by addressing the social and economic, as well as the ecological, aspects of sustainable practices. Unfortunately, however, what is missing from this discourse is informed comment on the way in which human psychology both contributes to, and hinders, the search for more adaptive forms of problem solving. In what follows, therefore, I attempt to redress the balance by affording a greater role to psychological analysis than one normally encounters in discussions of environmental management (Figure 1.1). The added emphasis on the psychological elements of collective problem solving is consistent with an older line of thinking exemplified by Sir Julian Huxley's notion of "psychosocial ecology"[38] (p. 132):

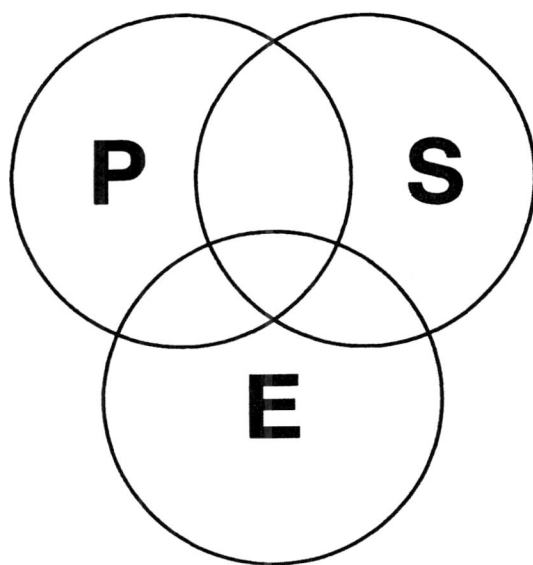

FIGURE 1.1 Problem domains (P, psychological; S, social; E, environmental). (Modified from Craik, K. "The personality research paradigm," in Wapner, S., et al. (1976) *Experiencing the environment.* Used by permission of Plenum Publishing Corp., New York.)

Man lives in a triple tier of environments, material, social and psychological. *Ecology* in the customary sense deals with man's relations with the forces and resources of external nature. *Social ecology* deals with man's social relations, both within and between societies. And what we may call *psychological ecology* is concerned with man's individual and collective relations with the forces and resources of his inner nature and the environment of ideas, beliefs, and values which he has created and with which he has surrounded himself.

Although Sir Julian's language would raise feminist eyebrows today, his ideas are prescient, predating current concerns with sustainability by many decades. In his view, psychosocial ecology forms the basis for a new ecology that would seek to develop "a new pattern of our relations with each other and with the rest of our environment, including the mental environment which we both create and inhabit" (p. 118). This conception of a triple tier of environments was not invented by Huxley but has a venerable history in many cultures. It is resurrected from time to time in philosophical and religious discourse[39] and has been used in more recent times to explain the complexities of environmental and other forms of problem solving.[40,41] Psychosocial ecology (PSE), therefore, provides an interdisciplinary framework that allows the structure and dynamics of environmental problems to be ex-

plored in a comprehensive and systematic way. The implication of a PSE approach is that environmental problem solving is potentially most effective at the intersection of the three domains (Figure 1.1). Conversely, research and actions conducted at the margins, as it were, uninformed by the insights of the other domains, are likely to be ineffective. A further implication is that adaptive problem solving requires unprecedented cooperation, far beyond what one normally sees in our competitive society.[42]

While few would argue with these sentiments about the need for effective interdisciplinary collaboration, the implementation of such a goal is fraught with interpersonal and political problems. A genuinely interdisciplinary strategy runs counter to the prevailing technical emphasis in orthodox environmental management, although practitioners may disagree. After all, environmental managers deal regularly with both technical and socioeconomic data in developing plans that must be implemented in a political context.[43] They might be inclined to say that they are only too aware of the need to attend to psychosocial matters. My own experience of the use of psychosocial analyses in environmental management, however, is that such information is either resisted or used superficially, more as a token gesture than a fundamental reappraisal of the decision process. Francis and Regier[44] (pp. 284–85) offer much the same view in commenting on management of the Great Lakes:

The narrow range of professional expertise that dominates individual organizations (e.g. engineers and lawyers in environmental regulatory agencies, or biologists in fish management agencies) can became a major source of rigidity . . . preventing organizational learning needed at times of change. Such organizations have a strong interest in making certain that new situations are not perceived or defined in ways that reduce the importance of their established expertise. . . . Evidence of this rigidity associated with a narrow range of expertise [can be seen in] recurring struggles for control over management agendas involving engineers and biologists, and more recently, public health specialists. . . . Research agencies have been focused by governmental funding agencies largely on biophysical questions, and increasingly, ones that would support regulatory agencies and/or government policy initiatives. . . . "Safe" social science . . . is acceptable. Research that could raise contentious issues about policies, institutional arrangements, information and communication systems, and decision processes . . . is deemed not to be scientific.

Safe social science, therefore, is tolerated. Because it questions neither the economic inequities and social injustices embedded within the sociopolitical status quo nor the conventional wisdom of professional practice, it tends not to disrupt orthodox decision processes. For those who believe that current professional and political orthodoxies are part of the problem, however, this use of safe social science is unacceptable. Instead, they argue for a more critical social science used analytically to reappraise prevailing orthodoxies and to develop new forms of environmental management.[45] This book provides an introduction to the way in which psychosocial ideas can be used in this more critical way to understand and, hopefully, improve environmental problem solving.

Organization of the Book

Because a major premise of this book is that orthodox problem-solving systems are inadequate, it would seem reasonable to begin our discussion by trying to demonstrate this point. Accordingly, chapters 2 and 3 cover some of the basic psychosocial ideas needed for such an analysis. Readers who have little background in psychology or political sociology might prefer to skim these chapters rather than getting bogged down in the detail, and after reading the rest of the book, come back to these theory chapters to clarify specific points. In chapter 4, I use a case study to evaluate the conventional problem solving commonly found in the natural resource sector prior to 1970. This is followed in chapter 5 by an analysis of more recent developments following the upsurge of interest-group politics in the 1970s. Although there is a great deal of literature on the shortcomings of orthodox problem solving, the basic message still bears repeating because it has yet to percolate down through the ranks, as it were, where orthodoxy is alive and well. Following the dissection of orthodox approaches to problems, we turn to the many reforms that have been proposed but that suffer from the myopia of single vision, a view of problem solving constrained by ideology and disciplinary affiliation (chapter 6). The remedy to these compartmentalized visions is an integrated perspective that combines the various reforms into a more adaptive form of problem solving (chapter 7). While fantasies about the ideal approach to problem solving are useful as goals to which we can aspire, their practical usefulness is limited by a wide range of psychological and sociopolitical barriers. The extent to which these barriers can be overcome is discussed in the final chapter (chapter 8), where I bring together previous discussions and make some (practical) recommendations for change in our environmental problem-solving systems. The book ends with a discussion of the prospects for adaptive problem solving in light of intractable nature of the psychosocial and ecological barriers to adaptive change.

References

1 Ponting, C. 1991. *A green history of the world.* London: Penguin Books.

2 Union of Concerned Scientists, eds. 1993. *World scientists' warning briefing book.* Cambridge, Mass.: Union of Concerned Scientists.

3 Myers, N. 1993. *Ultimate security: The environmental basis of political stability.* New York: Norton.

4 Smil, V. 1993. *Global Ecology: Environmental change and social flexibility.* London: Routledge.

5 Ludwig, D., R. Hilborn, and C. Walters. 1993. Uncertainty, resource exploitation and conservation: Lessons from history. *Ecological Applications* 3: 547–49.

6 Dryzek, J. 1987. *Rational ecology: Environment and political economy.* New York: Blackwell.

7 Arney, W. 1991. *Experts in the age of systems.* Albuquerque: University of New Mexico Press.

8 Clark, T., A. Curlee, and R. Reading. 1996. Crafting effective solutions to the large carnivore conservation problem. *Conservation Biology* 10: 940–48.

9 Allen, G., and E. Gould. 1986. Complexity, wickedness, and public forests. *Journal of Forestry* 84: 20–23.

10 Reckmeyer, W. 1990. Paradigms and progress: Integrating knowledge and education for the twenty-first century. In *Rethinking the curriculum.* Edited by M. Clark and S. Wawrytko, 53–64. New York: Greenwood Press.

11 Winsor, D. 1988. Communication failures contributing to the *Challenger* accident: An example for technical communicators. *IEEE Transactions on Professional Communication* 31: 101–107.

12 Presidential Commission. 1986. Report of the Presidential Commission on the space shuttle *Challenger* accident. Washington, D.C.: U.S. Government Printing Office.

13 Boisjoly, R., E. Curtis, and E. Mellican. 1989. Roger Boisjoly and the *Challenger* disaster: The ethical dimension. *Journal of Business Ethics*, 8: 217–230.

14 Couch, S., and J. Kroll-Smith. 1985. The chronic technical disaster: Toward a social scientific perspective. *Social Science Quarterly* 66: 564–75.

15 Collingridge, D., and C. Reeve. 1986. *Science speaks to power: The role of experts in policy making.* London: Francis Pinter.

16 Deyle, R. 1994. Conflict, uncertainty, and the role of planning and analysis in public policy innovation. *Policy Studies Journal* 22: 457–73.

17 Magill, A. 1988. Natural resource professionals: The reluctant public servants. *The Environmental Professional* 10: 295–303.

18 Schon, D. 1983. *The reflective practitioner: How professionals think in action.* New York: Basic Books.

19 Schon, D. 1987. *Educating the reflective practitioner.* San Francisco: Jossey-Bass.

20 Waddington, C. 1977. *Tools for thought.* St. Albans, U.K.: Paladin.

21 Morgenthau, H. 1946. *Scientific man vs. power politics.* Chicago: Phoenix Books.

22 Luten, D. 1980. Ecological optimism in the social sciences. *American Behavioral Scientist* 24: 125–51.

23 Ehrlich, P., and A. Ehrlich. 1996. *Betrayal of science and reason: How anti-environmental rhetoric threatens our future.* Washington, D.C.: Island Press.

24 Sessions, G. 1995. Political correctness, ecological realities and the future of the ecology movement. *The Trumpeter* 12: 191–96.

25 Malone, T. 1986. Mission to planet earth: Integrating studies of global change. *Environment* 28: 6–11, 39–42.

26 Malone, T., and R. Corell. 1989. Mission to planet earth revisited: An update on studies of global change. *Environment* 31: 6–11, 31–35.

27 Lubchenko, J., et al. 1991. The sustainable biosphere intiative: An ecological research agenda. *Ecology* 72: 371–412.

28 Costanza, R. 1993. Developing ecological research that is relevant for achieving sustainability. *Ecological Applications* 3: 579–81.

29 Merry, U. 1995. *Coping with uncertainty: Insights from the new sciences of chaos, self-organization, and complexity.* Westport, Conn.: Praeger.

30 Ludwig, D., R. Hilborn, and C. Walters. 1993. Uncertainty, resource exploitation, and conservation: Lessons from history. *Science* 260: 17, 36.

31 Montgomery, J. 1990. Environmental management as behavioural policy. *Canadian Public Administration* 33: 1–16.

32 Paehlke, R. 1989. *Environmentalism and the future of progressive politics.* New Haven: Yale University Press.

33 Bugliarello, G. 1988. Toward hyperintelligence. *Knowledge: Creation, Diffusion, Utilization* 10: 67–89.

34 Lee, K. 1993. *Compass and gyroscope: Integrating science and politics for the environment.* Washington, D.C.: Island Press.

35 Mitchell, B., ed. 1995. *Resource and environmental management in Canada: Addressing conflict and uncertainty.* Toronto: Oxford University Press.

36 Ophuls, W., and A. Boyan. 1992. *Ecology and the politics of scarcity revisited: The unraveling of the American dream.* New York: W. H. Freeman.

37 Ehrenfeld, D. 1993. *Beginning again: People and nature in the new millenium.* New York: Oxford University Press.

38 Huxley, J. 1964. *Essays of a humanist.* Hammondsworth: Penguin.

39 Roszak, T. 1992. *The voice of the earth: An exploration of ecopsychology.* New York: Simon & Schuster.

40 Craik, K. 1976. The Personality Research Paradigm in Environmental Psychology. In *Experiencing the environment.* Edited by S. Wapner, S. Cohen, and B. Kaplan. New York: Plenum.

41 Linstone, H. 1984. *Multiple perspectives for decision making: Bridging the gap between analysis and action.* New York: North-Holland.

42 World Commission on Environment and Development. 1987. *Our common future.* Oxford: Oxford University Press.

43 Dorney, R. 1989. *The professional practice of environmental management.* New York: Springer-Verlag.

44 Francis, G., and H. Regier. 1995. Barriers and bridges to the restoration of the Great Lakes Basin ecosystem. In *Barriers and bridges to the renewal of ecosystems and institutions.* Edited by L. Gunderson, C. Holling, and S. Light, 239–91. New York: Columbia University Press.

45 Costanza, R., B. Norton, and B. Haskell. 1992. *Ecosystem health: New goals for environmental management.* Washington, D.C.: Island Press.

2

Psychological Processes

In pursuit of more adaptive forms of problem solving, we are faced with the challenge of finding more effective ways of grappling with the complexity of environmental problems, as well as keeping the conflict surrounding them within reasonable bounds.[1,2] Both aims are subject to the vagaries of human psychology. For instance, individual human beings have a very limited capacity to process information, a cognitive impediment to the management of complexity,[3] while the highly emotional nature of human values provides a fertile substrate for conflict over both the means and ends of problem solving. If we are to overcome these barriers, it is essential that we develop a better understanding of the psychological frailties that influence our collective behavior, not the least of which are our own personal biases and foibles.

Competing Rationalities

One major source of conflict between environmental protagonists is their incompatible worldviews, environmental paradigms, or, as I prefer to say, the competing "rationalities" at war with one another in environmental disputes.[4,26] A rationality is simply a mental model composed of two broad sets of ideas: *what* people believe and value (their ideology)

and *how* they seek to achieve their valued goals (their preferred mode of reasoning).[4,5]

Beliefs (what I believe to be true) and values (what I prize) become organized over the years into personal *ideologies* that act at a covert level in guiding behavior. Often, we are unaware of the ways in which we are influenced by our ideology since the ideas involved have become habitual modes of thinking that we take for granted. For instance, natural resource professionals often dismiss their environmental antagonists as being excessively "ideological," the implication being that environmentalists' actions are biased by a set of rigid beliefs and an underlying political agenda. In contrast, many professionals view themselves as ideologically neutral, impartial scientists trying to provide a sound technical basis for responsible environmental management. However, this is not a tenable position given the ideological assumptions and other biases that riddle the scientific enterprise.[6,7] In the viewpoint being offered here, everyone, without exception, is "ideological." Most environmental scientists and professionals abide by a version of what is referred to below as the Imperial mode, whether they realize it or not.[8]

Modes of reasoning, or cognitive styles, are preferred ways of thinking that become so habitual and ingrained that they are summoned automatically in response to problem situations. The most commonly observed difference

in cognitive styles is between analytic and holistic modes. Analytic thinking is reductive in
the sense that it is assumed problems are best
understood by the study of their component
parts. Problems are dismembered into pieces
that are studied in detail before attempts are
made to reassemble Humpty Dumpty into
some semblance of a coherent whole. Holistic
thinkers, however, are inclined to argue that
just as the King's men were unable to put
Humpty Dumpty back together again, so the
emergent properties of complex problems are
lost in the analytic process. They prefer, instead, to use a more intuitive, experientially
based form of practical judgment, one that
takes into account the surrounding context of
the problem being studied. When coupled
with different ideologies, these two cognitive
styles contribute to distinctive and contrasting
rationalities.

There is a tendency, in a society that is
greatly influenced by scientific thinking, to believe that only scientific and positivistic modes
of thought are truly rational and all else is either inferior or irrational. Thus, the "rational
choice model" of decision making is frequently
held up as the ideal toward which we should
aspire. In other words, when confronted with
an environmental problem, it is argued that
we should develop a comprehensive understanding of the problem based on impartial scientific data; explore all possible alternative solutions; engage in logical, data-based decision
making; and seek evaluative feedback on the
consequences of our actions.[3,9-11] Mistakes in
policy are usually attributed to a lack of adequate information, and it is assumed that rationality will be achieved when this deficiency
is remedied.[12] Little attention is paid in this
model to the possibility that it is quite rational
at times to ignore "scientific" information. For
instance, if a public interest group suspects that
the available information on some local pollution issue is biased in favor of vested interest,
they may well be justified in dismissing it in
favor of their own experiential knowledge.[13]

In practice, "rationality" can take many
forms.[4] Being rational simply means that one
takes orderly steps toward achieving a reasonably coherent goal. Often charges of irration-

ality stem from a lack of understanding of the
reasoning behind the behavior in question. Or
one may simply disapprove of the behavior
and seek to dismiss it with a disparaging epithet. Thus, a "rationality" is a preference for a
specific mode of reasoning or problem solving
in pursuit of a desired goal. For example, a politician may seek to maintain his position in the
establishment by making judgments intui-
tively using his political instincts. This is very
different from a scientist who searches for an
effective malaria vaccine using analytic methods. However, while the politician and the scientist may use different rationalities, they are
similar in that their behavior is embedded in a
set of values and beliefs. It follows that all
problem-solving behavior is subjective; it cannot be "objective" in the sense of being totally
detached from personal and cultural values.
This makes it difficult to assert that any one
form of rationality, such as the scientific mode,
should be afforded higher priority in problem
solving than any other form. Currently, the
presence of multiple rationalities in environmental problem solving hinders communication and cooperation between protagonists
who cannot see why their preferred "rationality" is unacceptable to others. Finding ways
of overcoming this tendency to reject the rationality of others is a crucial step in improving
environmental problem solving.[12,14]

Two disparate rationalities are commonly at
loggerheads in environmental disputes. Worster[15] refers to these as the Imperial and Arcadian traditions and the conflict between them
as "a struggle between rival views of the relationship between humans and nature: one
view devoted to the discovery of intrinsic
value and its preservation (Arcadian), the
other to the creation of an instrumentalized
world and its exploitation (Imperial)" (p. xi).
This is not to say that there are *only* two rationalities or that combatants can be neatly divided into two separate and homogeneous
groups. No one uses a single rationality to the
exclusion of all others. Rather, individuals prefer one kind of rationality but use others as the
situation demands.[16] It is convenient for descriptive purposes, however, to identify polar
opposites while recognizing that many of

those involved in environmental problem solving will subscribe to some aspects of both viewpoints. It is those at the extremes, who hold to their beliefs with passionate intensity, that present an intractable barrier to the kind of compromises required in adaptive problem solving. Although the Imperial and Arcadian ways of thinking have been described in various ways, there is some convergence on the following outline.[17-26]

Imperial Rationalities

Ideology

All of us hold beliefs about *human nature:* what people are like in general. In the Imperial way of thinking, the core belief is that of "individualism" (Table 2.1).[17,23] Human beings, especially males, are seen as being aggressive, competitive creatures, intent on their own self-interest and material well-being through mastery over both the natural and social worlds. In the more extreme versions of this ideology, masculinity becomes associated with power and violence and a preference for a paternalistic society in which the most brutal men are often revered.[20] Individualism is also associated with emotional detachment from others and the natural world that facilitates freedom to pursue self-interest untroubled by feelings of compassion for the less fortunate. Thus, there may be some concern for others, but it is limited, because the ideology of individualism assumes that people are supposed to take care of themselves. At times, this competitive struggle can slip out of control, becoming almost Darwinian in its intensity, fueled as it is by human passions.[27] However, there is the concomitant belief that we have sufficient willpower to overcome these base emotions and impose rational control over our behavior. Many scientists, for instance, believe that by dint of training and self-control they can eliminate personal biases from, and suppress emotional influences on, their thinking, thereby attaining scientific "objectivity." Hence, the common belief among professionals that it is possible to separate fact from value and think-

ing from feeling.[19] In this ideology, therefore, objective rationality is assumed to be the best way to approach problems, one that involves a certain detachment from the matter at hand in pursuit of mastery and control.

Associated with the underlying preference for individualism that pervades N. American society[28] is the belief that *society* should be organized in such a way as to allow individual freedom to pursue self-interest.[18] This is best achieved if individuals are trained in specific marketable skills that they sell on the open market, their freedom to do so being protected by the state.[23] It is recognized that, to achieve this, some degree of centralization of power is needed. However, beyond this minimal level, state regulation should not interfere too much in the economy but, rather, allow market forces to operate. As a result, it is reasonable to expect unequal distribution of rewards (wealth) in society that will become stratified into high and low achievers.[22] This is both a reflection of the underlying dynamic of society (individual striving) as well as being a social necessity. It is essential that the best (most competitive) people should rise to the top and fill the important decision-making positions. Because individual self-interest is couched primarily in terms of material well-being, economic growth is seen as the foundation of a stable society and is best achieved through the stimulation of market forces and the promotion of efficiencies of scale. Thus, centralized, large-scale operations provide the best opportunity for society to survive in the global market place, thereby ensuring individual security through personal advancement. It follows that the hierarchical structure of society, which reflects personal striving and merit, should be protected. Where the status quo is threatened in any fundamental way, it is legitimate for those in authority to use the power at their disposal to neutralize the threat.

The natural world is perceived as a hostile, dangerous place where only the fittest survive and competition rather than cooperation is the ruling principle.[15] Although human beings are part of nature, our higher cognitive powers have helped us transcend the limits imposed on other species. This "exemptionalist" belief

TABLE 2.1. A comparison of Imperial and Arcadian idealogies.

Imperial	Arcadian
Human Nature (P)	
1. Humans are naturally aggressive and competitive.	1. Humans are naturally cooperative.
2. We are capable of only limited caring.	2. We are capable of widespread concern.
3. Rational and objective thought is more important than intuition and emotion.	3. Emotion and intuition are at least as important as any other knowledge.
4. Fact and reasoning can and should be separated from value and feeling.	4. You cannot, nor should, separate them.
5. We should take risks to attain reward.	5. We should err on the side of caution.
6. Individuals should look after themselves.	6. We should take care of each other.
Nature of society (S)	
1. Human societies are naturally hierarchical.	1. Social hierarchies are unnatural.
2. Decisions should be made by experts: politicians advised by scientists.	2. We should all be involved in decisions making.
3. The way forward is through representative (parliamentary) democracy.	3. The way forward is through direct, participatory democracy.
4. Material acquisition underlies social progress.	4. Spiritual quality of life and loving relationships are more important.
5. Economic growth is desirable and can go on forever without harming the environment.	5. Indiscriminate economic growth is bad and should not continue.
6. Trade should be increased to further economic development.	6. Trade should be reduced to foster local self-sufficiency.
7. Large-scale production and central control are desirable.	7. Small-scale production locally controlled is more desirable.
8. Current sociopolitical arrangements are OK.	8. There needs to be fundamental sociopolitical change.
Nature (E)	
1. Nature is hostile and neutral.	1. Nature is benign.
2. The natural world contains ample reserves.	2. Earth's resources are limited.
3. The environment is resilient.	3. Ecosystems are delicately balanced.
4. Humans are separate from nature.	4. Humans are part of nature.
5. Nature should be exploited for human material benefit.	5. We must respect and protect nature.
6. Environmental problems can be solved by analytic/scientific reasoning.	6. Environmental problems can only be solved by holistic approaches.

(Based on material contained in Stephen Cotgrove's *Catastrophe or Cornucopia*, John Wiley, by permission of the author, and David Pepper's *Modern Environmentalism* by permission of Routledge, London.)

assumes that we are not subject to the same laws as other species and can, therefore, continue to grow economically, making increased demands on the environment with relative impunity.[29,30] We can overcome problems that might arise with this strategy with our superior intellect. Resource substitution and more efficient, less polluting technology will save the day. The exemptionalist belief is further supported by the additional assumption that nature is both bountiful, a cornucopia of natural resources, and wonderfully forgiving in the sense that it is exceptionally resilient to human interference.[26] Although human actions may cause changes and ancient ecosystems may disappear, newer forms of stable state emerge to provide for us. Change is a prominent feature of the natural world[31] and should not be viewed with moralistic dismay. Rather, we should seek to manage the changes that occur to our satisfaction and benefit. We certainly have the capacity to do so.

In sum, the Imperial ideology emphasizes a philosophy of the rugged individualist, surviving in a competitive world through dint of superior powers and preferring a society in which the most powerful and competitive individuals rise to the top. The resulting control by elites ensures the improvement of material well-being through economic growth and continuing technological innovation. The basic principles are the apparently contradictory ones of individualism and centralization.[23]

Modes of Reasoning

To reiterate a point made earlier, a "rationality" is a form of thinking that involves a preferred style of reasoning embedded in a set of environmental and social values and beliefs or ideology.[4] When the Imperial ideology is combined with an analytical form of reasoning, the result is what is often referred to as the "technocratic rationality" commonly found among scientific and managerial professionals. In this way of thinking, the primary value is mastery of the natural and social worlds, an optimistic view that stems from an enthusiasm for technical virtuosity and technical fixes.[32] The form of reasoning preferred in this mode is instrumental analytic, in that it is more concerned with method and technique than with the normative (value) issues surrounding the selection of ends and is prone to decomposing problems into their constituent parts for each of which a separate solution is sought.[5] A problem solution, in this way of thinking, is the sum total of the separate part solutions. When applied to the economic sphere, technical rationality becomes "the dominant form of reason in contemporary industrial societies, [one that is] grounded in utility calculations."[4] In other words, decisions are considered to be rational to the extent that they follow utilitarian values and promise to enhance the material well-being of the greater number. Thus, the primary value is efficiency in both production and consumption with human relationships often being reduced to a matter of economic exchange.[4]

In contrast, when adherents of Imperial values prefer to use a more intuitive, practical form of reasoning, various kinds of professional judgment result. An example in the political sphere, for instance, is "political rationality," a heuristic way of thinking used by politicians to navigate through the messy complexity of public life.[4] Effective political rationality requires extensive personal experience with professional and political systems. This intuitive knowledge is a form of social intelligence quite different from the knowledge and abilities inherent in more technical ways of thinking.[5,33] It is certainly not the kind of knowledge amassed during formal education and, as a result, is in short supply.

In sum, the Imperial mode combines the master values of dominance, utilitarianism, and individual self-interest with two modes of reasoning: social (leading to political rationality) and technical (leading to economic and technical rationality). The net result of these ways of thinking is a hierarchical society bent on economic growth and personal advancement.

Arcadian Rationalities

Ideology

In contrast to the competitive individualism at the core of Imperial beliefs, the Arcadian way of thinking emphasizes the more cooperative aspects of *human nature* (Table 2.1). Although self-preservation and competitiveness may be a part of human nature, they are not considered to be the greater part.[15,20] Rather, it is believed that human beings are capable of behaving in a much more harmonious, nonaggressive way that, under appropriate circumstances, can lead to cooperative living. In addition, humans are thought to be capable of great compassion and empathy, something that would become more evident if it were not for the destructive influences of the patriarchal society imposed upon us by the dominant elites.[20,34] Human beings, therefore, have an innate capacity to form cooperative relationships because of their ability to extend their compassion beyond their own self-interest. Indeed, if we are to overcome the environmental problems that confront us, it is believed that we must encourage this social compassion, and the biophilia of which it is a part.[22] Thus, rather than material self-interest, it is essential to adopt voluntary simplicity as a lifestyle, replacing material self-seeking with nonmaterial (spiritual) forms of self-development. In pursuit of these goals, we are aided by our ability to make moral judgments, a form of thinking that relies on our emotional reactions to events. Because moral (value) judgments are so intimately connected with emotion, it is unwise, indeed impossible, to try to exclude emotion from decision making. Such an ambition is part of the compartmentalization of life under the Imperial way of thinking that has resulted in so much alienation and anomie. The unfettered pursuit of economic self-interest, which is at the root of our environmental problems, is possible only in circumstances where moral concern is dismissed or attenuated. It must be replaced, Arcadians argue, by a less materialistic vision of life, one that recognizes the cooperative and compassionate side of humans and promotes spiritual, rather than economic, development.

Thus, the Imperial society of competitive individualism needs to be replaced by a *partnership society* that emphasizes harmonious cooperation.[20] In such a society, the role of individualism and centralization would change. Because material self-aggrandizement would be discouraged, eventually there would be no need for a centralized state to regulate behavior. Likewise, in the absence of material acquisition, there would be no need for centralized industry producing massive quantities of consumer goods. Instead, it would be possible, and desirable, to return to a simpler, more decentralized form of social structure organized around small, self-sufficient communities. These would cater to the basic needs of local citizens enjoying lives of voluntary simplicity that made relatively few demands on natural resources. It might even be possible to view work as gratifying, to be engaged in for pleasure rather than the frenetic drudgery that characterizes many work situations today.[20,23,35] Under such conditions, income could be related to need rather than to competitive merit. In small communities, the primary group would be the basic unit of society, with the result that individual aspirations would take second place to the well-being of the group. This should not be problematic, it is argued, because humans are innately capable of generosity and compassion under the right conditions. It follows that in a world of small-scale communities, hierarchies would

gradually fall away as a genuine participatory democracy replaces the present representative "democracy".[22,23]

In light of the notions of cooperation and harmony that pervade the Arcadian worldview, it is not unexpected to find that *nature* is seen as a benign, hospitable place.[36] While nature is beneficent, however, it is also delicately balanced. In its natural state, the environment can withstand natural disruption and maintain steady states for extended periods of time, a view that reflects an older conception of ecological climax dating back to the 1930s that has been rejected by the majority of modern ecologists for some time.[31] Nevertheless, it continues to pervade the Arcadian way of thinking and underlies the belief that nature is ephemeral, easily disrupted for the worse by human meddling.[26] This fundamental fragility can only be dealt with by drastically reducing human interference, by "treading lightly on the earth." Our technical cleverness does not exempt us from the laws of nature, as Imperial advocates claim, but, instead, may actually create more problems for us. Thus, the idea of continuing economic growth, with its implication of even more human intrusion into the natural world, is seen as a form of insanity, because we can never be exempt from natural laws. It is only through establishing a more harmonious relationship with nature that we will survive and allow the biosphere to recover from the depredations we have already visited on it.

In sum, the Arcadian view is that human beings are capable of a much more harmonious existence both with themselves and with nature but only if we reject the assumptions and institutions of the dominant Imperial elites. This would require a basic shift toward a partnership society based on a less materialistic lifestyle of voluntary simplicity and frugality that made less demands on a fragile nature. We are capable of and must adhere to this lifestyle if we are to achieve some semblance of a sustainable society.

Modes of Reasoning

Like their Imperial counterparts, proponents of the Arcadian worldview share a common set of values but differ over their preferred modes of reasoning. Thus, the primary value is the wish for a more harmonious, cooperative life of voluntary simplicity. A central feature of this value system is the belief that all other values should be subordinated to the goal of ecological integrity, the "preservation and promotion of the integrity of the ecological and material underpinning of society"[4](p. 58). Without a healthy natural environment, it is argued, all human pretensions to sustainability and growth are fatuous. The means for achieving this goal of ecological sanity differ, however. Some environmentalists seek to achieve ecological restraint and harmonious human-environment interactions through social interventions. That is, self-restraint can be achieved by the creation of appropriate living conditions (small communities) that will result in less ecologically destructive lifestyles. The kind of reasoning involved need not be technically informed but can be based on common sense and an intuitive, indigenous knowledge resulting from long experience with local conditions. It may also be more emotionally flavored than one sees in the reasoning of the technical professional. This social route to ecological rationality is prevalent, as one might expect, among ecosocialists.[37] In contrast, a variant of ecological rationality, involving a more analytical form of reasoning, is found among deep ecologists and ecopsychologists, those who favor a more psychological route to restraint.[38,39] Whereas ecosocialists maintain that changes in our social arrangements necessarily precede psychological change, ecopsychologists argue the reverse. If we are to manage the environmental crises we have brought down upon ourselves, we must first change our destructive belief systems through introspective analysis of our decidedly warped mentalities. Thus, ecological rationality is to be achieved through a cognitive route, changing the way we think that, in turn, will radically alter the way we behave (see chapter 6).

In sum, the Arcadian worldview is based on the master values of nonintrusive harmony and cooperation that is to be achieved through either social or psychological reforms. Unless ecological rationality is promoted, it is assumed that Imperial value systems and rationalities will result in the irreversible degradation of the natural environment and the collapse of industrial civilization. It is believed that there is little time to achieve the necessary transition to a more Arcadian mode of living.

The Conflict in a Historical Context

Conflict between proponents of the two major rationalities is commonplace in both environmental and social policy. In western societies, the Imperial way of thinking dominates our lives, whereas proponents of Arcadian views are confined to the margins where they cast moral judgments on the material excesses of the industrial world. Yet adaptive problem solving requires the achievement of much greater levels of cooperation between the two rationalities than has hitherto been accomplished.

It is interesting to find that Imperial and Arcadian ways of thinking have been influencing debates in natural history for centuries.[15] Indeed, some of the basic elements of Arcadian thinking were evident among the Gnostic Christians in the first century A.D.[40] In addition, anguish about deforestation, overgrazing and soil erosion from the hillsides of ancient Greece was being expressed by the great reformer Solon in 590 B.P.E.[41] Likewise, protestations about mindless forest destruction in Nova Scotia dates back to the earliest days of European settlement in the eighteenth century.[42] Although these dissenting opinions were always in the minority, unease about our treatment of the environment does seem to be more widespread today among the general public than in previous decades.[43] Thus, the Arcadian challenge to the dominant Imperial worldview has been active for centuries, part of an endless battle between two fundamentally different personal styles, the implication being that the kinds of people who subscribe to each viewpoint are appearing on the scene generation after generation. It is also evident that the social conditions that maintain the relative imbalance between each tradition are also reproduced over time. One has to be cautious, therefore, in talking about paradigm shifts. While some changes in environmental policy and management have occurred over the years, stubborn resistance to any fundamental change by those of the dominant Imperial persuasion calls into question whether this might be termed a paradigm shift.[44] An old adage comes to mind, The more things change, the more they stay the same.

Given the ancient lineage of this conflict, it is difficult to be optimistic about its resolution any time soon. It seems reasonable to conclude that the balance of power will remain, as it has done for centuries, in the hands of the dominant Imperial tradition. Thus, the current problem-solving mode, in which proponents of the dominant rationality control environmental policy, with the marginalized minority snapping at their heels, is likely to persist. The question, then, is how might adaptive problem solving be achieved under these circumstances?

The historical persistence of ideological conflict suggests that individual preferences for particular paradigms are not easily amenable to change through persuasion or rational argument; otherwise, these conflicts would have been resolved eons ago. Instead, the core beliefs and values that form the basis of ideologies are, in fact, tenaciously held and intensely felt. They are a part of one's identity as a person and cannot be changed or replaced as if one were inserting a new battery. Many adults will resist any fundamental change in their core values, even if the consequences are dire.[45] How, then, can one achieve some accommodation between antagonists in environmental conflicts when their differences appear to be so deeply rooted? The most common way of dealing with conflict situations is to avoid psychological analysis altogether and concentrate on various forms of bargaining and negotiation, none of which require the participants to explore one another's psychological processes. Used in a cynical way, psychology may enter the scene only as another tool with which to outmaneuver one's opponents by providing insight into their motives and weaknesses.[46] However, the assumption made here is that a willingness to compromise is contin-

gent on developing a genuine understanding of one's own biases and the recognition that all "truths" are relative, so that no single viewpoint, including one's own, is absolutely correct. Unfortunately, resistance to this form of self-analysis is commonplace among professionals, just as it is among the general public.[47] It is also evident that those at the extremes of the ideological spectrum are markedly unwilling to engage in this kind of endeavor, being so emotionally involved in their ways of thinking. Because compromise is not in their lexicon, the chances are they will continue to wreak havoc on environmental problem solving and will have to be contained by political-legal means, as they are now. I recognize, therefore, that my advocacy of psychological self-analysis, as a route to environmental compromise, is directed at the middle ground occupied by those who are only too aware of their human fallibility. Thus, if we are to understand the roots of environmental conflict, the antagonism between proponents of the two paradigms, and the prospect for compromise or change, then we need to look more closely at the psychology of human personality. We shall begin with the psychological roots of ideology and then proceed to the question of differences in modes of reasoning. In what follows, some readers may become impatient with the broad generalizations offered. They will see a myriad of exceptions and complications that negate the arguments being propounded. Patience is needed in this regard. The purpose in sketching the links between personality and environmental behavior in broad brush strokes is that it provides a framework within which a more complex picture can be developed as we proceed through the book.

The Objective-Subjective Personality Dimension

A variety of explanations have been offered to account for ideological preferences. One perspective, for example, suggests that the rather gloomy, apocalyptic forecasts associated with

Arcadian thinking serve an ulterior motive. According to Douglas,[48] environmentalists' claims of cosmic doom serve the function of drawing together groups of like-minded people who have little else to sustain their group cohesion. In contrast, other scholars argue that adherence to one or another of the two main ideologies is determined, primarily, by place of employment.[49,50] Thus, professionals working in government agencies, such as the Environmental Protection Agency, are likely to have a much stronger adherence to Arcadian values than those working in the private sector[50] (p. 46). In addition, ideological differences, especially over matters of risk, appear to be related to disciplinary specialization, with life scientists exhibiting a more Arcadian orientation than those in the physical and engineering fields.[49] Although each of these explanations offers an insight into some of the factors involved in choice of ideology, they don't tell us why a particular idea is so attractive to someone that they are willing to join an environmental group or, in contrast, accept a job in an organization that operates within the Imperial worldview. Nor do they tell us why it is that a person may defend his or her ideological beliefs so strenuously. Explanations of these matters requires a more psychological route to understanding.

Two quite different psychological orientations underlie the major rationalities: the contrast between a tough-minded, impersonal, detached control (Imperial) and a more tender-hearted, personal, nonintrusive attachment (Arcadian). It is interesting that both of these orientations, or more precisely "motivations," are thought to exist in all of us from birth.[51] On the one hand, we feel compelled to assert our individuality, to separate ourselves from others, and to curtail our dependence on them. Psychologists commonly refer to this as the *agency* drive or, more simply, egoism. On the other hand, we experience the urge to join with others in cooperative, intimate, and, often, dependent relationships. This *communion* drive is perhaps more recognizable as altruism.[52] In other words, human behavior is believed to result from a continuing conflict between egoistic and altruistic drives, both of which operate at an unconscious

level. Whether these basic drives are genetically programmed into our nature is a contentious issue,[53] but the presence of both in very young children does imply a genetic origin.[54]

Subsequent socialization of a child will, of course, shape the relative balance between the two orientations, causing some of us to favor one over the other. Unfortunately, not all of us experience the kind of upbringing that leads to a happy compromise between egoism and altruism.[54] Problems often arise in early childhood during the crucial phase when emotional bonding with others takes place.[45] This is a very important turning point in a child's development, the consequences of which can persist into adulthood, coloring our interactions with feelings of trust or distrust. A child who, for whatever reason, grows up with the feeling that he cannot entirely trust, nor care about, other people is likely to see the world as a fundamentally hostile place. One way of dealing with this is to emphasize one's own self-interest by attempting to control one's hostile surroundings. In contrast, a trusting child may well develop feelings of love and concern for others and so look out on a more benign world. There would be little need for excessive self-assertiveness because she exists in a relatively safe, friendly world. When these differences in trust are taken to an extreme and one set of drives is suppressed, an unbalanced personality is produced. Thus, the result of this socialization process is that people develop different accommodations between their competing drives. Some show extreme forms of either agency (egoism) or communion (altruism), whereas most of us arrive at a compromise that leaves us somewhere between the two extremes. It is possible to find in adults, therefore, many examples of self-assertiveness unconstrained by feelings of belonging or concern for others.[55] While, in contrast, the suppression or underdevelopment of self-assertiveness is evident in those who are so tender-hearted that they are unwilling to make the hard choices sometimes thrust upon us. The tough-minded masculinity that is associated with the Imperial ideology can be interpreted as an expression of egoism and agency taken, in some cases, to a self-centered extreme. In contrast, the more tender-hearted femininity associated with the Arcadian position could be seen as an expression of altruism and communion taken, in some cases, to a romantic, unrealistic extreme. What this means is that environmental ideologies are rooted in psychological processes that have been shaping our personality since childhood and about which we may have little awareness. As a result, the contrasting personalities exhibit quite distinctive sets of problem-solving biases,[56] sometimes referred to as tough-minded versus tender-minded or hard-nosed versus soft-nosed. However, I prefer the more neutral terms: objective-subjective.

The objective-subjective personality dimension is a convenient way of talking about the effect of the agency and communion drives on personality development. The term "personality dimension" simply refers to the fact that it is possible to arrange people along a continuum on which a few show extremely objective or subjective behavior most of the time while the rest of us exhibit a more moderate combination of both qualities in our daily lives. In discussing the dimension, I shall focus on the polar opposites, contrasting extremes on the dimension as a descriptive convenience. Thus, extremes of objectivity are evident in those who are egoistic, power-seeking, emotionally detached, and prone to imposing their frame of reference on events. In contrast, those who are more subjective seek less controlling, more empathetic relationships with others, an orientation aided by their sensitivity to the beliefs and feelings of others. The implications of these qualities for problem solving are discussed below.

The Objective Personality

Power

One of the most characteristic features of Imperial behavior is the overriding concern with mastery and control, as well as a willingness to take risks in achieving this control. The attractiveness of megaprojects, the preoccupation with faster and more powerful technologies, as well as the age-old attempts to "tame"

nature all stem from the same underlying psychological theme. In moderation, there is nothing wrong with this technological fascination, as it is quite normal to enjoy the feelings of power involved in mastering a task or solving a problem. There is an inherent satisfaction in extending and exercising one's capabilities.[57] A sense of competence results from being a *cause*, being able to have an effect on the world around us and seeing ourselves as *potent* beings. Without these aspirations we would not have developed much beyond the Stone Age, because curiosity and mastery underlie innovation. It is not unexpected, therefore, to find that power seeking is also correlated with the need for excitement, an attraction to risk, and the seeking of stimulation and challenge.[56,58] These correlations might help explain some of the behaviors associated with the Imperial orientation, such as the willingness of adherents of this paradigm to tolerate high levels of risk in the use of technology.[59]

The need for excitement, stimulation, and challenge experienced by the "objective" individual may be at the root of what Pacey[32] refers to as the "technological imperative," a master value that extols technical virtuosity, cleverness, power, and other "existential joys of technology." One of the characteristics of this way of thinking is the excitement at being at the cutting edge of technological innovation and a contrasting disinterest in the more mundane tasks of servicing older technology. It is unfortunate that the feelings of mastery and potency associated with this power-seeking behavior can, under some circumstances, lead to the "illusion of control," the belief that one has more control over events than is actually possible. Gamblers exhibit this in their belief that they can control the roll of dice, while certain kinds of holistic medical treatments are based on the assumption that it is possible to control the growth of cancer cells through positive thoughts.[60] At the same time, the obsession with data and information seen in many western countries is associated with an overconfidence in the prospect for rational solutions to complex problems.[61,62] It is no wonder, therefore, that Ehrenfeld[63] refers to the present era as the Age of Hubris, one pervaded by the aura of magic and power surrounding science.

While critics of the Imperial way take this kind of behavior to task, their concerns deepen when we turn to more extreme forms of power seeking, the psychological origins of which we need to understand and ponder. When power seeking becomes tinged with aggressiveness and cruelty, we are beginning to enter a psychological terrain that is different from the simple pleasures of being able to solve a problem or to build something valuable. For instance, consider the contrasting forestry myths of Paul Bunyon (the mighty logger) and Johnny Appleseed (the tree planter). Both involve exerting power to change their environment, but the former introduces an image of masculine dominance, whereas the latter is a more gentle, even feminine conception of change. An example of the Bunyon mentality is provided by two loggers, whose conversation was overheard by a young forestry scientist on a ferry wending its way among the gulf islands of British Columbia. Upon seeing an enormous tree on a promontory, one turned to the other and said, " Someday I'm going to get that tree."

There is little or no reliable information on the frequency of extremely objective personalities among professionals, but feminist critics of science are often appalled at the aggressiveness and misogyny they perceive in both the behavior and language of some scientific and technical experts who appear to exhibit several of the less endearing personal qualities associated with objectivity.[64-68] However, while not all adherents to the Imperial way of thinking are beset by aggressive power striving, the point is that some are. These individuals can make the attainment of ideological compromise difficult by creating turbulence in the decision-making environment because of their overweening need for power and control. The disruptive nature of such behavior is easy to find in such examples as the following: the willingness of powerful elites to manipulate information,[69-71] the occurrence of scientific fraud,[72-75] a descent into pathological science,[67,76,77] and the use of environmental

terrorism as a political weapon.[78] I include the last example to underscore the point that aggressive power seeking is not limited to elites but can be seen in all walks of life, including certain kinds of "environmentalism."

Egoism and power seeking, therefore, range from relatively moderate forms of knowledge seeking in pursuit of solutions to environmental problems to the more aggressive forms of egoism that result in crime and disruption of "rational" problem solving. The problem is how to promote the former while controlling the latter, a matter to which we shall return later.

Emotional Detachment

One of the most common criticisms leveled at Imperial (imperious) professionals is their apparent coldness in responding to environmentalists' and the public's concerns. Levine observed this in the way that some officials from New York State's Department of Health dealt with citizens of Love Canal[79] (p. 98):

The person who conducted the meeting . . . mentioned later that he felt proud he had been able to remain cool, 'to talk like a machine,' despite the anger the Love Canal residents displayed when he read the Department of Health announcement. Task force members expressed similar feelings more than once when they conducted or even were present at meetings with residents who seemed so unreasonable. Privately, the officials congratulated each other on not giving anything away, on not conceding anything to the residents . . . and argued that the people were given answers, but, 'like spoiled children,' just did not like the answers they had received. . . . Simply living through a heated meeting, getting it over and done, without 'loosing cool' and without departing from an official position became one more mark of the professional.

This professional manner, with its impression of emotional detachment, greatly offends members of the public. At the same time, professionals who subscribe to the dominant Imperial mode are perceived by their Arcadian critics as lacking any genuine concern for the environment or for the people living in resource-dependent or contaminated communities. However, the very idea that they are not concerned about the environment or the people who depend upon it is incomprehensible to the majority of professionals. As Petulla[80] notes, most environmental professionals have an "ideal image of themselves as selfless, dedicated, tireless, competent, humble" (p. 137) individuals who have fought to protect the environment for the benefit of all. Indeed, the model of the environmental steward, one who preserves the natural environment for future generations, "provides a shining image of how to act and be the environmental professional for all seasons" (p. 139).

What we have, then, are two broad groups, each of which believes it is they who have a genuine concern for the environment in contrast to the other group who do not. Each is exasperated by the emotional behavior of the other. Environmentalists see professional and other members of the dominant mode as emotionally unresponsive and cold, whereas the latter see environmentalists and other dissenters as overemotional, irrational, and uninformed. In addition, the posture of detached "objectivity" preferred by many professionals in problem-solving situations is perceived as a way of covering up wrongdoing, ignorance, or lack of concern.[79] The net result is difficulty in communication and the continuation of conflict, the question being: What, if anything, can be done about this? If "professional manner" is a style imposed on professionals by the demands of their job, then it might be possible to change job requirements and so initiate a change in professional behavior. However, if there are deeper, more psychological, reasons underlying the apparent emotional detachment of professional adherents of the dominant paradigm, then the situation is less promising. It is possible, for instance, that those who are attracted to technical jobs are, indeed, more emotionally detached, less empathic individuals. What does psychology have to say about this?

Emotional detachment involves the psychological process of dissociation, which allows thoughts to be separated from feelings and the suppression of those feelings that may cause anxiety.[81] Dissociation is used by everyone at some point to diminish the impact of stressful

situations, such as the death of a loved one or some other shock. We suppress the feelings involved until we feel strong enough to acknowledge and deal with them. In moderation, this is adaptive. For instance, surgeons and pathologists have to do things to their patients that would make most of us rather bilious. To be effective, however, they have to suppress "normal" reactions to distress; otherwise, they would be useless, even dangerous. Thus, professional training in medicine, as well as other science-based professions, promotes objective detachment as a feature of professional competence. Several studies of professional education in natural resources have found, for example, that "thing-orientation becomes pervasive among resource professionals because this attitude is encouraged rather than suppressed during their education"[82] (p. 297). Deviation from this technical preoccupation is, apparently, not tolerated.

While the varying degrees of emotional detachment seen in professional life is certainly a reflection of training, its roots are established long before such training begins.[56] For instance, those who are attracted to scientific and technical disciplines as adults actually begin to develop specialized personality structures very early in life. The foundation for their analytical thinking styles, as well as their tendency to turn away from people toward the world of things, can take place as early as three or four years of age.[83,84] Thus, incoming student foresters have been described as tough-minded and conservative, with a primary interest in the production and management of things,[82] whereas engineering students are considered by Hyde and Mclean[85] to lack an interest in people, preferring, instead, to deal with objects rather than humans, thereby exhibiting the "typical detachment of the engineer from human relationships" (p. 7).

Although, to this point, I've talked about emotional detachment in professionals, such behavior is not confined to them but is, instead, ubiquitous in the human population.[55] Certain kinds of "environmentalist," share this capacity for objectification. One finds, for example, some radical environmentalists espousing the misanthropic view that the world would be better off if there were a massive culling of the human population.[78]

Extraception

One further component of the objective orientation is *extraception*, a preoccupation with external reality rather than the internal world of private experience and emotion. In this sense, then, "objectivity" becomes a preference for dealing with the immediately obvious "facts" of physical appearance, rather than the murky world of opinions, feelings, and values. One engineer, for instance, explained that, when he was a student, he "didn't want to be bothered with the fuzzy features of the rest of life."[85] (p. 11) Thus, facts are seen as the only legitimate basis for knowledge, while intellect and reason must be kept separate from that of emotion and speculation. When taken to an extreme, the lack of interest in introspection takes on a more active anti-introspective attitude, one that involves an intense dislike for dealing with those feelings and impulses that lurk beneath the facade of rationality.[86] The distaste for the introspective world is commonplace among environmental professionals,[82] as it is in all walks of life. Baum, for instance, was faced repeatedly with resistance to psychoanalytic knowledge in his study of problem solving by planners.[87] Apparently, such knowledge unsettles the planner "by making him aware of two domains which he has attempted to ignore: the external organizational and political environment and the internal unconscious realm. Both worlds exist and affect his efforts to solve problems." If he were required, or forced, to engage in introspection, "these insights may help him take more conscious control of his actions, (but) they immediately challenge his self-esteem and may discourage him from learning more." (p.170) Thus, experts in organizational change, who seek to encourage more constructive introspection by both managers and workers, find that the majority put up enormous resistance to such demands and engage in defensive avoidance, a kind of flight from the unpalatable truths that often result from introspection.[3,88,89] My own experience in trying to foster introspection among forestry and agriculture students led me to conclude that it is unwise, to say the least, to push the process too far. A substantial minority of these students quickly become angry, even violent, if they are expected to engage in an activity that seems to threaten their sense of identity. The implication of this for environmental problem solving is rather disconcerting, because an anti-introspective attitude precludes any systematic reflection on one's own goals and methods. This is made even more troubling by the occurrence of these same attitudes among some "environmentalists" who are similarly intransigent in their adoption of particular positions on environmental issues.

One unfortunate consequence of anti-introspection is that many professionals lack insight about the way in which their personal biases influence their work. As a result they are inclined to take their own perspective as "immediately objective and absolute"[66] (p. 117) and have few qualms about the imposition of their way of seeing things on others. Perhaps the best example of this in environmental problem solving is the persistent dismissal of lay knowledge by professionals and other elites, even when such knowledge is based on many years of experience with local environmental conditions[90,91] Thus, several studies of resource professionals have found that they perceive the general public as largely uninformed on resource issues and believe that they know best what the public wants. As a consequence, "professionals often undertake public information programs more from a desire to shape public opinion than to incorporate public opinion into policy decisions."[82] (p. 298) It follows that such professionals are also inclined to be sympathetic to interest groups whose ideas correspond with their own.[82]

Summary

In sum, the objective personality is one that combines power seeking with emotional and cognitive detachment. The reason for this is simple enough. If one seeks to exert one's authority and control over others, or over the environment, it is easier to do so if one adopts a degree of emotional detachment; otherwise, one is troubled by pangs of empathy and other unwelcome emotions. Likewise, taking other people's views into account tends to raise troublesome doubts about one's own viewpoint, something that is unsettling and should be avoided. I hasten to add that, under some circumstances, such as emergency situations requiring decisive action, "objective" qualities are crucial. However, when dealing with long-term environmental problems that require years of sustained cooperation, the more extreme forms of objectivity can be disastrous.

The Subjective Personality

Love

As an expression of the communion drive and the subjective orientation to life, "love" is in direct contrast to the power-seeking behavior of the objective personality. "Love" is not a topic that is often included in discussions of environmental problem solving. When I broach the matter with students and colleagues, their reaction is raised eyebrows and a certain discomfort, as if "love" was an unscientific concept that has no place in scientific discourse. However, if we are to understand paradigm conflicts, we need to explore the way in which "love" underlies the subjective personality and shows itself in Arcadian ways of thinking. One of the more amenable conceptualizations of "love" is that of biophilia, the love of life[92] (p. 406):

Biophilia is the passionate love of life and all that is alive . . . The biophilous person . . . wants to mold and to influence by love, reason and example, not by force, by cutting things apart, by the bureaucratic manner of administering people as if they were things . . . he enjoys life and all its manifestations.

In this view, biophilia is genetically programmed in all of us, its survival into adulthood depending on how we are socialized. Under unfortunate conditions, Fromm[92] argues, this natural love of life can be overshadowed by the development of necrophilia, a form of malignant aggression that sometimes arises in response to life's exigencies. Necrophilia, literally the love of death and dead things, is an extreme form of the kind of "objectivity" described earlier. While Fromm's view of biophilia implies a love of all living things, including people, more recent conceptions appear to emphasize its biocentric, love-of-nature aspects.[36] For instance, Wilson[93] speaks of biophilia as an innate urge to be near living things and natural stimuli, a genetically programmed need for "deep and intimate association with the natural environment, particularly its living biota."[53] Similarly, Roszak[39] talks of an ecological unconscious that generates a spontaneous rapport with the natural environment.

History tells us that there is a lot of confusion over the meaning of "love." For instance, in making recommendations about the way in which science in tropical countries might be developed, Wilson[93] suggests that progress will be made if those who truly love the land for its own sake study their local regions by, among other things, collecting and preparing specimens of local flora and fauna. Yet how can one love something and then proceed to kill it for the purpose of study? He also argues that innovation in science has its roots in someone who truly loves a subject to the point of obsession. Does he, I wonder, mean an obsessive fascination with, rather than the love of, some subject matter? In a similar vein, he comments, in passing, that "mechanophilia, the love of machines, is but a special case of biophilia" (p. 116). To psychologists such as Fromm, and others,[94,95] however, this would seem a rather bizarre suggestion as they prefer to recognize the love of machines as being quite the opposite—a form of necrophilia. The confusion is compounded when one turns to Aldo Leopold,[96] who is often held up as an example of someone who took a loving attitude to the land. Yet, in his classical book, he describes his hunting experiences with some relish, especially the occasion of his first kill (of a duck): "I cannot remember the shot; I remember only my unspeakable delight when my first duck hit the snowy ice with a thud and lay their, belly up, red legs kicking" (p. 129). Once again, what form of love is being expressed here? It is clearly unlike that shown by Rachel Carson who was known to take great pains in returning her specimens from whence they came, ensuring that while she studied them, she did not kill them.[56] Finally, Magill suggests that "young people enter natural resource disciplines because they 'love the outdoors'; they like to hunt, fish, camp and generally participate in outdoor activities"[82] (p. 296). However, it might be better to say that they love *being* outdoors rather than love *the* outdoors. We need to be skeptical, therefore, when faced with protestations of love. Perhaps the confusion can be disentangled by looking more closely at two defining characteristics of "love": intimacy and altruism.

Intimacy is the wish for a deep union in which one can share oneself with another without egoistic demands and manipulation. An intimate relationship is egalitarian, therefore, one in which there is no desire to dominate the other person or creature.[97] Some of the characteristics of an intimate relationship with nature are well illustrated by Wordsworth's youthful experiences[98] (p. 276):

As a boy, he communed with all that he saw as something not apart but inherent in his own 'immaterial nature'; many times while going to school, he would grasp at a wall or a tree to recall himself from 'this abyss of idealism to the reality'. As an adult, he was still able to enjoy moments of heightened awareness when all nature appeared an organic living whole. He experienced a 'visionary power' when looking at a flower, a tree or some other object in the countryside. A kind of light flowed from his mind and bestowed a splendor upon the object, which then lost its identity and became a presence, an energy or a force . . . This visionary power resembles primitive animism felt by many children and preliterate peoples. It is also a kind of pantheism, feeling a divine presence in all things. The joyous apprehension of the whole brings a deep sense of unity, harmony, tranquillity and love.

Intimacy with nature, therefore, is a form of "oceanic feeling," a mystical union in which the separation between oneself and the natural world is blurred—a kind of animistic mode of experience.[39] Ecopsychologists speak of an ecological self that reflects their "experience of connecting, of being in touch with the planet, of feeling earth's pain, hearing earth's cries"[38] (p. 246).

The second aspect of biophilia, altruism, involves empowering others, enabling them to cope more adequately with the demands of life. These nurturing, prosocial behaviors are evident, along with aggressiveness, in toddlers. As mentioned repeatedly throughout this section, how they develop into adulthood depends on subsequent socialization. It is not surprising, therefore, to find that prosocial behavior develops through loving contact with nurturing parents.[54,99] On the other hand, the origins of proenvironmental values and behavior have received less attention than the equivalent prosocial behaviors. Clearly, a sense of compassion for the natural world and the wish to care for it has its roots in childhood experiences,[100] but evidence in support of a genetic basis for this biocentric urge is sketchy,[53,101] and the jury appears to be out, as it were, on its credibility.[102] What is most likely is that a form of attribution error is occurring here. It is well known among psychologists that theorists often base their theoretical formulations, albeit unconsciously, on their own experiences. They tend to believe that what is true of their own lives generalizes to others.[45] Such may be the case in discussions of biophilia. Those who take great delight in the natural world may wish to believe not only that *some* others share their passion but that it is built into the species as a whole. Such hypotheses, however, are not testable, and one is left, in discussing biophilia, with speculation about its origins and extent.

However, a good example of an Arcadian behavior that appears to be based on a combination of both prosocial and proenvironmental values is that of voluntary simplicity, the willingness to restrain one's material consumption so as to make fewer demands, and have less impact, on the natural environment.[103] It involves the adoption of material frugality and a more cooperative lifestyle than one normally sees in our individualistic society. Such drastic steps would seem to require a deep concern for the way in which material overconsumption affects the natural world and a willingness to forego some material comforts for the collective benefit of all. Smil[62] has calculated, for instance, that a more equitable sharing of global energy supplies, which would help in the development of third-world countries, would require the affluent nations to reduce their energy consumption to the levels of utilization seen in France in the 1960s. This would entail a reduction in energy use in the United States by about two-thirds to three-fourths of present consumption. The prospects for such a radical shift in consumption patterns are rather dismal, as Worster[104] (p. 174) suggests:

Despite (evidence) that people have begun to rediscover the moral ideal of self-imposed restraint, the obstacles to a wholesale move in that direction remain very imposing. It may well turn out that human appetite will not be, cannot be, moderated. On a planet teeming with four or eight or twelve billion people, examples of restraint may remain what they are now: a series of minor, isolated gestures, unthinkable to the starving, unacceptable to the aspiring, unappealing to the affluent.

Emotional Involvement

A crucial element in the attainment of intimacy (communion) is one's capacity to empathize with others.[105] This contact with another person can be at an emotional or cognitive level: You can feel what others feel and share their joys and sorrows, or you can understand and sympathize with what they are going through. Emotional involvement or empathy, therefore, implies an attempt to make contact with the subjective experience of another person or creature. It also implies a genuine compassion, caring, and concern for that other being. How this develops in children is well documented, and, as with all other valuable characteristics, caring parents tend to produce caring children, although there are exceptions of course.[106]

When one turns to emotional involvement and feelings of compassion for the natural world and its creatures, less is known about the origins of these feelings. There is every reason to believe that the capacity for empathy

shown by young children for other children and pets can be nurtured by exposure to adults who provide a model of environmental concern.[100] This process of transcending the self, moving beyond one's own selfish concerns by expanding one's sense of identity to include other beings and things, is the central concern of ecopsychologists and environmental educators who adopt a deep ecology viewpoint.[38,107] However, the extent to which it is possible to promote such proenvironmental feelings in adults is an open question. While it may be possible to generate some *sympathy for* the creatures around us, a genuine *empathy with* the natural world is more difficult to achieve.[108] For instance, as mentioned earlier, it is easy to arouse public concerns about charismatic species, such as the larger predators, but immensely more difficult to arouse passions about slime molds, which play an equally important role in ecological processes.[62] Similarly, how much genuine empathy can one expect to develop for HIV or other pathogenic organisms? Being genuinely concerned for the immensity of nature, with all of its troublesome species, is perhaps too much to expect. Possibly, the kind of selflessness involved is restricted to those steeped in a religious tradition that promotes self-abnegation, but, for most people, this monastic self-denial is impossible in large doses. While one must accept that there are definite limits to caring for the natural world,[109] it would seem that the subjective personality has the best prospect of transcending such limits. Once again, I should point out that attaining this level of sensitivity is not an unmixed blessing. There are times when empathy may stand in the way of making tragic choices, particularly those surrounding the allocation of scarce resources or making difficult medical decisions.

Intraception

The Romantic tradition argues that attaining a real understanding of nature is an introspective process, one that cannot be achieved in the absence of love or sympathy based on a bond of kinship with other creatures. This intraceptive orientation is characteristic of the subjective personality in which there is a clear preference for looking inward toward the world of personal feelings rather than outward toward the objective world of "facts" and physical events.[56] Thus, the intraceptive person is interested in learning about not only their own thoughts, feelings, and motives but also about the subjective experiences of others. This kind of orientation helps develop a sensitivity to others that supports or enables the growth of empathy and attachment. Reflecting on what is going on in one's own head is sometimes referred to as metacognition. That is, one takes a step back and, in a sense, observes oneself thinking about some issue or other. This introspective and reflective process results, hopefully, in an understanding of one's biases and defenses that is a prerequisite for the most advanced forms of thinking sometimes referred to as "wisdom."[47] How to get professionals to become "reflective practitioners," who engage in sustained introspection, is a puzzle with which educators continue to struggle,[110,111] as we shall see in chapter 8.

Summary

In sum, the objective and subjective personalities are profoundly different human beings, showing characteristics that are both genetically determined as well as being the product of quite different relationships with, and upbringing by, their parents. While relatively few of us are at the extremes of this personality dimension, it is unlikely that those who are will be amenable to change in their adult years. They are likely, as mentioned earlier, to approach environmental problem solving as they always have done. Thus, the prospects for adaptive change are much greater among those who occupy the mid-range of the dimension, those who are less intensely committed to extreme beliefs and values. My point is that when one is considering strategies for attempting to resolve environmental conflicts, it is essential to bear in mind the differences that have just been outlined; otherwise, one is simply groping in the dark.

The Cognitive Dimension of Personality

Different modes of reasoning are adopted within the Imperial and Arcadian traditions. To understand how these ways of reasoning influence problem solving, we need to turn to the cognitive dimension of personality.[56] Cognition has to do with the way in which we perceive the world around us, store information in memory, and use such information to reason and think (Figure 2.1). It is well known that when we look out at the world, we perceive only a tiny fraction of the information available to us. Our perceptual processes are highly selective, carefully filtering out all but a tiny fragment of "reality," so that perceptual biases distort the information we store in permanent memory and use, subsequently, in thinking about problems. It follows that all claims to intellectual objectivity should be treated with some skepticism. The particular set of filters we construct for ourselves leads to individual differences in how we "see" problems that, in turn, contribute to the difficulties we have in communicating with those who have radically different cognitive styles. Two of the most commonly occurring styles are described in the following, along with their impacts on environmental problem solving.

Cognitive Styles

There has been a great deal of discussion in recent years about the need for a more "holistic" approach to environmental management based on nonreductive modes of planning and decision making.[112-116] Underlying this debate is the recognition of a distinction between *analytic* and *holistic* modes of thinking, or cognitive styles. An analytic approach to a problem implies the reduction of the problem to a "manageable" size and the further study of its individual components. It is assumed that an understanding of the problem can be attained by integrating information derived from the separate study of these individual parts. This process of reduction, compartmentalization,

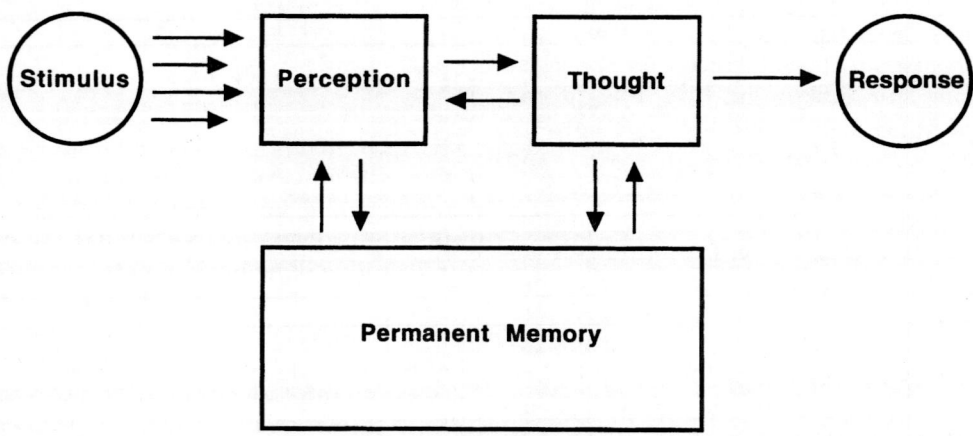

FIGURE 2.1. Cognitive processes. (Modified and reprinted with permission of Simon & Schuster from *Vital Lies, Simple Truths* by Daniel Goleman, Ph.D. Copyright 1985 by Daniel Goleman, Ph.D.)

and simplification has troubled scholars since ancient times[117] and has been rejected passionately by cultural critics in recent years.[95] The feeling of loss, of missing something important, when engaged in scientific analysis is captured in the following reflections of an eminent scientist[118] (p. 163):

I came on my first Military orchid, a species I had long wanted to encounter but hitherto had never seen outside a book. I fell on my knees before it in a way that all botanists will know . . . I measured, I photographed, I worked out where I was on the map, for future reference. I was excited, very happy, one always remembers one's 'firsts' of rarer species. Yet five minutes after my wife had finally . . . torn me away, I suffered a strange feeling. I realized that I had not actually *seen* the three plants in the little colony we had found.

Thus, advocates of a more holistic style argue that it is not possible to understand problems solely on the basis of analysis, for there are emergent properties of any system over and above the sum of its parts. Instead of putting strict boundaries around a problem and proceeding "inward" to a detailed analysis of its subcomponents, the holistic approach places a problem in its broader context, seeking understanding of the problem through its relations with a larger whole. Carvajal[119] offers an interesting example of the two different approaches to problem solving in the context of waste disposal at a petrochemical complex.

The company's primarily analytic approach to waste pollution identified major wastes as belonging to gaseous, liquid, and solid categories that, in turn, were further subdivided into numerous subproblems. Because each of the latter has a preferred treatment, it follows that the analytic "solution" to the problem of petrochemicals effluents, therefore, was the aggregate of all the part solutions. A more holistic approach, had it been tried, would have sought an understanding of the problem by linking wastes to the systems that produce and receive them. By going "outside" the problem as it is defined analytically, the holistic approach allows some rather unusual possibilities. For instance, Carvajal notes that because the analytical approach does not include the petrochemical processes themselves in the problem definition, then research on reducing waste production by improving the basic chemical processes involved was not, at that time, being contemplated. However, switching attention from waste treatment to waste prevention is possible with a more holistic orientation. Similarly, the possibility of reducing wastes by changing consumer habits is a legitimate concern for holistic thinkers but quite irrelevant to those of the analytical persuasion. Although holistic thinking can result in baroque schemes seemingly far removed from the original concerns of petrochemical wastes, it does succeed in breaking problems open, thereby exposing conventional wisdom to imaginative shocks. Conceptual mayhem of this kind is not always welcome of course. The extent to which it is may depend on how well it is combined with analytical thinking. Clearly, what is needed, in adaptive environmental management, is the ability to combine the precision and clarity of analytical thinking with the breadth and imagination of holistic thought. While this may seem simple enough in theory, it is difficult in practice.

First, it is not at all clear what "holistic" thought is supposed to be. Frequently, "holistic" and "systems" thinking are used synonymously, thereby confusing two modes of thought that may have little in common. Although systems thinking often involves an holistic-integrative element, it can also be primarily analytic. A systems analysis can, as the name implies, proceed analytically by fitting together the pieces of a jigsaw puzzle. There may be little holistic thinking involved. It is better to restrict the term "holistic" to refer to imaginative, intuitive thinking. Kolb,[117] for instance, makes the distinction between a very direct experience of one's environment, which he calls "apprehension," and the more detached, abstract analysis of that environment using concepts and ideas or "comprehension." All of us engage in both modes when we subject our personal experiences to a more refined analysis, drawing lessons from them in the process. For instance, you can stand in front of an artistic masterpiece simply absorbing its splendor (apprehension) or engage in a more active analysis of the techniques used in producing it and the meaning it intends to convey (comprehension).

Cognitive styles arise because some of us tend to spend a lot of time in developing the analytic skills involved in comprehension, whereas others prefer to live in a world of unanalyzed personal experience and interpersonal anecdotes.[117] It is difficult to convey our personal experiences (apprehensions) to others because they remain at a tacit or intuitive level in our minds until we subject them to some kind of analysis (comprehension). In the professional world, tacit knowledge gained over a lifetime of professional experience is,

likewise, difficult to explain to others. It can only be demonstrated in practice. This causes problems in decision-making situations where experienced individuals may have great difficulty in explaining the basis of their judgments or justifying their opinions.

Thus, in both the lay public and among professionals, there occur two kinds of thinking, analytic (comprehension) and holistic (apprehension). All of us switch back and forth between the two as we work our way through problems,[120] but there is a tendency in some individuals to favor one over the other. The role of holistic thought (tacit, intuitive, personal experience) is to provide the hunches, the professional sense of a problem, the intuitive framework that can guide analysis. Problems arise when the two modes are not used in combination, when problem solving is based on purely analytic rationality in the absence of the guidance provided by tacit knowledge and, the reverse, when personal experience and anecdote are not subjected to analysis. Stereotypically, the former is characteristic of scientists whereas the latter is common among environmentalists and members of the lay public. More specifically, the preoccupation of many professionals with technical rationality, and their rejection of lay knowledge and experience, is seen as being problematic.[121,122]

In addition to problems with understanding what "holistic" thinking is, any attempt to encourage a synthesis of cognitive styles is handicapped by the fact that many people find it difficult to change their preferred cognitive styles. The tenacity with which habitual ways of thinking are protected is best understood by looking at the way in which these analytic-holistic differences are rooted in every aspect of cognition. We shall look at the influence of analytic and holistic styles on perception, memory, and reasoning.

Perceptual Styles

Often, we are puzzled and irritated when other people do not perceive events as we do. However, as mentioned earlier, what we perceive of the world around us is so filtered by our perceptual processes that we see only a highly distorted and selective version of events. There are two main reasons why we need this filtering process. First, there is so much information out there that we simply cannot cope. Some means has to be found for reducing the cognitive load. This is achieved, in part, by a form of unconscious screening in which much of the information in one's perceptual field simply does not register as being significant. Second, selective attention is a convenient way of avoiding the many upsetting aspects of our world that we would prefer not to recognize. Human beings are skilled in using selective attention to avoid discomfort, a trick that helps us get through the day. However, if taken to extremes, we drift into delusional fantasies and become incapable of dealing with real-world problems.

People differ in how prone they are to engage in selective attention. The two perceptual styles that result have been referred to as *field independence* and *field dependence*. One characteristic difference between these styles is the relative ease with which field-independent individuals can "disembed" a relevant detail from a confusing context. In contrast, the field-dependent person finds this task difficult and is apt to be overwhelmed with "irrelevant" detail. Thus, the extremely field-independent person is good at attending to detail, whereas the more field-dependent individual may miss important details, being inclined, instead, to think in broad generalities. A case in point occurred recently during a heated discussion on CBC Radio between a forest ecologist who supported the judicious use of clear-cutting and an environmental activist who did not. At one point, the activist made the kind of statement that one would expect to hear from a field-dependent person, that clear-cutting was leading to the extinction of numerous indigenous species. The forest ecologist responded in a typical field-independent manner and demanded that the activist name one such species. Unfortunately, the activist was unable to do so.

From a psychological perspective, what seems to be happening is that field-independent person's attention is grabbed by differ-ences rather than similarities, thereby predisposing him or her to analytical-reductive thinking.[56] In contrast, the inclination of the field-dependent person is to avoid such fine-grained distinctions and to see similarities between widely disparate events. The result is a less discriminating perspective that is prone to information overload. This is not to say that the field-dependent style is, in some sense, inferior. There are times when sensitivity to "irrelevancies" is a crucial prerequisite for solving a problem. The information that was originally ignored, the surrounding context that was dismissed, and the emotional atmosphere of a debate that is avoided may all prove to be the most important elements in making progress in resolving a problem. In a similar vein, the focused style of the field independent can at times result in a preoccupation with trivia,[123] a descent into obsessive-compulsive futility. Thus, the exclusive adoption of one or the other styles is likely to be maladaptive. Clearly, adaptive problem solving requires some judicious balance between the two styles. For instance, the inclination of field-independent technical experts to reduce and compartmentalize problems, and to seek to ignore the broader political context, infuriates citizen groups who interpret this behavior as an attempt to keep control of events in elite hands or cover up wrongdoing.[124] In contrast, the woolly vagueness and irrelevancies of holistic arguments (field dependent) raise the ire of more precise minds.[125]

Problem solving would be facilitated if those with differing perceptual styles were able to understand the strengths and weaknesses of their contrasting styles and understood how each can contribute to the problem-solving process. This may mean learning to change one's own style, or adopting a different perceptual style when it is appropriate. However, this is not easy. Even if someone wanted to incorporate more of another style in their thinking, they may not have the skills to do so.[126,127] In addition, there is considerable evidence that the styles in question start developing early in childhood and may even have a genetic basis.[56] Thus, by the time a young person enters higher education, they have well-established perceptual (and cognitive) styles. They also tend to select majors that are consistent with their styles. Field-independent students gravitate toward engineering and science, while field-dependent students are attracted to liberal arts.[128] The reinforcement of these preferred styles by the selective training that takes place in these faculties makes subsequent change difficult. The prospects for a more holistic environmental management look decidedly remote in light of these traditional educational arrangements.

Memory Styles

One of the most common scenarios in environmental disputes is the conflict between experts and the lay public over the interpretation of data. Typically, a group of citizens perceive that their neighbors are dying of cancer in excessive numbers or that their are too many birth defects or some other problem only to find that the scientific experts called in by the government can find no evidence of this. The public's reaction is incredulity and anger when the experts dismiss lay and anecdotal "evidence" in favor of the more credible information gained through scientific analysis. What is involved here is a fundamental difference in the way in which the lay public and experts organize their knowledge and, as a result, what kind of information is more compelling to each side in the argument. To understand this we need to examine how human memory works.

Psychologists frequently distinguish between two different kinds of memory. *Episodic* memory is where we store our personal experiences in a relatively undigested form, a sort of verbatim repository for memories of the events that fill our daily lives. In contrast, we also have a more abstract kind of memory that stores the "lessons" we have drawn from this great mass of personal experience. These lessons take the form of abstractions or conceptions that provide a symbolic means of understanding what has happened to us. For instance, novice cross-country skiers may be bewildered by the variations in snow conditions that affect the efficiency of their skis.

However, as time elapses, they may begin to draw lessons from their experiences by categorizing and labeling snow conditions, an analytic process that leads to further development of the skiing part of their *semantic* memory. We can also amass these conceptual abstractions through academic study, where the experiences of others are transmitted to us through the use of language. It is interesting to find that people differ in the kind of memory on which they rely when thinking about environmental (and other) issues. Scientifically and technically trained professionals have been educated to make more use of their semantic memory, whereas lay people are inclined to rely on the more holistic, but less abstract, episodic memory in recalling information about the matter at hand. It is no wonder that there is conflict between the two groups.

It is unfortunate that there is so much misunderstanding about the use of these two different kinds of knowledge and the memory structures that underlie them, for each can contribute to the problem-solving process. Combining indigenous and scientific (positivist) knowledge is difficult, however, because they are organized so differently.[129] For example, semantic memory is arranged in a hierarchical, "logical" manner. Each idea or concept is defined by a set of distinctive characteristics linked to other concepts in higher-order groupings similar to biological taxonomies in which species are grouped into genera, families, and so on. Of course, people differ in how systematic these structures are, as we shall see shortly, but all are alike in one fundamental way: the abstract nature of the memories. For instance, when someone refers to a certain snow type as that of "corn snow," the category tells us only about a particular combination of temperature, moisture, and graininess of snow and that klister wax is probably the appropriate choice for one's skis. It tells us little else about the specific circumstances in which that snow is found. The advantage of such ideas is that they can be applied in a variety of situations, that is, they are generalizable. However, it is easy to see how they might lead to a rather detached, abstract way of thinking about problems, a quality of technical experts that annoys lay groups.

In contrast, the organization of indigenous or lay knowledge is much more difficult to understand and communicate. In all cases, we are talking about tacit knowledge gained over years of experience in a particular job or situation. A case in point is that of "grouters" who work on maintaining and repairing dams.[91] Grout is a mixture of cement and water that is pumped into holes in the structure to maintain an underground barrier to seepage that might erode the foundation of the dam. The exact nature of the erosion that has occurred, as well as the amount and composition of the grout needed, is not known in most cases. Thus, grouting is an art that depends for its success on the skills of the grouters (workers) who, after years of experience, develop a "feel for the (w)hole" and other tacit knowledge. Schmidt[91] argues that "over time grouters build up a repertoire of strategies for treating various kinds of rocks in specific situations and acquire a kind of general knowledge. Because . . . each site, each hole, and even different stages of a single hole are unique, they can never rely upon formal models or general rules of thumb or recipes. If they settle into a mindless routine, it will jeopardize the quality of the work. They must constantly be alert to the 'back talk' of the specific situation" (p. 526). Because it is so difficult to communicate this knowledge in the abstract terms more familiar to professionals, and possibly because of the lowly status of the grouters, such knowledge is often ignored by the professionals involved in dam maintenance. "For it is science that dismisses knowledge expressed as feelings, engineering that scorns the knowledge of uneducated laborers, and bureaucracy that disaggregates such knowledge"[91] (p. 527).

Part of the problem in communicating lay knowledge is that it may not be organized into the systematic, logical networks one finds in semantic memory. Instead, it is organized on a different principle, one that makes use of stories and cultural traditions. For instance, indigenous peoples in Bolivia can identify many different varieties of sweet and bitter potatoes that they themselves cultivate.[129] They also know that they do so to reduce the risk of re-

liance on a few varieties. But if one were to push them to explain why they planted a certain variety in a particular place at a special time in the season, one would be met with a mixture of stories based on cultural and social tradition. Outsiders trying to understand agricultural practices and knowledge in such places, therefore, would have to understand local social-cultural history, something that is difficult even for trained anthropologists.[129]

Often, research on problem solving reaches the conclusion that experts are more competent than novice or lay people in a variety of problem-solving situations.[130–133] One explanation for this is that experts have amassed a large amount of information about a subject area and, through years of experience, have learned to organize this into sophisticated memory structures and problem-solving strategies. In contrast, novices are depicted as suffering from a lack of relevant information, a poorly organized memory, and a limited repertoire of strategies for approaching problems.[134,135] However, on closer inspection, research on novices and experts is usually conducted in an artificial laboratory situation that favors the abstract thinking of experts. If, instead, experts were taken out into the field and their problem-solving competence was compared with indigenous people in a particular location over time, then a different picture might emerge. Something of the sort has been done, albeit inadvertently, as part of the developed world's technical assistance programs to the developing world. The results of this attempt to transfer technical expertise to nonwestern cultures and environments have been equivocal to say the least, raising fundamental questions about the competence of "experts" and the assumed lack of competence in indigenous people.

In sum, there is a fundamental difference between the knowledge stored in semantic and episodic memory. All of us use both kinds of memory to some extent but may rely on one or the other as a matter of preference. It is clear, however, that both kinds of knowledge are important in problem solving. Professional or bureaucratic abstractions in the absence of local contextual knowledge can be dangerous, just as indigenous knowledge may not generalize beyond local conditions. Adaptive problem solving would seem to require some balance between the two.[136,137]

Reasoning Styles

"Reasoning" is the cognitive process in which information is used to think through a problem. It involves the retrieval of information from memory and the surrounding context, which is then used in making inferences and judgments.[138] People differ in how they go about this; that is, they adopt different reasoning styles. For instance, Allen[139] has detected two harvesting strategies among fishing-boat captains: the Cartesians who, in their search for fish, rely on the available technical information and the Stochasts who rely on their hunches. The former is logical and systematic, whereas the latter uses intuition, a feel for the fishing grounds. Allen believes that both have a role to play in fishing policies. The technical fisher is likely to achieve a steady but unspectacular return, whereas the hunches of the Stochasts may lead to new sources of fish. It is also possible of course that the latter may also lead to expensive wild goose (or fish) chases. One would think that a hunch played out in light of the available technical information might be a better prospect. The more systematic reasoning of the Cartesian skippers is typical of *convergent (analytic)* reasoning, whereas the Stochastic skippers show a more intuitive form of reasoning seen in *divergent (holistic)* ways of thinking.

Convergent reasoning is a narrowly focused kind of thinking in which attempts are made to fashion arguments that follow a clear, logical sequence. Each step in the argument is verified before moving on to the next step in the chain of logic. As a result, convergent reasoning tends to look for the single correct answer and does not encourage flights of fancy. Even so, a preference for convergent thinking does not necessarily imply that exponents are good at it. Several studies have shown, for instance, that many scientists cannot distinguish between logical and illogical arguments.[7,140] In addition, convergent thinking also has a num-

ber of weaknesses that limit its usefulness when facing complex problems. Chief among these is the way in which convergent thinkers, especially technically trained professionals, place a great deal of salience on only certain kinds of information, notably scientific "data." This *data conservation*, a preoccupation with the quality of data and caution in its interpretation, is a tendency to become obsessed with the kinds of inference that can be made "legitimately" from a data set. The endless wrangling of scientists over data interpretation is a case in point.[8,141] Convergent reasoning, therefore, is a narrow, highly focused kind of thinking, concerned with making tight, "logical" inferences in the construction of arguments. There is a preference for staying close to the data while avoiding speculation and other forms of intellectual looseness.

Divergent thinking, on the other hand, involves a much more vague, indeterminate kind of argumentation, almost as if the person is stumbling around with no particular aim in mind. It is as if the diverger ignores strictly logical connections between ideas and "free associates," spewing forth notions that are only distantly related to the question at hand.[83,84] The result is the possibility of a number of cognitive errors. Since the memories of divergers are embedded in a much more emotional and situational context than one finds among convergers, the former are influenced by *recency* and *vividness* effects. Thus, for the diverger, recent exposure to a vivid event makes it more retrievable from memory and more likely to influence judgment. For example, seeing a TV program on nuclear hazards may increase their perception of nuclear risks, even though available scientific evidence suggests otherwise. Thus, simple exposure to the "discussion of a low-probability hazard may increase its memorability . . . and hence its perceived riskiness, regardless of what the evidence indicates"[142] (p. 465). It follows that divergent thinking is influenced strongly by the emotional impact of information, rather than its objective content. This can be an advantage when imaginative fantasy is at a premium, but the fuzzy looseness of divergent thinking can create problems where precision is needed. For instance, divergent reasoners are inclined to see illusory correlations and misperceive patterns in events.[143] Thus, links are frequently made by environmental activists between exposure to some toxic material and a public health effect, the assumption being that the two are correlated. Subsequent research often concludes that the correlation is spurious and that no link can be found. Similarly, "cancer clusters" occur from time to time, the causes of which are difficult, if not impossible, to determine. The possibility that they may be random occurrences is not acceptable to local residents who perceive meaningful patterns where there are none.[144] In other words, the judgments of divergent thinkers are troubled by a tendency to ignore such things as sample size and base-rate information, something that biases their thinking toward the immediately salient, emotionally vivid memory.

In sum, convergent reasoning, also referred to as instrumental-analytical thinking, is the mode of choice in dealing with the evaluation and testing of hypotheses. Where one wants a critical evaluation of a single, well-defined issue, then the precision and rigor of convergent thinking is at a premium. However, convergent reasoning does not provide a useful basis for generating innovative ideas, nor is it helpful in the early stages of coping with ill-defined problems. What is needed in this latter case is a more imaginative kind of thinking, one that is less constrained by the demands of data and "logic" and can draw together previously disparate aspects of the problem, rather than subjecting it to the analytic scalpel of convergent thinking. Thus, the intuitive freedom of divergent, or normative and holistic, thinking makes it valuable during some, but not all, phases of problem solving.

As with all of the styles discussed in this chapter, convergers and divergers tend to irritate one another. The narrow, nitpicky, persnicketiness of the converger incenses the diverger who, in turn, is seen as undisciplined, befuddled, and specious by convergers. It is crucial, however, that the two forms of reasoning should be combined in both individual and group problem solving, no matter how dif-

ficult this might be.[33,121] In environmental management, for instance, the preference for convergent thinking among many scientific and technical professionals can lead to tunnel vision and other cognitive pathologies.[145] On the other hand, by relying too much on divergent reasoning, one risks falling "into the fallacy of replacing empirical science with metaphysical guesswork"[36] (p. 106).

Personality Types

Earlier in the chapter, I pointed out that the persistent conflict between proponents of different environmental viewpoints is best understood by examining the underlying psychological differences between them. So far, we have explored this in terms of two separate personality dimensions: objective-subjective motivations and analytic-holistic cognitive styles. However, in real life, these personality characteristics interact in complex ways to produce different personality types, each of which takes a different approach to problem solving.[19,56] In closing the chapter, therefore, four main personality prototypes (Fig. 2.2) are outlined along with a description of how each prototype approaches environmental problems. In doing so, I do not mean to imply that there are only four different kinds of people in this world but, rather, offer that these type descriptions are useful heuristic devices that help us understand the psychological basis of common problem-solving styles. This personality classification is based in part on the ideas of Cotgrove.[19]

The Objective-Analytic Prototype

The objective-analytic (OA) type adopts a tough-minded, analytical orientation to life in which the person's central concern is the achievement of a sense of power and control. Strenuous efforts are made to establish control over their emotional and cognitive interactions with other people. Typically, this is achieved by a process of emotional distancing

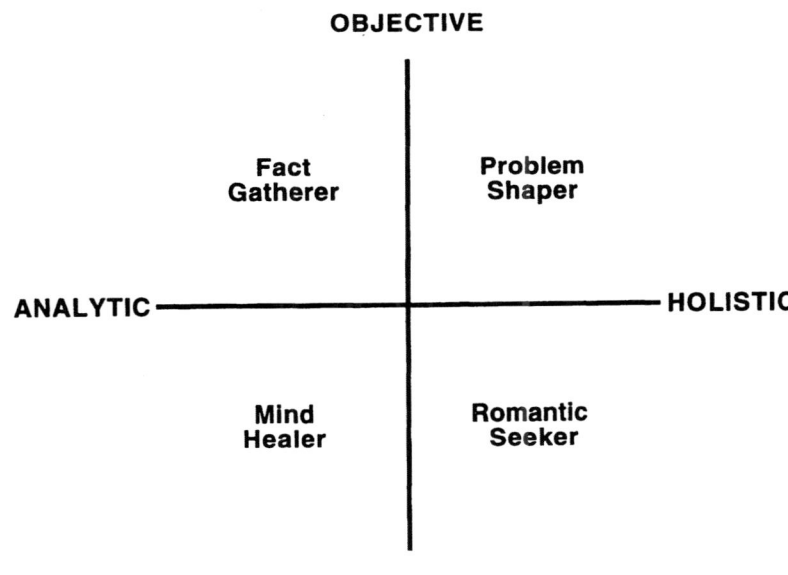

OBJECTIVE

**Fact
Gatherer**

**Problem
Shaper**

ANALYTIC ———————————————————— **HOLISTIC**

**Mind
Healer**

**Romantic
Seeker**

SUBJECTIVE

FIGURE 2.2. Personality prototypes.

or detachment, the purpose of which is to limit emotional involvement with others so that one will not be bothered by the vagaries of one's emotional reaction to or the emotional pressure exerted by others. Similarly, cognitive control is sought through giving preference to one's own interpretation of events over the perspectives of others. The potential confusion engendered by paying undue concern to another person's viewpoint is thereby avoided, as are the upsetting consequences of introspection, the exploration of one's subjective world. An illusion of control is sustained, therefore, by focusing on the exterior world of objective certainty, the world of outward appearance and physical reality. As a consequence, a mechanistic worldview is developed in which simple cause-effect relationships are prized as a means of understanding and control. One way to ensure the control of events is to pay painstaking attention to the minutiae of external experience. This goal is facilitated by an analytical style in which factual detail is sought in a relatively circumscribed field. Thus, "objective" facts are thought to provide the key to understanding and control, whereas intuitions and impressions are not. What is most threatening to this kind of person is loss of control. For instance, any implication that their ideas or methods are insufficient to the task would challenge their carefully constructed sense of competence, a crucial element in maintaining their emotional stability. Similarly, emotional situations that threatened to disrupt their finely crafted facade of objectivity and impartiality would be profoundly aggravating. A typical response would be to put some emotional distance between themselves and the threat by, for instance, reasserting their objectivity and rationality. A threatened scientist will claim the high ground, asserting his moral and technical competence in contrast to the emotional bias and technical fumbling of his opponent. The OA type, therefore, is empirical, reductionist, impersonal, and obsessive. It follows that members of this category might be labeled "fact gatherers," a prototypical example being those natural and social scientists who seek understanding and control through the collection of "objective data" in a narrowly defined segment of the physical and psychosocial worlds. Thus, technical rationality is the preferred approach to problem solving of the OA type, a search for technical fixes using instrumental-analytic ways of thinking.[5,32]

The Objective-Holistic Prototype

The objective-holistic (OH) and OA types share the same objective, impersonal manipulative orientation, but power or control is achieved using a different strategy. Rather than an attempt to understand and control reality by seeking detailed, factual information (as one finds in the OA type), the illusion of control is achieved by the development of schemes, theories, systems of thought, and fantasies that are imposed on "reality." In contrast to the OA type, who seeks to document reality, the OH type may seek to impose a system of order on to it, or seek to force the surrounding environment to comply with and fit into his or her model of how things should be. Thus, empirical data are often ignored in favor of an intuitive judgment of events. Indeed, by attending too much to data, there is the risk of having one's conceptual schemes and fantasies deflated, a most threatening prospect for the OH type. When the mismatch between fantasy and reality becomes too great and the OH type's conceptual schemes are threatened, then psychological defenses, such as denial and repression, may be used to suppress this unpleasant truth. Thus, this personality type takes a broader, less data-bound approach to problems that one sees in the OA type, one that emphasizes intuition, professional judgment, and contextual knowledge.[112] Typical examples are the system scientists, model builders, conceptual theorists, managers, and policy analysts who adopt a holistic and systems form of technical rationality. A similar emphasis on professional judgment and intuition is also found among senior bureaucrats and politicians whose approach to problem solving involves political rationality. In all these examples, there is an attempt to impose a solution on a problem situation by using one's best guess about the nature of a complex situation. It is for this reason that the OH type might be termed a "problem shaper."

The Subjective-Holistic Prototype

The primary concern for subjective-holistic (SH) types is with establishing intimate, nurturing relationships with other people and with the surrounding environment. The SH types seek to join with others in what the more objective types derisively refer to as "dependent" relationships. Regardless of how one might label this behavior, it does seem that SH types, rather than seeking power over others, strive to empower others through nurturing behavior. This is facilitated by a well-developed sense of empathy, which implies a main interest in subjective experience, be that of one's own inner reality or the inner, psychological world of others. Feelings and personal impressions are given priority over the details of "objective reality." When subjectivity is coupled with a holistic style, there is a lack of interest in the analysis of personal experience, or in imposing emotional control on one's thought processes. The result is that personal reactions intrude into thought, making the latter evaluative, emotionally tinged, and intensely subjective. Thus, the *romantic* lives in an impressionist, often imaginative, world of personal anecdote and unanalyzed subjective experience. One can see why such personality types find the detached, impersonal manner of scientific professionals so disagreeable, as the orientation here is toward feeling, not logic; cooperation, not control; and, involvement, not detachment. In its most constructive manifestation, SH types are to be found among local helpers, facilitators, community developers, and activists who seek to reestablish a "healthy" relationship with nature by forging bonds, fostering networks of human relationships, and empowering others. Here too are the organic farmers and holistic foresters seeking to recreate what they consider to be a more humane form of living with the land. At worst, this type of person shows a form of unexamined rejection of the technical world of "them," the people who run things in industrial society. In other words, this is the stereotypical "overemotional" and "uninformed" environmentalist, the bane of bureaucrats and politicians.

Ecosocial rationality is the preferred approach to problem solving, one that emphasizes community self-development rather than the elite control favored by the more objective types. Given the difficulty in implementing such problem-solving strategies in a world dominated by objectivity,[55] it is reasonable to label this type "romantic seekers."

The Subjective-Analytic Prototype

The main agenda of the subjective-analytic (SA) types is to understand themselves and others. They are much given to self-analysis and contemplation of the psychological world, especially their spiritual relationship with nature. There isn't much interest in the technical world or in trying to gain control over the physical processes of life. As a consequence, there is a tendency to withdraw into a reflective, self-preoccupied world of introspective thought at the expense of engagement in the broader world. It is as if the person truly believes in mind over matter, that problems can be solved by pure thought. Because all environmental problems lie in the head and are due to distorted values and wrong thinking, then a prerequisite for any solution is critical self-analysis and the development of more benign environmental beliefs and values. How such self-awareness might be translated into practical action on the world stage is not a concern of this personality type, who would see such questions as matters of technical trivia. When objective reality intrudes into the introspective landscape of the mind, then the SA type has a particular problem. Of all four types, the SA types have the greatest difficulty in summoning effective defensive coping because of their inclination to introspective analysis. Thus, the SA types have difficulty in protecting themselves from what Smail[55] calls the horrors of psychological honesty, with the possible consequence of descent into ruminative self-doubt and despair. It is for this reason that the recipe of self-analysis offered by this personality type as a solution to environmental problems is particularly unpalatable to the rest of society. Typical examples of the SA type are

to be found among ecopsychologists, environmental philosophers, naturalists, novelists, poets, hermits, psychotherapists, New Age mystics, and so on. The preferred approach to problem solving is a psychological variant of ecological rationality, one that emphasizes self-awareness and transcendence.[39] It is for this reason that they may be referred to as "mind healers."

Personality Types and Environmental Problem Solving

The problem-solving strategies adopted by the four personality types are so dissimilar that they appear to be incompatible, the resulting conflict between them diminishing the prospect for effective cooperation. How this unfolds in practice can be illustrated with a brief case study. Before doing so, I need to comment on how such psychological analyses should be interpreted and used. In what follows, note that certain ways of approaching the problem in question are typical of a certain personality type; however, one who adopts such a problem-solving style does not necessarily belong to that personality type, that is, does not show *all* of the *other* characteristics associated with that type. Such reasoning is not logical for it is tantamount to saying that because all birds have two legs, anything with two legs must be a bird. Thus, it is inappropriate to infer that someone belongs to a personality type simply on the grounds that they seem to have adopted a specific problem-solving style in approaching this one particular problem.

In recent years there has been controversy over the use of bovine growth hormone (bGH) as a means of increasing milk yields in dairy herds. According to proponents, injections of bGH can increase productivity by as much as 25 percent and is simply another tool that can be used to make milk production more efficient. Critics worry about the health effects of synthetic hormones on both cattle and consumers, as well as being dubious about the ecosocial effects of greater centralization of milk production. Thus, the decision to allow the use of bGH could be treated as a very simple matter, merely a question of determining whether or not there are deleterious health effects associated with its use and how cost-effective it might be. On the other hand, the problem could be defined in a much more sweeping manner, one that viewed bGH as a symptom of a more general malaise pervading industrial agriculture itself. Needless to say, the four personality types differ in the way in which they define the problem, as well as the strategies they offer to deal with the situation.

A case in point is the debate surrounding a study conducted by the Sustainable Agriculture Research and Education Program (SAREP) at the University of California. The SAREP study compared the bGH system of milk production (in which feed is brought to cattle housed in a central facility) with an alternative system based on rotational grazing (cattle are dispersed on pasture).[146] The basic conclusion reached by the multidisciplinary team that conducted the three-year study was that the rotational grazing system has a number of significant health and environmental advantages over the bGH/confinement feeding system. This controversial finding was unwelcome in some circles but embraced by others as evident in the lively debate that ensued in the pages of the spring and summer (1993) issues of the UC Davis Magazine. Of interest to us is not the validity, or otherwise, of these arguments but, rather, the kind of problem-solving style that the four hypothetical personality types might adopt in approaching this controversial problem.

One of the main criticisms leveled at the study, or more precisely, at a nontechnical summary of it,[147] was that it violated scientific standards. Considerable emphasis was placed by critics on the need to conduct the debate in an objective manner, sticking closely to the "facts" and avoiding speculation and emotionality. This concern for scientific rigor, empirical detail, and a distaste for "speculation" is a typical objective-analytic characteristic. It follows that, from the perspective of an OA type of person, the most appropriate problem-solving strategy would most likely be to have experts and specialists conduct rigorous empirical research within the confines of their specializations and to limit their public pronouncements

to the essentials of their findings. Thus, it would be appropriate to have studies done by dairy and public health scientists on the effect bGH has on bovine and consumer health, much as the Federal Drug Administration has been doing for the last eight years.[148] Economists could also conduct cost-benefit analyses of bGH use, while social scientists might determine consumer attitudes to milk produced in this manner. All of these studies would be highly circumscribed and conceptualized within the language of each disciplinary specialization, so that suitable levels of rigor could be maintained. Most likely they would be conducted in a piecemeal manner with little or no coordination between the various researchers, the assumption being that the role of the scientific specialist is to provide valid data within his or her field that would, in some unspecified way, contribute to policy decisions. How the subsequent integration of widely disparate studies might be achieved is not the responsibility of the specialist but a problem for policymakers whose decisions should be based on objective data produced by impartial experts.

In contrast, the objective-holistic personality type would see the research problem as that of understanding a system of agricultural production and its many ramifications, rather than its isolated parts. In other words, a systems approach would be taken. Research efforts might be conceived from the outset as part of an integrated effort to elucidate relevant aspects of the system. As a result, the research questions asked would extend beyond matters of physical health to those of social and environmental health and even to questions about the viability of industrial agriculture itself. The SAREP study appears to have taken this holistic or systems approach, within a primarily positivist (empirical) framework. One sees, for instance, an examination of the way in which bGH use might increase centralization of milk production in extremely large farms to the detriment of smaller operations and the sustainability of community life.[149] (This is a good place to reiterate that because someone appears to be using an OH research style in this one instance, it does not necessarily mean that they belong to that personality category. Thus, I am not saying that the SAREP researchers necessarily belong to the OH type and must, therefore, show *all* the characteristics of this type). The danger faced by a holistic or systems approach is that the research questions included in a systems analysis may be too broad, outside the scope of any research team. Those who *are* OH personalities are unlikely to be disturbed by this, however, because their main objective is to develop a comprehensive picture consistent with their overall view of the situation. If this requires informed judgment that goes beyond the available information, so be it. Some of the subsequent support for the SAREP study in the local community appeared to adopt an OH form of thinking, emphasizing the importance of taking a broad, systems view of the problem that includes more than mere technical detail. On the other hand, it is this broad, holistic strategy that seems to incense the OA personality types, as we have seen.

Conspicuously absent from this limited example of the bGH debate are viewpoints representative of the subjective personality types. Such viewpoints turn the problem around and direct solutions toward changes in people and social relationships, not toward the more objective world of technical data. Elements of subjectivity, in the sense we have used it, are present in some of the disputants, but an overtly subjective viewpoint is missing. There is an occasional reference to the need to establish human networks, but this is limited to a few respondents. Thus, the first inclination of the SH type is not to seek data but to focus on establishing linkages and relationships both among people and between people and nature. Although technical data are useful, they are secondary to the process of developing networks and communities. In a sense, technical studies and the disputes surrounding them are irrelevant because they are conducted in a sociopolitical context that is dominated by experts and elites operating at some distance from the life of producers. In contrast, one of the first steps, in an SH approach, would be to try to establish what might be called a "learning community." Rather than an ad hoc group or research team brought together for the oc-

casion and then disbanded when the research is completed, a learning community would be composed of an actual community of producers, consumers, researchers, administrators, managers, and policymakers who planned on being together for a very long time. Establishing this network of relationships, together with the necessary commitment, would take precedence over the collection of technical data. Whatever research was eventually conducted would be generated by the perceived needs of the community itself. In other words, the SH type is likely to be attracted to what is referred to as "participatory" research and policymaking, a form of lay-expert cooperation that has been attracting interest for some years now.[5,150]

The subjective-analytic approach to problem solving, one that emphasizes self-analysis, occupies a minority role in environmental controversies. A step in this direction was taken by several commentators on the bGH debate who tried to introduce a more psychological slant by pointing to the need to question some of the assumptions underlying the positions taken by protagonists. The main agenda for the SA type, then, is to establish the psychological conditions under which there can be genuine communication. Technical debates and yet more data collection are seen as futile in the absence of a willingness to tolerate others' perspectives. Instead, some effort must be made not only to understand and tolerate the views of others but to turn inward and seek out one's own biases. One reason why SH and SA approaches to problem solving are uncommon is that both question the distribution of power in the prevailing political system. Each has an egalitarian thrust in which elites are required to share decision-making power, as well as exposing their viewpoints and psychological motives to grassroots scrutiny. It is no wonder, therefore, that one sees so much conflict between proponents of the subjective and objective modes.

The previous example illustrates that the approach we take to any problem is influenced by our personality structure. Our ideologies and cognitive styles act as filters in determining which "problems" we are willing to acknowledge and the problem-solving style we adopt. We are all biased, therefore, not in the pejorative sense of the word but rather in the sense that our thinking is embedded in a system of values and beliefs, directed by preferred cognitive modes, and pummeled by our anxieties. Under these circumstances, better cooperation in environmental problem solving might be possible if the various players involved were willing to recognize the psychological burden under which they labor.

Conclusions

The main point of this chapter is that psychological conflicts underlie environmental controversies and hinder efforts to achieve working solutions to environmental problems. Examples of the four personality types are ubiquitous in environmental debates. Each sees the problem in a different way, uses different means to reach disparate goals, and has great difficulty in comprehending the arguments of other personality types. None, however, have a monopoly on the truth, while, on the other hand, everyone has something useful to offer. The problem is how to generate constructive dialogue. Fortunately, a few rare souls appear to have developed the kind of integrated personality structure that combines the best elements of all four types. The crucial role of these unique individuals in environmental problem solving is discussed in a later chapter. It is reasonable to conclude that while psychological differences among people are not the sole cause of destructive conflict, they create an unstable foundation on which to build cooperative problem-solving efforts. To date, there have been few systematic efforts, in any profession, to deal with the kinds of psychological barriers to cooperation described earlier, let alone in environmental management and planning. The evident distaste for attending to such personal matters must be overcome if adaptive problem solving is to be achieved.

References

1 Lee, K. 1993. *Compass and gyroscope: Integrating science and politics for the environment.* Washington, D.C.: Island Press.

2 Mitchell, B., ed. 1995. *Resource and environmental management in Canada: Addressing conflict and uncertainty.* Toronto: Oxford University Press.

3 Janis, I. 1989. *Crucial decisions: Leadership and policymaking in crisis management.* New York: The Free Press.

4 Dryzek, J. 1987. *Rational ecology: Environment and political economy.* New York: Blackwell.

5 Fischer, F. 1990. *Technocracy and the politics of expertise.* Newbury Park, Calif.: Sage.

6 Dickson, D. 1988. *The new politics of science.* Chicago: University of Chicago Press.

7 Mahoney, M. 1976. *Scientist as subject: The psychological imperative.* Cambridge, Mass.: Ballinger.

8 Hays, S. 1987. *Beauty, health and permanence: Environmental politics in the United States, 1955–1985.* Cambridge: Cambridge University Press.

9 Clarke, L. 1989. *Acceptable risk: Making decisions in a toxic environment.* Berkeley: University of California Press.

10 Deyle, R. 1994. Conflict, uncertainty, and the role of planning and analysis in public policy innovation. *Policy Studies Journal* 22: 457–73.

11 Mayer, R. 1985. *Policy and program planning: A developmental perspective.* Englewood Cliffs, N.J.: Prentice-Hall.

12 Ulen, T. 1990. The theory of rational choice, its shortcomings, and the implications for public policy decision making. *Knowledge: Creation, Diffusion, Utilization* 12: 170–98.

13 Wandersman, A., and W. Hallman. 1993. Are people acting irrationally? Understanding public concerns about environmental threats. *American Psychologist* 48: 681–86.

14 Cavanaugh, J., and H. Stafford. 1989. Being aware of issues and biases: Directions for research on postformal thought. *Adult development: Comparisons and applications of developmental models.* Edited by M. Commons et al., 279–92. New York: Praeger.

15 Worster, D. 1977. *Nature's economy: A history of ecological ideas.* Cambridge: Cambridge University Press.

16 Keraudren, P. 1996. In search of culture: Lessons from the past to find a role for the study of administrative culture. *Governance: An International Journal of Policy and Administration* 9: 71–98.

17 Bowers, C. 1993. *Education, cultural myths, and the ecological crisis.* Albany: State University of New York Press.

18 Cowen, H. 1994. *The human nature debate: Social theory, social policy and the caring professions.* Boulder, Colo.: Pluto Press.

19 Cotgrove, S. 1982. *Catastrophe or cornucopia: The environment, politics and the future.* Chichester: Wiley.

20 Milbrath, L. 1989. *Envisioning a sustainable society.* Albany: State University of New York Press.

21 Pepper, D. 1996. *Modern environmentalism: An introduction.* London: Routledge.

22 Roussopoulos, D. 1993. *Political Ecology.* Montreal: Black Rose Books.

23 Smith, G. 1992. *Education and the environment: Learning to live with limits.* Albany: State University of New York Press.

24 Stokols, D. 1990. Instrumental and spiritual views of people-environment relations. *American Psychologist* 45: 641–46.

25 Taylor, D. 1992. Disagreeing on the basics: Environmental debates reflect competing world views. *Alternatives* 18: 26, 28–33.

26 Thompson, M. 1990. Plural Rationalities: The rudiments of a practical science of the inchoate. Paper presented at a conference on *The Environmental Edge of My Profession.* September 17–21. Elsinore: Danish Academy of Technical Sciences.

27 Degler, C. 1991. *In search of human nature: The decline and revival of Darwinism in American social thought.* New York: Oxford University Press.

28 Robottom, I., and P. Hart. 1996. Behaviorist EE research: Environmentalism and individualism. *Journal of Environmental Education* 27: 5–9.

29 Dunlap, R. 1980. Paradigmatic change in social science. *American Behavioral Scientist* 24: 5–14.

30 Dunlap, R. 1983. Ecologist versus exemptionalist: The Ehrlich-Simon debate. *Social Science Quarterly* 64: 200–203.

31 Botkin, D. 1990. *Discordant harmonies: A new ecology for the twenty-first century.* New York: Oxford University Press.

32 Pacey, A. 1983. *The culture of technology.* Cambridge, Mass.: MIT Press.

33 Sternberg, R. 1990. *Wisdom: Its nature, origins, and development.* Cambridge: Cambridge University Press.

34 Collard, A., and J. Contrucci. 1988. *Rape of the wild.* London: The Women's Press.

35 Ewens, W. 1984. *Becoming free: The struggle for human development.* Wilmington, Del.: Scholarly Resources Inc.

36 Oates, D. 1989. *Earth rising: Ecological belief in an age of science.* Corvallis, Oreg.: Oregon State University Press.

37 Eckersley, R. 1992. *Environmentalism and political theory: Toward an ecocentric approach.* Albany: State University of New York Press.

38 Reser, J. 1995. Whither environmental psychology?: The transpersonal ecopsychology crossroads. *Journal of Environmental Psychology* 15: 235–57.

39 Roszak, T. 1992. *The voice of the earth: An exploration of ecopsychology.* New York: Simon and Schuster.

40 Pagels, E. 1981. *The gnostic gospels.* New York: Vintage Books.

41 Ponting, C. 1991. *A green history of the world.* London: Penguin Books.

42 Webster, P. 1991. Pining for the trees: The history of dissent against forest destruction in Nova Scotia 1749–1991. Master's thesis, Dalhousie University: Halifax, Nova Scotia.

43 Arcury, T., and E. Christianson. 1990. Environmental worldview in response to environmental problems: Kentucky 1984 and 1988 compared. *Environment and Behavior* 22: 387–407.

44 Lertzman, K., J. Rayner, and J. Wilson. 1996. Learning and change in the British Columbia forest policy sector: A consideration of Sabatier's advocacy coalition framework. *Canadian Journal of Political Science* 29: 111–33.

45 McMartin, J. 1995. *Personality psychology: A student-centered approach.* Thousand Oaks, Calif.: Sage.

46 Saul, J. 1992. *Voltaire's bastards: The dictatorship of reason in the West.* Toronto: Penguin.

47 Orwoll, L., and M. Perlmutter. 1990. The study of wise persons: integrating a personality perspective. In *Wisdom: its nature, origins, and development.* Edited by R. Sternberg, 160–77. Cambridge: Cambridge University Press.

48 Douglas, M. 1985. *Risk acceptability according to the social sciences.* New York: Russell Sage Foundation.

49 Barke, R., and H. Jenkins-Smith. 1993. Politics and scientific expertise: Scientists, risk perception, and nuclear waste policy. *Risk Analysis* 13: 425–39.

50 Dietz, T., and R. Rycroft. 1987. *The risk professionals.* New York: Russell Sage Foundation.

51 Maddi, S. 1976. *Personality theories: A comparative analysis.* Homewood, Ill.: Dorsey Press.

52 Penner, L., et al. 1995. Measuring the prosocial personality. In *Advances in personality assessment.* Edited by J. Butcher and C. Speilberger, 147–63. Hillsdale, N.J.: Lawrence Erlbaum Associates.

53 Gardner, G., and P. Stern. 1996. *Environmental problems and human behavior.* Boston: Allyn and Bacon.

54 Cummings, E., et al. 1986. Early organization of

altruism and aggression: Developmental patterns and individual differences. *Altruism and aggression: Biological and social origins*. Edited by C. Zahn-Waxler, E. Cummings, and R. Iannotti, 165–188. Cambridge: Cambridge University Press.

55 Smail, D. 1984. *Illusion and reality: The meaning of anxiety*. London: Dent and Sons.

56 Miller, A. 1991. *Personality types: A modern synthesis*. Calgary: University of Calgary Press.

57 White, R. 1976. *The enterprise of living: A view of personal growth*. New York: Holt, Rinehart and Winston.

58 Miller, A. 1985. Ideology and environmental risk. *The Environmentalist* 5: 21–30.

59 Buss, D., K. Craik, and K. Dake. 1986. Contemporary worldviews and perception of the technological system. In *Risk evaluation and management*. Edited by V. Covello, J. Menkes, and J. Mumpower, 93–130. New York: Plenum Publishers.

60 McAdams, D. 1990. *The person: An introduction to personality psychology*. San Diego: Harcourt Brace Javanovich.

61 Carver, J. 1976. Energy, information, and public policy. *American Behavioral Scientist* 19: 279–85.

62 Smil, V. 1993. *Global Ecology: Environmental change and social flexibility*. London: Routledge.

63 Ehrenfeld, D. 1993. *Beginning again: People and nature in the new millenium*. New York: Oxford University Press.

64 Bowling, J., and B. Martin. 1985. Science: A masculine disorder? *Science and Public Policy* 12: 308–16.

65 Cohn, C. 1987. Sex and death in the rational world of defense intellectuals: Signs. *Journal of Women in Culture and Society* 12: 687–718.

66 Keller, E. 1985. *Reflections on gender and science*. New Haven: Yale University Press.

67 Russell, D., ed. 1989. *Exposing nuclear phallacies*. New York: Pergamon.

68 Tripp-Knowles, P. 1994. Androcentric bias in science? An exploration of the discipline of forest genetics. *Women's Studies International Forum* 17: 1–8.

69 Chomsky, N. 1989. *Necessary illusions: Thought control in democratic societies*. Montreal: Canadian Broadcasting Corporation.

70 Forester, J. 1989. *Planning in the face of power*. Berkeley: University of California Press.

71 Nelson, J. 1989. *Sultans of sleaze: Public relations and the media*. Toronto: Between the Lines.

72 Bechtel, H. and W. Pearson. 1985. Deviant scientists and scientific deviance. *Deviant Behavior* 6: 237–52.

73 Ben-Yehuda, N. 1986. Deviance in science: Towards a criminology of science. *The British Journal of Criminology* 26: 1–27.

74 Billig, M. 1988. Selling science to the Devil. *The Psychologist* 1: 475–76.

75 Rosenzweig, R. 1989. Public policy issues in scientific fraud and misconduct. *BioScience* 39: 552–54.

76 Müller-Hill, B. 1988. *Murderous science: Elimination by scientific selection of Jews, Gypsies, and others, Germany 1933–1945*. Oxford: Oxford University Press.

77 Rousseau. 1992. Case studies in pathological science. *American Scientist* 80: 54–63.

78 Lewis, M. 1992. *Green delusions: An environmentalist critique of radical environmentalism*. Durham, N.C.: Duke University Press.

79 Levine, A. 1982. *Love Canal: Science, politics, and people*. Lexington, Mass.: Lexington Books.

80 Petulla, J. 1987. *Environmental protection in the United States*. San Francisco: San Francisco Study Center.

81 Nandy, A. 1983. The pathology of objectivity. *Ecologist* 13: 202–7.

82 Magill, A. 1988. Natural resource professionals: The reluctant public servants. *The Environmental Professional* 10: 295–303.

83 Hudson, L. 1968. *Contrary imaginations*. Harmondsworth: Penguin.

84 Hudson, L. 1970. *Frames of mind*. Harmondsworth: Penguin.

85 Hyde, R., and G. McLean. 1992. Alienated engineers: Part of the problem. Paper presented at a conference of the Canadian Sociology and Anthropology Association, 30 May, at University of Prince Edward Island, Charlottetown, P.E.I., Canada.

86 Levinson, D., M. Sharaf, and D. Gilbert. 1966. Intraception: The evolution of a concept. In *Concepts, theory, and explanation in the behavioral sciences*. Edited by G. DiRenzo, 116–35. New York: Random House.

87 Baum, H. 1987. *The invisible bureaucracy: The unconscious in organizational problem solving*. New York: Oxford University Press.

88 Hirschhorn, L. 1988. *The workplace within: The psychodynamics of organizational life*. Cambridge, Mass.: MIT Press.

89 Trist, E. 1993. The assumptions of ordinariness as a denial mechanism: Innovation and conflict in a coal mine. In *The psychodynamics of organizations*. Edited by L. Hirschhorn and C. Barnett, 165–75. Philadelphia: Temple University Press.

90 Neis, B. 1992. Fishers' ecological knowledge

and stock assessment in Newfoundland. *New-foundland Studies* 8: 155–78.

91 Schmidt, M. 1993. Grout: Alternative kinds of knowledge and why they are ignored. *Public Administration Review* 53: 525–30.

92 Fromm, E. 1973. *The anatomy of human destructiveness*. Greenwich, Conn.: Fawcett Publications.

93 Wilson, E. 1984. *Biophilia*. Cambridge, Mass.: Harvard University Press.

94 Mumford, L. 1967. *The myth of the machine*. New York: Harcourt, Brace, and Jovanovich.

95 Roszak, T. 1973. *Where the wasteland ends: Politics and transcendence in postindustrial society*. New York: Doubleday and Company.

96 Leopold, A. 1970. *A Sand County almanac*. San Francisco and New York: Sierra Club/Ballantine Books.

97 McAdams, D. 1988. *Power, intimacy and the life story: Personological inquiries into identity*. New York: Guilford Press.

98 Marshall, P. 1994. *Nature's web: Rethinking our place on earth*. New York: Paragon House.

99 Clary, E., and J. Miller. 1986. Socialization and situational influences on sustained altruism. *Child Development* 57: 1358–69.

100 Horwitz, W. 1996. Developmental origins of environmental ethics: The life experiences of activists. *Ethics and Behavior* 6: 29–54.

101 Wright, R. 1994. *The moral animal: Evolutionary psychology and everyday life*. New York: Vintage Books.

102 Kaplan, S. 1995. Book review: The biophilia hypothesis. *Environment and Behavior* 27: 801–3.

103 Elgin, D. 1981. *Voluntary simplicity: Toward a way of life that is outwardly simple, inwardly rich*. New York: Morrow.

104 Worster, D. 1983. Water and the flow of power. *The Ecologist* 13: 168–74.

105 Feshbach, S., and N. Feshbach. 1986. Aggression and altruism: A personality perspective. In *Altruism and aggression: Biological and social origins*. Edited by C. Zahn-Waxler, E. Cummings, and R. Iannotti, 165–88. Cambridge: Cambridge University Press.

106 Noddings, N. 1987. Do we really want to produce good people? *Journal of Moral Education* 16: 177–88.

107 Thomashow, M. 1995. *Ecological identity: Becoming a reflective environmentalist*. Cambridge, Mass.: MIT Press.

108 Fisher, J. 1987. Taking sympathy seriously: A defense of our moral psychology toward animals. *Environmental Ethics* 9: 197–215.

109 Robinson, J. 1993. The limits to caring: Sustainable living and the loss of biodiversity. *Conservation Biology* 7: 20–28.

110 Fien, J., and R. Rawling. 1996. Reflective practice: A case study of professional development for environmental education. *Journal of Environmental Education* 27: 11–20.

111 Zundel, P., T. Needham, and J. Kershaw. 1994. Designing and implementing a learning system in forestry to create reflective practitioners. In *Fifth National Conference on Problem Solving across the Curriculum*. Geneva, N.Y.: Hobart and William Smith Colleges.

112 Dorney, R. 1989. *The professional practice of environmental management*. New York: Springer-Verlag.

113 Maser, C. 1990. *The redesigned forest*. Toronto: Stoddart.

114 Naveh, Z., and A. Lieberman. 1984. *Landscape ecology: Theory and application*. New York: Springer-Verlag.

115 Risser, P. 1985. Toward a holistic management perspective. *BioScience* 35: 414–18.

116 Savory, A. 1988. *Holistic resource management*. Washington, D.C.: Island Press.

117 Kolb, D. 1984. *Experiential learning: Experience as the source of learning and development*. Englewood Cliffs, N.J.: Prentice-Hall.

118 O'Neill, J. 1993. *Ecology, policy & politics: Human well-being & the natural world*. New York: Routledge.

119 Carvajal, R. 1982. The dialectics of analytical and synthetic approaches. *European Journal of Operational Research* 10: 361–72.

120 Hamm, R. 1989. Moment-by-moment variation in 'experts' analytic and intuitive cognitive activity. *IEEE Transactions on Systems, Man, and Cybernetics* 18: 757–76.

121 Baum, H. 1982. Policy analysis: special cognitive style needed. *Administration & Society* 14: 213–36.

122 Brown, P., and E. Mikkelsen. 1990. *No safe place: Toxic waste, leukemia, and community action*. Berkeley, Calif.: University of California Press.

123 Entwistle, N. 1981. *Styles of learning and teaching*. Chichester, U.K.: Wiley.

124 Richardson, M., J. Sherman, and M. Gismondi. 1993. *Winning back the words: Confronting experts in an environmental public hearing*. Toronto: Garamond Press.

125 Efron, E. 1984. *The Apocalyptics: Cancer and the big lie: How environmental politics controls what we know about cancer*. New York: Simon and Schuster.

126 Cooper, L. 1985. Strategies for visual comparison and representation: Individual differences. In *Advances in the psychology of human intelligence*. Edited by R. Sternberg, 77–124. Hillsdale, N.J.: Erlbaum.

127 Cooper, L., and R. Mumaw. 1985. Spatial aptitude. In *Individual differences in cognition*. Edited by R. Dillon. Orlando: Academic Press.

128 Witkin, H., et al. 1977. Role of the field dependent and field independent cognitive styles in academic evolution: A longitudinal study. *Journal of Educational Psychology* 69: 197–211.

129 Redclift, M. 1987. *Sustainable development: Exploring contradictions*. London: Routledge.

130 Chi, M., R. Glaser, and E. Rees. 1982. Expertise in problem solving. In *Advances in the psychology of human intelligence*. Vol. 1. Edited by R. Sternberg, 7–75. Hillsdale, N.J.: Lawrence Erlbaum Assocates.

131 Costigan, K. 1985. How top executives think: They link their decisions to form multidimensional strategies. *Science Digest* 93(1): 26

132 Hamm, R. 1988. Clinical intuition and clinical analysis: Expertise and the cognitive continuum. In *Professional judgment: A reader in clinical decision making*. Edited by J. Dowie and A. Elstein, 78–105. Cambridge: Cambridge University Press.

133 Gick, M. 1986. Problem-solving strategies. *Educational Psychologist* 27: 99–120.

134 Voss, J., S. Tyler, and L. Yengo. 1983. Individual differences in the solving of social science problems. In *Individual differences in cognition*. Vol. 1. Edited by R. Dillon and R. Schmeck, 205–32. New York: Academic Press.

135 Voss, J., et al. 1983. Problem-solving skill in the social sciences. In *The psychology of learning and motivation*. Vol. 17. Edited by G. Bower, 165–213. New York: Academic Press.

136 Hammond, K. 1986. A theoretically-based review of theory and research in judgment and decision making. Report No. 260. Center for Research on Judgment and Policy, Institute of Cognitive Science, University of Colorado, Boulder, Colorado.

137 Hoyer, W., J. Rybash, and P. Roodin. 1989. Cognitive change as a function of knowledge access. In *Adult development: Comparisons and applications of developmental models*. Edited by M. Commons, et al., 291–305. New York: Praeger.

138 Evans, J. 1989. Some causes of bias in expert opinion. *The Psychologist* 2: 112–14.

139 Allen, P. 1989. Towards a new science of human systems. *International Social Science Journal* XLI: 81–92.

140 Kern, L., H. Mirels, and V. Hinshaw. 1983. Scientists' understanding of propositional logic: an experimental investigation. *Social Studies of Science* 13: 131–46.

141 Collingridge, D., and C. Reeve. 1986. *Science speaks to power: The role of experts in policy making*. London: Francis Pinter.

142 Slovic, P., B. Fischhoff, and S. Lichtenstein. 1982. Facts versus fears: Understanding perceived risk. In *Judgement under uncertainty: Heuristics and biases*. Edited by D. Kahneman, P. Slovic, and A. Tversky, 463–551. Cambridge: Cambridge University Press.

143 Miller, A. 1985. Psychological biases in environmental judgements. *Journal of Environmental Management* 20: 231–43.

144 Black, D. 1985. Sellafield: the nuclear legacy. *New Scientist* 105: 12–13.

145 Miller, A. 1985. Technological thinking: Its impact on environmental management. *Environmental Management* 9: 179–90.

146 Liebhardt, W., ed. 1993. *The dairy debate: Consequences of bovine growth hormone and rotational grazing technologies*. Davis, Calif.: University of California Sustainable Agriculture Research and Education Program.

147 Liebhardt, W. 1992. Hormones, grass and milk. *UC Davis Magazine*, Winter, 12–13.

148 Halprin, L. 1993. New book takes hard look at controversial dairy technology. *Sustainable Agriculture* 5: 4–5.

149 Campbell, D. 1993. The economic and social viability of rural communities: BGH vs. rotational grazing. In *The dairy debate: Consequences of bovine growth hormone and rotational grazing technologies*. Edited by W. Liebhardt, 277–316. Davis, Calif.: University of California Sustainable Agriculture Research and Education Program.

150 Merrifield, J. 1989. *Putting the scientists in their place: Participatory research in environmental and occupational health*. New Market, Tennessee: Highlander Research and Education Center.

151 Goleman, D. 1985. Vital lies, simple truths: The psychology of self-deception. New York: Simon and Schuster.

3

Sociopolitical Dynamics

Environmental problem solving is a collective effort requiring sustained cooperation between many different kinds of people. Sometimes this cooperation is achieved without incident, but more commonly there is conflict and confrontation between protagonists who "face one another in a spirit of exasperation, talking past each other in mutual incomprehension . . . a dialogue of the blind talking to the deaf"[1] (p. 33). As we have seen, one of the main reasons for this maladaptive behavior is the collision of incompatible personality types. However, this is only part of the story because individual behavior is, in turn, controlled by a variety of social pressures that, at times, compel us to behave in ways that are inconsistent with our wishes or personal inclinations. Thus, we are embedded in a nested series of social groups, each of which imposes a set of constraints on our behavior. While these social rules and obligations may constrain our individual agency, they serve to structure our lives in such a way as to make collective action, such as problem solving, possible. The influence of these social institutions on problem solving is, therefore, a crucial feature in understanding the barriers to adaptive change. In what follows, I discuss the effect of sociopolitical factors on problem solving in work groups, organizations, and, finally, within the broader society. Much of the discussion focuses on conflict within each of these levels of social organization, especially power struggles among those who wish to control events. Thus, this chapter deals with the structure and dynamics of problem-solving systems, particularly conflict and cooperation at the intragroup, organizational, and intergroup levels.

Work Groups

The various problem-solving and decision-making activities involved in environmental management are conducted by working groups of one kind or another. Unfortunately, the effectiveness of working groups leaves much to be desired. One reason for this is that professionals, for instance, are inclined to treat group activity as a simple extension of their work as individuals. They assume that, intellectually, they can persist with their disciplinary habits and that, interpersonally, they are capable of working together in a rational, objective way. Neither of these assumptions necessarily holds, for working groups have their own set of demands that require a change in behavior on the part of those involved, particularly a willingness to cope with the more irrational aspects of behavior. Unfortunately, few of us receive any significant training in collaborative problem solving. Instead, we enter the workplace after many years of education that stresses individual accomplishment and personal career advancement in narrowly specialized areas.[2,3] The net result is a persistent ineffectiveness in group problem solving that has long been recognized but has proven difficult to ameliorate (Table 3.1). It is not surprising, therefore, to hear leading researchers[4] (p. 46) conclude:

The few general findings that have emerged from the literature do not encourage the use of groups to perform important tasks. Research has shown, for example, that for many tasks the pooled output of noninteracting individuals is better than that of an interacting group . . . It is tempting . . . to recommend to decision-makers . . . that they use groups as infrequently as possible.

Despite this unfortunate record, there are those who are confident about the usefulness of groups when managed shrewdly.[12,13] For instance, all proponents of participatory democracy assume a crucial role for decision-making groups.[14–16] In any case, we have no choice but to use groups. Given the complexity of environmental problems and the interdisciplinary nature of solutions, collective efforts are essential, regardless of the difficulties involved. Indeed, many corporate[17] and bureaucratic organizations[18–20] have come to similar conclusions. We are faced, therefore, with the task of trying to understand both the psychological and sociopolitical barriers to effective group problem solving as well as the challenge of finding ways in which these might be overcome. The first question is discussed here; the second, in chapters 7 and 8. A convenient way of talking about how groups work (or don't work) is to distinguish between *group structure*, how groups are organized, and *dynamics*, what actually happens in a group.

Group Structure

Group Composition

The kinds of people included in a group will, of course, determine its behavior. Groups that are homogeneous in the personal and professional qualities of their members are likely to behave quite differently from those that are more heterogeneous. The latter are more prone to experiencing conflict but may also

3.1. Intragroup difficulties in group problem solving.

1958. "The interdisciplinary experience often results in certain strains on the individual and brings out a variety of personal defenses, many of which may be expressed in disciplinary terms. A disciplinary viewpoint may be used as a defense, and an individual, when challenged, may retreat to a highly orthodox position within his own field in order to protect his role on the team"[5] (p. 267)

1972. "One of the initial hang-ups not predicted by those enthusiastically involved in developing these large-scale biome programs was the 'personal factor.' In each biome there seems to have been realized, in initial efforts, an unforeseen magnitude of psychological hang-ups of participants towards working together"[6] (p. 129)

1975. "I should not like to minimize the effort needed to have the different disciplines work together. . . . In general, the researchers are either unaware of, indifferent to, or hostile to the work of others in very different disciplines"[7] (p. F16)

1980. "In the forest project, the research board had underestimated the social and psychological factors such as interests, cognitive styles and academic career. . . . The consequences were social problems among the scientists"[8] (p. 234)

1983. "The records of the Ecosystem Group show (that) the will to proceed of the group as a whole was lacking. The remaining records suggest a realization that the Project was not going to be successful. . . . The remaining period was characterized by personal animosities"[9] (p. 283)

1988. "Other scientists, encountering parallel frustrations, had also concluded that we often aggravated problems in resource management by approaching them from the perspective of narrow disciplines. . . . Where I had accumulated knowledge in several fields and had teamed up with other experts, others also formed interdisciplinary teams of various kinds but fared no better. . . Why these teams (my own included) did not work deserves a close look"[10] (p. 29)

1994. "Groups have been subjected to a number of weaknesses and problems that interfere with their effectiveness. One notable group tendency has been labeled 'group think'—a defective decision-making process afflicting highly cohesive and conforming group. One contemporary type of group that appears particularly vulnerable to group think is the self-managing or self-directing team"[11] (p. 929)

contain the seeds of innovative action.[17] This paradox is particularly relevant in light of the demographic and cultural changes occurring in western societies, especially in North America. In decades past, natural resource decision making came under the purview of a very narrow range of people. As Kennedy[21] (p. 165) puts it:

In the 1960s federal resource management agencies in the United States, such as the Soil Conservation Service or Forest Service, were proud, cohesive, and successful organizations . . . They were also professional and gender monocultures with utilitarian conservation ethics, soon to be confronted with the social policies and environmental values of an urban, postindustrial U.S. society.

That is, natural resource agencies have been staffed largely by white men of rural working and lower-middle-class origins.[21,22] The so-called challenge to this arrangement came as part of a broader cultural change that has been taking place in North America in which previously excluded groups, such as women and ethnic minorities, are demanding a more meaningful role, not only in environmental matters but in all aspects of life. In the United States, the federal government has responded with legislation that sought to diversify the natural resource planning process by using interdisciplinary teams and affirmative action policies to diversify the professional workforce.[21]

The effect on disciplinary diversity in environmental management has been significant. In addition to the traditional natural resource areas of agriculture, forestry, fisheries, and wildlife management, one now sees a much more diverse array of environmental specialities drawn from the natural and social sciences, humanities, medicine, and law.[23,24] While the integration of these nontraditional professionals into, for instance, the U.S. Forest Service has been a difficult experience for many of those involved, significant disciplinary diversification has taken place.[21] Similar efforts to encourage greater representation of women and ethnic minorities in natural resource management has had more mixed results, however. The extent to which this is due to resistance from within the traditional professions themselves remains an open question. Women may shy away from careers in natural resources because of male-dominated work environments in which it is possible for them

to experience isolation, male-insensitivity, and sexual harassment that "can overwhelm and exhaust the individual, especially an isolated woman"[25] (p. 86). Whatever the reasons, the fact remains that only 10 percent of members of the Society of American Foresters are women,[26] roughly the same proportion as that found in the Canadian province of British Columbia. Across all traditional resource fields, women typically "hold less than 6% of upper-level management positions, and approximately 21% of professional positions" in the same province[25] (p. 85). On the other hand, the role of women in environmental management is not limited to the professional sphere, as they appear to have a significant impact on decision making as members of nongovernmental organizations and other grassroots activities.[25]

Ethnic minorities have not, until recent years, had any significant impact on mainstream environmental management. Their presence in the professional ranks is minimal, although there are efforts underway in the United States and elsewhere to redress the racial and ethnic imbalance.[27] At the grassroots, however, one does see indigenous people seeking to regain control of their ancestral lands, an activity that has significant consequences for natural resource policy.[28,29] In addition, there is a growing belief among the poor and minorities that they suffer disproportionately from the environmental risks posed by toxic dumps and industrial effluents that are frequently located in their neighborhoods. The result has been grassroots activism in search of "environmental justice."[30,31] Thus, formerly marginalized sectors of the community are seeking a more prominent role in environmental decision making.

The growing heterogeneity of stakeholders in environmental management has resulted in an increase in both the intellectual and interpersonal complexity of decision making. This complexity is least evident in professional teams where technical experts remain in the majority[18] but is hard to ignore in public participation exercises. When we add to this complex brew the differences (and conflicts) generated by the personality types described in chapter 2, we begin to understand why deci-sion-making groups have such a checkered history. The mere presence of heterogeneity in working groups, however, does not necessarily imply that some political revolution is under-way in environmental management. Environ-mental decision making is still very much in the hands of the elites who have dominated environmental management for so long. For instance, technically trained professionals, steeped in positivist epistemology, are much in evidence, whereas the social sciences and hu-manities are usually allocated a token role.[32] Meanwhile, environmental and other interest groups often complain of being marginalized and ignored during deliberations. Schnaiberg and Gould, for instance, argue that the com-position and administration of working groups is manipulated by powerful political-corporate interests to make it difficult for those with dis-senting opinions to participate meaningfully.[33] In a similar vein, one disillusioned participant in a policy workshop concluded[34] (p. 14)

Resolving public conflict over forest management policies does not seem to be a matter of finding bet-ter policies; it seems to to be a matter of quieting the voice of opposition. In workshop sessions, make sure they're in the minority (its important that they always feel that way). . . . use the majority to sand-bag them when necessary. Blunt their attack, wear them down. Most of all, try to keep them talking in forums like this one—away from the public eye.

Thus, while the composition of groups cer-tainly implies an increasingly difficult task of dealing with the complexity being generated, it is the way in which this complexity is han-dled that creates so many barriers to adaptive problem solving. In turn, this depends very much on how groups are organized.

Group Organization

In its initial deliberations, a group may abide by its formal organization, but this doesn't last long. That is, a leader may have been selected and team members assigned to specific roles, so that the relative power of each person in the group has already been decided at the be-ginning. In most instances, however, an infor-mal restructuring process takes place as group members seek to exert some influence on the proceedings. This informal reorganization commonly supersedes whatever structure ex-isted on paper and reflects, in part, the political context within which the group operates. Thus, power relations in the group often echo the distribution of power in the broader soci-ety.[35] Particularly troublesome is the develop-ment of status differentials or pecking orders within groups.[35,36] Those who represent pow-erful outside interests may assume a dominant role in the group, often because of their access to resources, while the politically and eco-nomically disadvantaged, such as women and minorities, occupy marginal, low-status roles in the proceedings. This is especially unfortu-nate when low-status members have infor-mation that is crucial to the success of the group but are simply not given a hearing.[36] Of course, in specific cases, a powerful personality can override these more general socioeco-nomic effects, and ostensibly low-status indi-viduals can become major players in group decision making; for example, there are char-ismatic popular leaders such as Chico Mendez, deceased spokesman for the Brazilian rubber-tappers, and Ovid Mercredi, Grand Chief of the First Nations, who has played a major role in negotiating aboriginal land claims in Canada.

The pecking orders that arise due to discipli-nary chauvinism are also troubling in interdis-ciplinary teams. The predominance of technical professionals in environmental management reinforces the assumption that it is the so-called hard sciences that have most to contrib-ute to problem solving, while the so-called soft disciplines, such as social science and the hu-manities, are afforded a secondary role, one that supports the main, scientific thrust.[37] Thus, it is not unexpected to find a pecking order in the minds of hard scientists, one that places physical scientists at the top and social scientists at the bottom. One result of this dis-ciplinary chauvinism is a studied disinterest by natural scientists in what social science has to offer. A case in point is my experience as the token social scientist on a technical assessment team that was conducting an environmental impact assessment on a proposed nuclear power station. During my presentation of some of my reservations about the way in which the soft (psychosocial) side of the assessment was being conducted, one of the nuclear engineers on the panel simply went to sleep. His snores, while discreet, were a source of great amusement among some of the panel members who, in psychological terms, used the snoring as an opportunity to express their irritation, albeit covertly, at having to listen to my drivel about the subordination of the soft side of the project.

People handle these disciplinary antagonisms and other status differentials in different ways, seeking some role in the group that will allow them to contribute to the proceedings. In doing so, characteristic roles emerge in groups, each serving a different function. One important research conclusion is that effective groups are successful, in part, because they tolerate a diverse array of roles within their group structure. Groups that fail to do so by suppressing some of the more "troubling" roles are likely to develop the pathologies that commonly afflict group problem solving.[38] Two of the most difficult roles for groups to tolerate are those of the critic and the visionary. The hard-nosed, objective-analytic critic, whose incisive criticisms of the woolly minded nonsense that often afflicts group problem solving, tends not to generate feelings of warmth and complacency. In addition, the lateral thinker (objective-holistic, subjective-holistic) seems, so often, to be operating on such a totally different plane of being that he or she is frequently seen as an irritation, a hindrance to progress. Yet both of these discordant roles are crucial to the well-being of groups, the effectiveness of which depends on how well they are incorporated into group activities.[38]

The point is that working groups are cauldrons within which participants struggle to cope with the personal, social, and political conflicts that inevitably arise. Adaptive problem solving requires that these psychosocial tensions be recognized and that effective steps are taken to deal with them. My experience is, however, that environmental professionals are not equipped either by training or inclination to do so. Although psychosocial conflicts of this kind are ubiquitous in all work situations,

they are seldom discussed or even recognized in the environmental management literature. There seems to be a certain distaste for the topic, as if its mere existence reflects badly on the professional credibility of scientists, managers, and policymakers. It follows that professional education ignores the problems raised by personality and other psychological conflicts, as do the most commonly used decision-making and problem-solving techniques used by environmental professionals. The customary way of dealing with conflicts in work groups is to ignore them in the fond hope that they will go away or resolve themselves in some unspecified way. They don't, of course, but persist as background interference that reduces the effectiveness of group problem solving.[37]

Attempts to avoid personality conflicts by, for instance, including only compatible individuals in work groups may have the paradoxical effect of contributing to collective myopia or groupthink.[39] An alternative is to try to manage personality conflicts so that they play a useful role in environmental problem solving. Obviously, too much destructive conflict can hinder the problem solving process,[40] but, on the other hand, a degree of conflict stimulates problem solving by challenging those involved to reexamine their stance on the matter at hand. The trick, though extremely difficult, is to find the appropriate balance so that conflict can be used constructively.[41] Any efforts in this direction requires us to understand why conflicts arise and what can be done to promote a more constructive interaction between different personality types in problem-solving situations.

Group Processes

The dynamics of working groups are commonly described in terms of two interrelated processes: the way in which the group approaches the task at hand and the social or interpersonal climate that develops within the group. Thus, the *task* dimension has to do with the problem-solving activity of the group and focuses on the substantive issues being addressed. On the other hand, the *social* dimension refers to the interpersonal relationships that develop in the group, particularly the informal rules and roles that govern interactions.

Task Dimension

Attempts to work through the task at hand is influenced by the personality characteristics of each member, as well as by social pressures both from within and from without the group. For instance, disruptive conflicts between those differing in cognitive style are commonplace. Thus, at the beginning of a recent interdisciplinary project, one of the members, a natural scientist, pressed for a speedy and precise definition of the problem so that we could move on quickly to the important task of collecting data. He became very impatient when it was suggested by others that we didn't know what the problem was and that we would consider the research a success if, at the end of the project, we had managed to develop a better understanding of it. Of the many differences contributing to this interchange, the most striking was the contrast between a scientific preference for *converging* on an unequivocal definition of the problem at an early stage in the proceedings and the more *divergent* preference for playing with the problem in the hope that a structure might emerge in some unspecified way. The resulting tension between these disparate cognitive styles was not resolved but continued to resurface throughout the planning stage of the project. This same tension is also evident in current debates over the merits of Euclidean and non-Euclidean planning methods.[42,43]

In a similar vein, the conflict between empirical (analytic) and systems (holistic) scientists over appropriate research strategies afflicts many projects. In the Swedish Forest Ecology Project (SWECON), a 10-year interdisciplinary project involving up to 100 scientists, empirical and systems or theoretical biologists experienced persistent problems in attaining meaningful collaboration.[44] The theoretical biologists were primarily interested in developing quantitative models of the forest ecosystem, both as a summary of current knowledge and as a guide for further research. From the empirical biologists' viewpoint, however, this was an exercise

in premature speculation, one that should have awaited the collection of more data, the latter being the primary role of the empiricists. At an early stage in the planning of this project, it had been assumed that the various scientists involved would be able to reconcile their differing cognitive modes, but, unfortunately, this did not take place. "This led to few cases of cooperation during the project," and what did occur consisting mostly of isolated exchanges of data[44] (p. 234). Like so many multidisciplinary projects, before and since, these conflicts resulted in considerable difficulty both in coordinating ongoing research and in integrating research findings at the end of the project. Thus, cognitive style conflicts can be unfortunate if they remain unproductive because there is ample evidence that, when dealing with ill-structured problems, multiple perspectives are needed if problem solving is to be effective.[45,46]

As groups experience these early tribulations, intellectual conflicts may be resolved by dominant individuals or coalitions imposing their views on the proceedings.[47] More than likely, it is the objective personality types who assume this position of dominance, simply brushing aside the confusion by structuring group activities to their liking. This objective dominance can be imposed by force of will or, more insidiously, through disinterest in, and apathy toward, more subjective options. For instance, in a recent workshop in which I was involved, the "social" work group was required to identify and quantify the social component of the model being developed. However, any attempt to discuss "philosophical" (value) issues was resisted by the technical (objective) people present by the simple tactic of declining to engage in discussion. Only when more practical issues arose, such as calculation of pesticide drift, did they re-engage, at which point fingers flashed on pocket calculators and the atmosphere became quite animated. Thus, the choice of topics for discussion was controlled by noncompliance.

Personality differences between group members are accentuated, in professional teams and committees, by disciplinary affiliations. As mentioned in chapter 2, there is a link between personality and discipline in that the more objective-analytic personality types gravitate toward science and engineering while subjective-holistic types are more commonly found in the liberal arts.[48] Subsequently, personality tendencies are reinforced through selective education and professional socialization. The net result is disciplinary chauvinism and endless disciplinary rivalry that disrupts group efforts.[49] Part of the problem is the narrow perspective encouraged by professional training.[18] One can see this in the disparate conceptions of the same problem developed from different disciplinary perspectives. Barker,[50] for instance, asked a variety of budding professionals to elaborate their conception of air pollution. While some student groups were open to interdisciplinary cooperation, others were not. Prominent among the latter were students of economics and engineering, two disciplines that play an important role in resolving environmental problems:

Students in economics were . . . extremely inflexible . . . unwilling or unable to broaden from an economic approach to a multidisciplinary perspective. They seemed to be constrained by a rigidly defined framework of economic concepts. Economic solutions were considered to the exclusion of any others . . . the students in this group were reluctant to consider any information which lay beyond these bounds and (were) hostile to other professional contributions. . . . Engineering students . . . saw their role as . . . the technical expert who is consulted after others have decided on a course of action . . . Many engineering students were intolerant of multidisciplinary efforts, even to the extent of other specialists providing input into the design stage (pp. 369–370).

It is aggravating to realize how well these various disciplinary perspectives could contribute to a more complete problem conception if it were not for the presence of mutual hostility between at least some disciplinary groups.[18,51] Brown[52] (p. 56) makes the same point in explaining how anthropologists could make a substantial contribution to projects dominated by the hard disciplines such as engineering— were they given the opportunity:

What anthropology could add to this (engineering) perspective, from its position of holism and reflectivity, is a critical reflective consciousness and an awareness of the richness of cultural diversity within our own society. An awareness of cultural diversity represented by other ways to meaning-construction would be a signal service to a discipline currently blinkered by an aging dogma of positivist empiricism.

Disciplinary chauvinism is held in place, however, not only by personality and training but also by the professional reward systems within which we all labor. The difficulty in obtaining professional cooperation, therefore, may not be so much a reflection of individual intransigence but, rather, the way in which rewards are organized. For academics, participation in interdisciplinary groups may not be recognized by one's disciplinary peers so that *productivity panic,* the genuine fear that valuable time is being wasted, may push academics in research groups toward seeking some disciplinary payoff.[7]

There are various ways of dealing with these intellectual disputes in working groups, some more peremptory than others. For instance, Erickson[53] recommends the surgical removal of what he calls "high grade morons," those individuals who "hold the view that what they know they know very well indeed and what they do not know is irrelevant." However, a less radical approach requires that we understand something about the more social dimension of group problem solving.

Social Dimension

Although groups can at times deal constructively with conflict, there is a tendency in many cases to let intellectual and interpersonal conflict persist below the surface, to the detriment of the problem-solving process. This flight from conflict is commonplace in all manner of groups. For instance, Hirschhorn[54] describes how self-management teams can fail when team members as a whole, fearing the consequences of interpersonal conflict, refuse to confront the damaging behavior of one or two individuals. Similarly, Putnam[55] (p. 175) argues:

Perhaps no aspect of group decision-making inspires greater simultaneous attraction and avoidance than conflict. On the one hand, researchers acknowledge that conflict in group decision-making is not only inevitable but also beneficial to the group process. Disagreements lead to the reexamination of opinions, the sharing of diverse ideas, and the discovery of creative solutions . . . In fact, researchers concur that the presence of ideational conflict during group deliberations produces higher quality decisions than does the absence of controversy . . . On the other hand, controversy in groups often leads to hard feelings, personal apprehensions, and unpleasant experiences. To avoid strained interpersonal relationships, members often foster harmony by taking flight from controversial topics, placating or smoothing over difficult issues, or compromising to get disputes settled in a hurry.

Therefore, informal norms or rules often develop in working groups, supported at times by the group's leader, to avoid the seamier side of professional behavior.

While there are numerous reasons for the prevalence of conflict avoidance in professional working groups, one can trace the behavior back to underlying personality factors. In chapter 2, reference was made to a difference between field-independent and field-dependent individuals. The former have the capacity to restructure what they are seeing, ignore distracting stimulation, and maintain highly focused attention, whereas the latter do not seem to impose these constraints on their behavior to the same degree. It was also mentioned that scientists and technically trained professionals are more likely to be field independent, whereas those in some of the social sciences and humanities disciplines are more field dependent. Individuals who are field independent and dependent respond to group activity differently. Where the group's task is well structured, or capable of being structured, field-independent individuals (e.g., scientists and engineers) prosper. They are efficient at resolving the task at hand and appear to be comfortable with the proceedings.[56] In addition, under such conditions, they can be more receptive to information that discredits their viewpoint than are field-dependent individuals.[57] However, in less-structured situations, especially those that involve emotional conflict, their behavior changes. They begin to show signs of discomfort and may withdraw

psychologically from the proceedings by such behaviors as noncompliance with suggestions and other forms of psychological distancing. In contrast, field-dependent individuals seem to be able to tolerate the social-emotional confusion of unstructured situations; are more sensitive to the emotional climate; and seem willing to sustain a responsive, task-oriented demeanor.[58,59] Thus, scientific professionals may be predisposed to the avoidance of conflict in working groups, a predisposition that would be accentuated by their educational experiences. While scientific and technical training encourages intellectual debate, the latter is conducted, at least in most cases, in a formal, highly stylized manner. It is not the kind of training that prepares one for the emotional rough and tumble of working groups. Unfortunately, the consequence of conflict avoidance is groupthink, the descent into unwarranted complacency and narrowness of vision,[11] an unfortunate condition that some researchers have identified in natural resource agencies.[60]

Working groups are the workhorse of environmental problem solving. The fact that they are beset with so many problems is unfortunate to say the least. Overcoming such problems requires that participants become more skilled in group work. Unfortunately, neither the intellectual nor the interpersonal training needed for such a transformation is a part of traditional educational curricula, a deficit that continues to handicap effective collaboration.

Organizations

Problem-solving groups and teams usually operate under an organizational umbrella, an arrangement that imposes additional layers of social control on an individual's behavior. Not only are we constrained by the rules and norms of our working group but also by the demands for social conformity all organizations require from their members. However, the sociopolitical pressures we experience inside organizations are not imposed by some abstract entity called an *organization*, but rather by powerful groups within it. Thus, organizations are composed of competing interests in a state of perpetual conflict.[47] "Coalitions struggle for the control of resources and jockey for power. Division heads fight the corporate center for control over budgets; scientists don't want line managers to oversee their work . . .," and so on[61] (p. xiv). Organizational politics is, therefore, a fact of life for which few of us are prepared by formal education. While skills in maneuvering through the minefield of internal politics are crucial elements in adaptive problem solving,[62] the danger for any organization is that politics and self-protection become more pressing than dealing with environmental problems in an adaptive way.[63,64] As individuals and groups scramble to protect themselves from threats to their career or positions of influence, fear and rigidity may become so pervasive that the institution is incapable of the flexibility essential for coping with the major environmental (or other) problems facing it. Holling and Meffe[65] conclude, for instance, that this kind of rigidity is widespread in natural resource management where command and control strategies are causing pathological responses to environmental problems. In what follows, I outline the way in which organizational politics serve to distort the problem-solving process.

Organizational Politics

The structure of an organization is, in part, a result of its internal politics. For instance, the basic architecture of the U.S. Forest Service (its programs and policies, administrative units, budget system, professional staff, and reward systems) was put into place in an earlier era when it enjoyed more unequivocal public approval than it experiences today.[66] It was also a time when internal disputes over agency policies were more muted and political struggles within the agency less intense than they are now. According to Clarke and McCool,[67] the U.S. Forest Service, as well as other agencies such as the Army Corp of Engineers, was politically successful because it developed a clearly recognizable organizational culture founded on the values and competences of a

dominant professional group. In addition, it was able to draw on, and make itself useful to, a well-defined economic clientele. Thus, the Forest Service had adapted successfully to its sociopolitical context by assuming the role of a commodity-based organization that catered to the timber needs of society. In contrast, agencies with an interdisciplinary base and no specific economic constituency, such as the National Park Service, are less politically powerful and have experienced more internal discord.[67] However, increasing professional diversification is diluting the influence of the dominant profession in resource agencies, and this, together with their attempts to cope with changing social values, is creating internal conflict. Thus Brunson and Kennedy[68] (p. 156) observe:

Older practitioners of traditional specialties face a crisis in confidence as younger colleagues call for change and the agencies institute programs with names like 'New Perspectives' or 'Change on the Range,' implying that the old perspectives (and perhaps they personally) badly needed changing. Meanwhile, affirmative action programs have fostered resentment of women and minorities . . . further souring intra-agency relations.

For those who identify with, and have a vested interest in, older agency policies and viewpoints, current developments may be perceived as a threat to their well-being. After all, who wants to be told that their cherished beliefs are no longer valid and that the programs and policies on which they have been working for 30 years or so have been so economically and ecologically flawed that they must be replaced by some new system of management, a shift in emphasis for which they may not be professionally prepared. One reaction to such threats to self-esteem is to seek reaffirmation of their psychological and political position in the organization. For those objective personalities who subscribe to the Imperial mode, the most obvious route is to engage in *power seeking*, to gather unto themselves the reins of power within the organization so that they can control the agency's agenda. Hence, we find pressure toward greater centralization, hierarchy, and control that exists in many organizations. This sociopolitical pressure, which seeks to counteract and slow down change, is an important way of maintaining stability in organizational functioning; otherwise, there would be chaos. It is when this pressure takes the form of a social defense against needed change that one should become concerned.

Psychological defenses are used by individuals to cope with the anxiety produced by a complex and uncertain world. In seeking to protect our rather fragile egos, however, we do not stop at the individual level but join with one another in the development of *social* defenses.[54] These might be seen as collective efforts to stave off feelings of anxiety associated with (social) impotence and meaninglessness. For instance, one would like to think that the products of one's work are useful both to the organization and to the larger community. When this is reaffirmed regularly by those around you and by tangible rewards such as promotion, then such social confirmation is reassuring. Organizational life is replete with such social defenses as we seek to reassure one another that we are engaged in a worthwhile endeavor.[54] However, the role of social defenses is paradoxical. On the one hand, they serve to make worklife bearable and meaningful, thereby reducing the fear and anxiety that pervades most, if not all, institutions.[54] On the other hand, they create the conditions for delusional thinking, witch-hunts, purification rituals, and other irrational behaviors more associated with medieval times than with modern, scientific organizations.

Consequently, all organizations face a cruel paradox.[69] To engage in effective action, members of an organization have to be highly motivated in pursuing a particular goal. This is only possible when there is a sense of mission based on a set of shared assumptions or organizational ideology. The latter provides a basis for the social and intellectual organization of the institution, establishing the framework within which acceptable goals and behaviors are defined.[63] It can also be seen as a complex illusion, a so-called legitimating myth that is important in maintaining morale within the organization and its credibility in the eyes of the public, which ensures continued access to public funding.[64] It is struggles over the control of this myth that creates the most intense conflict in organizations.

While a strong ideological core within an agency may facilitate choices, it also tends to distort organizational thinking along lines congenial to the dominating group. Only those ideas that are consistent with the dominant group's ideology are welcomed, while divergent views are ignored or suppressed. For example, although natural resource agencies see themselves as managers of natural resource policies under legislative mandate, the programs that are actually put into place are screened through the agency's (i.e., dominant group's) ideology. Thus, "agency program managers, bureaucrats and administrators, team leaders, front-line biologists, technicians, and others . . . translate the original policy aims through their existing programs, modes of operation, and technical capabilities. In doing so, they make numerous professional and organizational interpretations, which are simultaneously both technical and political decisions. Thus the program's outcomes may diverge significantly from the original policy aims"[62] (p. 501). In other words, broad policy directives are filtered through and sometimes subverted by the dominant organizational ideology. Unfortunately, these organizational biases can reduce the organization's capacity to adapt to a changing world. The paradox, then, is between the need for sufficient motivation and commitment to overcome the many barriers to action in a difficult world and the rigidity introduced into organizational behavior by these very same motivations.

The most fundamental conflict in organizations is between proponents of the Imperial and Arcadian ideologies, each of whom seeks to impress their particular way of thinking on organizational operations. Because no organization is ideologically homogeneous, there is always an element of dissent against the dominant value system. In natural resource agencies, where the Imperial mode predominates, dissent usually takes an Arcadian form—proponents of which are spread ubiquitously throughout government agencies.[70] Dissent can range from relatively minor differences of opinion within the limits laid down by the dominant ideology (deviance), to the questioning of the very ideology itself (heresy).[71] Needless to say, of the two forms of dissent, it is heresy that is met with the most bitter disapproval in all organizations. Dissent, of course, is not limited to bureaucratic organizations but is ubiquitous in all collective activity. For instance, the internal political disputes that have troubled environmental groups such as Earth First!, Greenpeace, and the Sierra Club all involve divergent interpretations of the dominant organizational ideology that, in these instances, is more Arcadian in nature.[72–75]

According to Brown and Harris,[76] the dominant Imperial tradition of resource management in the U.S. Forest Service (Table 3.2) is most evident among senior staff with decision-making responsibility, such as *line officers* (district rangers, forest supervisors, regional foresters, chiefs, and deputy chiefs) and headquarters staff. The beliefs and ideals of this group derive from an earlier period in the Forest Service's history, mentioned previously, which produced a selective focus on anthropocentric values and an organizational culture that "reinforced in its employees a self-image of stewardship, of objective and scientific professionals assigned to guard the public forests. To such a proud professional monoculture, the interdisciplinary and public power-sharing imposed by (legislation) represented a direct rebuke"[68] (pp. 151–152). This Imperial tradition is being challenged by proponents of an alternative management paradigm, one that has Arcadian elements but stops well short of the radical views held by radical environmental groups. In a sense, the two management paradigms might be seen as hard and soft versions of the Imperial mode (Table 3.2). The alternative mode is deviant, therefore, rather than heretical.

The alternative resource management paradigm is being promoted by, among others, the Association of Forest Service Employees for Environmental Ethics (AFSEEE), an interest group within the U.S. Forest Service that sees itself as a "vigilant watchdog over the successes and failures" of agency policy. Composed of junior staff (including a high percentage of women professionals) and concerned citizens, it aims to "forge a socially responsible value system for the Forest Service based on a land ethic which ensures ecologically and economically sustainable resource management" (AFSEEE internet home page). The land ethic referred to is that of Aldo Leopold, a pioneer conservationist with whom this group identifies. The origins of AFSEEE appear to lie in the frustrations of junior staff with their experience of professional life within the U.S. Forest Service, graphically depicted by the Association's founder Jeff DeBonis, a timber manager[77] (p. 163):

Shock was my initial reaction to this forest's timber management; shock at the rate and ferocity of cutting, at the steepness of the cutting units, the resulting erosion and slope failures, and at the environmental damage so nonchalantly dismissed by my superiors.

TABLE 3.2. Contrasting Forest Service resource management paradigms.

New resource management paradigm	Dominant resource management paradigm
Amenity outputs have primary importance	Amenities are coincident to commodity production
Nature for its own sake	Nature to produce goods and services
Environmental protection over commodity outputs	Commodity outputs over environmental protection
Primary concerns for current and future generations	Primary concern for current generation
Less-intense forest management, such as "new foresty," less herbicides and slash burning	Intensive forest management—clear-cutting, herbicides, and slash burning
Limits to resource growth: emphasis on conservation for long-term sustainability	No resource shortages: emphasis on short-term production and consumption
Consultive and participatory decision making	Decision making by experts
Decentralized decision authority	Centralized decision

Frustrated by what he saw as the agency's "timber-dominated mindset," he struggled to deal with his own ethical dilemmas. "As the negative impacts of agency actions became more and more obvious, employees tried to pretend it was not happening. And yet, at some subconscious level, we knew that we were overcutting" (p. 165). Finally moved to drastic action, he founded the AFSEEE in 1989 to pursue public education, monitoring of Forest Service activity, and to help whistle-blowers or *clarion callers*, as they are now labeled. Subsequently, in 1993, DeBonis formed a broader dissenting group, the Public Employees for Environmental Responsibility (PEER), to extend the promotion of the alternative paradigm throughout U.S. public agencies.

Conflict within organizations is not limited to those who wish to change the organization's ideology. Even among those who subscribe to the dominant ideology, there is conflict over preferred forms of thinking and decision making. One can see this most clearly in the tension between agency scientists and managers. Scientific professionals prefer to see the fruits of their work used in rational decision-making rather than in the more political manner in which they believe managers often operate.[78] A case in point is the tension between the headquarters of the Environmental Protection Agency and its outlying, field offices. Villanueva[79] has found that some officers in the latter, who see themselves as more scientific in nature, chafe at what they consider to be the political orientation of headquarters. Similar disputes over the emphasis to be placed on science can be seen in resistance to the incursion of rational-scientific logic into decision making. Anderson,[80] for instance, describes the way in which an attempt by clinical scientists and computer experts to introduce computer diagnosis into a teaching hospital was opposed by physicians because it would reduce their control over a central feature of their work. Replacing their professional judgment with a computer not only threatened their self-esteem but degraded the craft knowledge they had worked so hard to develop.

Response of Dominant Coalitions

Not all organizations respond in the same way to dissent. Much depends on how heavy-handed the dominant coalition chooses to be. Responses range from a degree of tolerance to more harsh measures aimed at suppressing the deviant or heretic.[81] Strong reactions are most commonly observed when the ideology of the dominant group, who seek to monopolize the interpretation of the agency's legitimating myth, is under attack from a credible source in the organization.[64] For instance, opinion leaders in a Department of Natural Resources might wish to maintain the myth that it has a monopoly of technical expertise and knowledge about environmental management. To do so, it must discredit opinions from unqualified sources that conflict with departmental policy, select recruits who are willing to sustain the myth, and stifle dissent from within.[63] This organizational protection is conducted primarily by *standard bearers* and *dominant coalitions* drawn from upper management.[63,64] The bureaucrats who occupy these positions inhabit a stressful world to say the least. According to Williams et al.,[82] their professional life is one of maintaining outward cordiality while studiously pursuing the hidden games of self-protection and power seeking. Powerful men can hold on to power only with the help of loyal support, especially during periods of crisis. Hence, it is "little wonder that 'whistleblowers' pose such a threat to bureaucracy. Whatever their motives . . . they may be in a position to expose the inner workings of the organization which typically are protected from public scrutiny. If they are able to obtain support from the press or legislative bodies, whistleblowers can readily undermine certain high-level personnel as well as individuals loyal to them"[82] (p. 398). Thus, whistle-blowing is a form of bureaucratic opposition, an attempt to change an organization by those with little power or authority.[81] However, organizational protection is not restricted to the upper levels of management for everyone has a " personal stake in preserving the complicated fantasy of the organization, even though conditions in the organization are in fact unsatisfying to all but a few elite members"[83] (p. 383). Thus, dissent is seen as a threat when it promises to disrupt the complicated web within which professionals operate and with which they protect themselves. A major social defense is to take steps to reaffirm ones beliefs and shared illusions.[54] This can take many forms, such as trying to persuade doubters that one's assumptions are valid and useful, using unobtrusive control to shape perspectives and, when all else fails, the open suppression of dissent.

Ritual and Public Relations

Control of dissent and disaffection at the grass-roots level involves public relations exercises aimed both internally at organizational members and externally at public criticism. All skilled organizational politicians know that the majority of employees crave stability, meaningfulness, and a sense of moral rectitude. Powerful groups maintain their grip on organizations to the extent that they can instill these feelings in subordinates. Kennedy[60] refers to this as reinforcing the "illusion of morality." Upper management may engage, therefore, in bolstering the morale among subordinates by fostering an air of change and adaptation. This involves reformulating organizational goals in light of changing social norms and, for instance, renaming organizational units to divert attention away from discredited activity.[64] The new goals, of course, need not be more adaptive, they may simply afford a period of grace in which the institution may rebuild its credibility and morale. Similarly, public relations for external consumption serve to present, and reinforce, an air of reasonableness. Thus, organizations may deny harm from, and disclaim any responsibility for, the unfortunate consequences of institutional policies. In turn, they present new institutional policies, the intent of which is to create the appearance of bureaucratic efficiency and rationality and to reaffirm the need to leave matters in the hands of those best qualified to

deal with them.[64] Thus, public relations are meant to reassure while avoiding the trauma of fundamental change. The extent to which this is true of recent developments in natural resource agencies is a matter of opinion. Certainly, there has been widespread espousal of the need for change by upper-level agency personnel.[68,84] In the U.S. Forest Service, for example, the "New Perspectives" initiative is an overt attempt at critical self-examination, one that has resulted in the establishment of a set of principles for encouraging more adaptive forest management through, among other measures, the adoption of ecosystem management as the central operational agenda. Kessler and Salwasser[66] report that while one reaction to the New Perspectives initiative was to dismiss the whole thing as so much "smoke and mirrors," a final evaluation of the exercise must await the passage of time. Thus, it remains to be seen whether it will result in substantive change.

Rituals are solemn ceremonies and customary acts, the purpose of which is to reaffirm one's social identity, one's place in life. It is comforting to re-enact, once again, time-worn rites. In the workplace, while we may find daily routines boring at times, at another level they provide reassurance and confirmation of purpose. Whether it be the usual round of committee meetings, project evaluations, planning sessions, data collection, or whatever, all of this provides structure and, hopefully, meaning to one's daily life. According to Hilborn,[85] "every (fisheries) agency I know has its rituals: certain types of size and area regulations, specific data collection methods, definite forms of decision-making . . . few of their decisions have been replicated or controlled, and therefore they simply do not know if their successes are the result of their actions; they have adopted some strange modes of management as a result" (p. 9).

These activities become excessively ritualistic, however, when they serve to allay psychological fears rather than being directed at environmental problems. Under such circumstances, organizational life is, as Shakespeare reminds us, a tale told by an idiot, full of sound and fury and signifying nothing. Perhaps it is too cynical to suggest that, at times, it is to the benefit of ruling elites to have their subjects engage in rituals and time-consuming work, for it keeps the masses out of mischief and settles discordant emotions. My personal reaction to the enthusiastic data collection exercises, the esoteric modeling, and the obsessive planning engaged in by many government agencies is that they are time-wasting rituals encouraged by the dominant elites to submerge otherwise intelligent, potentially troublesome, professionals in ostensibly important, but actually trivial, activity. This could not be achieved, of course, if it were not for willing participation among the troops, something Springer[86] was astonished to find in the organizations he studied.

Shaping by Unobtrusive Control

While rituals and public relations operate by spreading a comforting miasma, more focused control methods are also used to contain dissent. Direct control and the open use of arbitrary power are still ubiquitous but tend to be less favored than more indirect, unobtrusive methods.[87] When the latter operate effectively, "members share the belief in the organizational mission and share common assumptions regarding organizational reality, and they can be counted on to behave in the best interest of the organization"[87] (p. 256). Unobtrusive control is maintained by peer pressure and managers who interpret events in line with organizational goals and ideology. In other words, there is constant pressure to think in a particular way, to be a team player. Mattson[78] is of the opinion that:

Virtually all scientists who work in natural resource agencies are exposed to management culture (which) values cooperation as well as loyalty and obedience to the 'group' and to agency leaders, and correspondingly emphasizes the importance of being a 'team player' . . . Most natural resource agencies, in fact, use these symbols (e.g. 'the good soldier' or 'family member'), along with selective hiring, advancement, and other enculturation, to promote homogeneity of outlook and purpose . . . Scientists who defy agency superiors in the name of ethical science not only risk disapproval and cen-

sure by 'the group,' but also, under more extreme circumstances, may put their paycheck, career, and research opportunities on the line.

In a similar vein, DeBonis[77] remembers that, during his years as a team leader in a natural resource agency, he believed that his duty was to "pressure unbelievers into submission, if not conversion." More specifically:

At one particular interdisciplinary team meeting, Ernie was being pressured to back down on his analysis of grizzly bear habitat needs that conflicted with timber cutting units proposed. Ernie was correct in his assessment—ecologically, morally, and professionally. I, on the other hand, was part of an effort to discredit and intimidate him into joining the 'team,' in order to proceed with the timber cutting at any cost. At that point, I realized I had become part of the agency's institutionalized culture without even knowing it (pp.160–161).

Bullis[87] has also observed several examples of what she believes to be unobtrusive control in the U.S. Forest Service. For instance, in one informational meeting a team member was welcomed back from a continuing education session with the announcement by the team leader that it would make the worker a better team member. Because everyone present already knew about the return, Bullis interprets the intent of this comment as one of underscoring the importance of being a team player, a quality held at a premium in an organization that is known for its integrity in public service. On another occasion, after a presentation by a representative of an outside special interest arguing for the protection of cultural resources during forestry operations, the supervisor offered an ideologically appropriate interpretation of what they had just heard:

We must be realists in looking at current economic conditions which will come first. We need to look at the significance of all the laws and do a quality job in all of them and it will take innovation and working relationships together. We must look at special interest groups. All want their two cents worth in management. The more we work together the better off we are in the long run. In confrontation we all lose. (p. 263)

Bullis suggests that the supervisor is taking the opportunity to guide his team toward interpreting the idea of multiple-use forestry (the dominant organizational premise at the time of this study) as one in which "economics comes before cultural resources." Other interests must take their place behind this overriding value and be evaluated in the context of competing values. The supervisor has thus managed to position team members as arbiters of competing interests trying to work for the common good. On another occasion, toward the end of a meeting during which a great deal of confusion was expressed about a new program, the team leader summarized the discussion by saying: "I sense from your comments that people think this is a good, valid program" (p. 265). These observations sought to resolve any doubt that was present and to interpret the new program in a positive light, although, in this case, the effort was not entirely successful. Bullis concludes that what is happening in many instances of unobtrusive control is an attempt to encourage thinking that conforms with agency goals, something she also sees in *participative decision making*[87] (p. 266):

Participative decision making provides an additional opportunity for members' active indoctrination . . . The forest supervisor decides to make a structural change by reassigning and relocating one work group . . . He then charges one of his staff members with developing a set of alternatives for implementing the change. The management team is brought into the decision making after this set is developed and the preferred alternative specified. The team is charged with making the decision whether to accept, modify or reject the alternative. During the meeting, the plan is presented to the team, after which the supervisor praises the report, comments on the need for generating community acceptance of the change, and calls for a round robin vote. Each team member compliments the staff officer on the plan, notes a concern with community acceptance, and registers an affirmative vote.

Bullis concludes that such meetings have less to do with decision making than with providing an opportunity for the team "leader to monitor the compliance of team members." Any disagreement with decisions that had been made prior to the meeting would have provided the leader with "the opportunity to

help the deviant reconsider." Thus, the meeting was a way of enlisting support for the planned action and a way of reinforcing acceptance of "the authority of the line officer to define decision-making situations" (p. 266).

The Suppression of Dissent

Unobtrusive methods of control work best when the dissenter holds deviant rather than heretical views. When there is some common ground between the deviant and the orthodox, and appeals can be made to organizational loyalty, then control methods can remain relatively benign. However, when dissenting views are more heretical and threaten both organizational policy and those identified with it, then the gloves come off, as it were.[81] The dissenter and management may become locked into an escalating response on both sides of the dispute in which continuing intransigence on the part of the dissident is met with increasingly severe retribution.[88] O'Day[83] refers to this sequence of events as "intimidation rituals" (Table 3.3). For instance, a controversy that occurred some years ago between some of the engineers and managers working on the BART system in San Francisco lends itself to interpretation along such lines. Perruci et al.[89] outline the central events as follows: "On March 2 and 3, 1973, three engineers were fired from the San Francisco Bay Area Rapid Transit District. These engineers were reported to be concerned that the automatic train control system . . . was not safe enough. They expressed their concerns to management in oral and written form over an extended period of time, and were not satisfied with the response. They thereupon carried the issue privately to certain members of the board of directors and outside professional consultants, who made the issue public. They were then fired. The furore surrounding this event became known as the BART incident . . ." (p. 151).

According to Perucci et al., prior to the escalation of the dispute, the three engineers had had an unusual degree of freedom in obtaining information about the project as a whole, largely because of their relatively undefined roles. Step 1 of the ritual occurred when, in-

3.3. Intimidation rituals.

Phase one: Indirect intimidation

Step 1. Nullification

There is reassurance by superiors that accusations and complaints are invalid, and further investigation would simply affirm this. Attemps are made to overawe the dissenter.

Step 2. Isolation

The isolation and separation of the dissenter from peers by closing communication links, restricting freedom of movement, reducing allocation of resources, or transfer.

Phase two: Direct intimidation

Step 3. Defamation

Should the dissenter refuse to remain silent, middle management will begin to impugn his or her character and motives, thereby cutting off potentially sympathetic support, by attributing attempts at reform to questionable motives, underlying psychopathology or gross incompetence.

Step 4. Expulsion

Taken as a last resort, for it injures the organization's carefully managed image of reasonableness. A voluntary withdrawl of the dissenter is preferred to dismissal proceedings.

Reproduced from Miller, A. "Professional dissent and environmental management," *The Environmentalist*, volume 4, p. 147, 1984 by courtesy of Chapman and Hall, Hants, England.

dependently of one another, they took their initial worries to their respective supervisors who gave tacit acknowledgment of their concerns but saw the matter as falling outside of their jurisdiction. However, one engineer was warned that he was criticizing too much for his own good. Subsequently, two of the three engineers attempted to bypass middle management in an effort to convey their disquiet to upper levels. This too failed and, coupled with a lack of peer support, further exacerbated their sense of isolation (step 2). The persistent failure to have their concerns effectively aired eventually led them to a more political form of protest. A memo was circulated among engineers and management expressing the need for a new, specialized unit to look into the problems they had uncovered. Management immediately hardened its position (step 3). "In early reactions to the engineers, management

was inclined to view them either as (a) narrow specialists who couldn't see the 'big picture,' (b) persons with personality traits that led to confrontation with authority figures, or (c) general troublemakers incapable of cooperative 'team' behavior"[89] (p. 160). With the advent of the memo, however, they came to be seen as both "self-interested and power-hungry." Finally, the three engineers brought in an outside consultant and approached a member of the BART board with their findings (step 4), a step that eventually resulted in events slipping from their control and the matter going public. Shortly thereafter they were fired. It is interesting that the consequences of their actions followed the dissenters beyond this point for they experienced "great difficulty in finding suitable employment, financial hardship (in one case, bankruptcy) and personal familial problems"[89] (p. 162).

Whistle-blowing and intimidation rituals appear to be universal phenomena cutting across all political and organizational boundaries. A case in point is the dismissal of a provincial forester in the employ of the Ontario Ministry of Natural Resources for allegedly breaching his oath of office.[90] At issue were timber-cutting rights to an area of forest on the shore of Lake Superior. Local small-scale loggers were of the opinion that the land had been reserved for their use. When the forester and his supervisor were instructed by central office to give priority to a particular local entrepreneur, they raised questions about the adequacy of the data on which decisions about cutting licenses were to be made. "The government's Forest Resource Inventory (FRI) for the area was, they felt, drastically overestimating the amount of available wood. If a cutting license were issued . . . on the basis of such an inventory, the principles of sustained yield forestry would be defeated"[90] (p. 232). The forester continued to resist pressure from middle management to issue the cutting license and refused on professional grounds to comply with a direct order to develop a licensing agreement. Feeling embattled, he decided that his only option was to seek the political support of his local member of parliament. At the same time he took the rather drastic step of filing charges of professional misconduct against upper-level administrators in the organization to which he belonged. Three months later he was fired for breaking his oath of office, but he was reinstated after a legal appeal. The grievance settlement board that dealt with the case ruled that he should be offered his job back.[90]

In sum, the pervasiveness of fear and defensiveness in many organizations makes it unlikely that it will be possible to break out of this cycle of dissent and intimidation. The problem is compounded by the need for both stability and change in organizations, a paradoxical state of affairs that arouses unbearable tensions within organizations. Yet, dissent is potentially a constructive element in organizational responses to a changing environment, if it could be used constructively.

Pathologies in Organizational Thinking

All organizations have an internal knowledge base from which members draw information in pursuit of their various tasks. This knowledge base defines organizational culture and gives a characteristic flavor to the daily round. Schon suggests that it is a repository of "principles and maxims of practice, images of mission and identity, facts about the task environment, techniques of operation, stories of past experience which serve as exemplars for future action"[62] (p. 513). In a sense, organizational culture is a kind of ongoing discourse shaped and also distorted by the competing interests described earlier.[91] The result is what Lee[92] refers to as "information flow pathologies," which occur when we react not to the real world but to our personal (biased) construction of the world around us.[93] In so doing we create and seek to sustain personal conceptions that are more comforting than realistic. When this becomes a collective phenomenon, a form of group delusion, then organizations or subgroups within them distort information to meet their needs.[94] Typically, what is involved is a narrowing of perception, a form of

cognitive constriction in which relevant information is avoided, alternatives limited, and decisions made on the basis of familiar collective experience.[93] One unfortunate result is that "in situations of conflict and uncertainty, natural resource scientists and the information they provide ... hold sway with agency leaders seemingly only to the extent that their results agree with traditional commodity interests and agency ideology"[78] (p. 768). For instance, research that has a negative effect on a project may be challenged; subjected to unusual scrutiny; contracted to other research groups; or in various ways explained away—a treatment that is seldom meted out to favorable evidence. A case in point, Bella[94] argues, is the way in which unfavorable information about the booster-rocket seals was filtered out of the decision process leading up to the *Challenger* launch. "Such distortions do not depend upon deliberate falsifications by individuals. Instead, people who are competent, hard-working, and honest sustain systematic distortions by merely carrying out their organizational roles" (p. 369).

Distortions of organizational thinking appear to be greatest in highly centralized organizations in which there is conflict between the constituent subgroups.[95] That is, when significant decisions are made at increasingly higher levels in the organization, less powerful groups are left scrambling for what little influence they may be able to exert. Thus, centralization and hierarchical arrangements tend to isolate individuals into self-contained groups. When this compartmentalization is formalized by rules and standard procedures, there is likely to be an increase in the chance that information flow will be distorted by cognitive biases. For instance, Kennedy[60] suggests that, under such circumstances, decisions are unduly influenced by shared stereotypes about each other that reinforce an in-group (us) and out-group (them) mentality. It is no wonder that friction between subgroups over such things as territory and budgets is commonplace within hierarchical or authoritarian systems. This is another way of saying that the more hierarchical and bureaucratic an organization, the more inflexible and fractious it becomes. To make matters worse, individuals are reduced to the role of functionaries, simply doing their assigned roles within their confined niche, largely unconcerned about, and even resistant to, information emanating from other parts of the organization.[94] Thus, compartmentalization and subsequent conflict between groups working on a project may lead to *knowledge disavowal*, the avoidance of available information to protect the status quo or to avoid a difficult choice.[95]

Knowledge disavowal may also result from what Trist[96] calls the "hatred of learning from experience." This is a common phenomenon that afflicts us all and is evident in the widespread resistance to the evaluation of what we are doing. For example, Trist concluded that the introduction of new work methods into a coal-mining operation in England foundered because both workers and management refused to face up to the fundamental changes that were required in their work habits. Rather than recognizing the early failure of the self-managing work teams that were at the core of the new procedures, all of those involved maintained the view that nothing out of the ordinary was happening, and, therefore, no extraordinary efforts were required to make the project a success. In a similar vein, Hilborn[85] argues that fisheries agencies have difficulty in learning from experience because they avoid evaluation of their activities and fail to engage in the kind of active experimentation needed to develop adaptive policies. One consequence is that they do not learn from their failures and, in some cases, cannot determine if a given policy is a failure or not. This problem might be referred to as a lack of institutional memory. For instance, "learning and thinking within organizations is affected by record keeping, the memories of people working in the organization and the patterns of communication between people within the organization"[91] (p. 211). However, most organizations make only feeble efforts to organize and integrate the enormous store of information they have generated. Consequently, rich sources of knowledge remain unused in long-forgotten files and databases.[85] The shifting, temporary quality of professional employment adds to this difficulty in learning from past ex-

perience. As professional managers move or are moved around the country, the latter policy being adopted by some agencies to discourage too close an identification with local interests,[97] it is difficult to acquire a long-term understanding (i.e., memory) of any one locale.[92]

Given these distortions in the flow and availability of information within organizations, it is not surprising to find that organizational decision making is far from the rational ideal. Indeed, it has been described as a chaotic mess, a kind of garbage can in which streams of activity within the organization coalesce at some point and a decision occurs. Such is the confusion that, often, it is not evident to those involved that a decision has actually been made or who made it.[98] Information may play a role in this process, or it may be manufactured after the event to justify the decision.[86] Under these conditions, it is understandable that policy making in organizations tends to follow an incremental pathway, one in which only minor changes in policy are tolerated. Thus, new policies are modified to fit existing procedures,[62] and management conceptions, such as the multiple-use paradigm, can be used to stifle debate over alternatives.[60] While their are those who support incrementalism as the most sensible response to environmental uncertainty,[99] others believe that the adoption of incremental policy making is simply a way of protecting the organizational status quo.[86] In this way of thinking, incrementalism is seen as a form of studied inaction, an avoidance of making significant decisions. For instance, Lyles and Mitroff,[100] in a study of management problem formulation, identified *inactivists*, managers who would only take a position on an issue if forced to do so:

At all organizational levels, there were individuals with fears caused by various pressures such as personal failure, threats of punishment, organization failure, or time constraints. Fear of punishment or failure could even cause a person to try to cover up the real problem . . . some managers obscured problems or even deliberately distorted information to protect their positions (p. 113).

Inactivists, it appears, are particularly terrified of retaliation by the politically powerful within their organization and, accordingly, take great pains not to expose their own past errors or those of their superiors.

In conclusion, the vagaries of organizational and bureaucratic politics are so entrenched and of such long-standing that they are unlikely to disappear any time soon. If so, adaptive problem solving is possible only to the extent that professionals and others are able to master the political game without losing sight of the more pressing ecological agenda.

Society

"Environmental governance is implemented through institutions, exercising jurisdiction that is conferred upon them by law or tacit social consent. In North America these agencies include municipalities, provincial and state governments and their subdivisions, public and private corporations, communal organizations . . . and private property owners . . . In all, a complex regime of mixed authorities governing the use of resources and the environment has emerged"[101] (pp. 27–28). Thus, environmental management involves a tangled web of institutions and jurisdictions, the resulting confusion and conflict acting as an obstacle to adaptive management. Caldwell,[101] for instance, estimates that there are at least 650 institutions involved in Great Lakes management, hardly a recipe for effective action. In his view, progress toward the officially declared goals of ecosystem management in the Great Lakes region is hindered because it is the system of governance itself that is the problem. In this last section of the chapter, I shall examine problems with the system of governance that has developed in North America, drawing heavily on John Dryzek's notion of social choice mechanisms.[102]

Societies differ in the extent to which decision making rests in the hands of elites or is accessible to organized elements among the general public. For instance, decisions (social choices) about the levels of risk posed by toxic chemicals differ in Canada and the United States because they are reached through different political decision-making procedures.[103] More specifically, the process in the United

States is more open to public scrutiny than that found in Canada, which has tended to follow the relatively closed British model of decision making. Of interest to us is the ecological adaptiveness of collective decision making and the conflict that swirls around the adoption of particular mechanisms of social choice. A variety of such mechanisms can be found in the world's societies, often combined into a complex mixture of decision-making arrangements. It is convenient, for our purposes, to examine some of these basic social choice mechanisms separately before looking at how they are used in practice. The following summary is based on Dryzek's[102] views on the matter.

Free Markets

Markets are forms of social organization that imply an open exchange of material goods and ideas among the populace. The regulation of such a system is through feedback from price systems that determines the direction of economic activity. As valued goods become more scarce, rising prices stimulate production and the search for alternatives. Advocates of free markets, therefore, are adamantly opposed to government regulation, which they see as an impediment to industrial growth and efficiency. From an environmental perspective, however, there are several problems with free-market economics. First, there is no mechanism for including the real cost of goods into the pricing system, because this would require inclusion of externalities such as the environmental damage caused by production. The result is unrealistically low prices and the continuing degradation of natural resources.[104] Second, markets need continued economic growth if they are to be politically stable. To be successful, markets depend on the continued stimulation of material acquisitiveness that, in turn, requires persistent inequalities in the distribution of wealth within society. Only if greed and avarice are promoted will the populace struggle to keep up with or outdo their neighbors. At the same time, sufficient growth has to take place so that enough material goods can trickle down to the poor and placate any stirring of political unrest. These strenuous efforts to sustain economic growth, as one might anticipate, have detrimental effects on the state of the natural environment. The pursuit of unrestrained self-interest within a market economy results in the tragedy of the commons and other public goods problems. Markets have few answers to overutilization of resources and pollution of common properties such as air and water. It follows that government intervention in a market economy is difficult because it may be seen as damaging to the profitability of business and results in reduced investment and greater unemployment.[102]

Administered Systems

In an administered system there is central control of decision making by an elite who operate through a subordinate bureaucratic structure.[102] One normally thinks of this kind of central control as being typical of the former Soviet Union and other totalitarian states. However, less draconian examples of central regulation are a common feature of western democracies, in the form of environmental management and regulatory agencies. As we shall see in chapter 6, liberal doses of central control have been proposed as a way of limiting the damage done to the environment by the material excesses of market systems. Dryzek, for instance, points out that Thomas Hobbes' belief in a powerful sovereign as the remedy for societal ills has been adopted by latter-day Hobbesians who speak of technocratic and scientific priesthoods, benevolent authoritarians who will impose some semblance of restraint on an otherwise feckless public. There are several problems with this avenue toward ecological sanity, however. First, central elites do not necessarily hold a monopoly on the knowledge and skills needed to resolve environmental problems, as explained earlier. In practice, Dryzek argues that highly centralized administered systems, such as those in eastern Europe, have had an extremely poor record of environmental protection and seem to have been at a loss in trying to cope with the devastation wrought by industrial production. The reason for this may

lie, in part, with the recalcitrance of bureaucracies that may give the appearance of complying with central directives but, in practice, subvert policies in pursuit of self-interest. The resulting insensitivity to negative feedback, the information pathologies, the compartmentalization of effort, and the rampant internal conflicts make centrally controlled bureaucratic states clumsy vehicles for attempting to solve environmental problems.[102]

Polyarchies

Polyarchies are systems of choice involving a large number of players or stakeholders, none of whom are able to dominate the proceedings. In its ideal form, a polyarchy is what most people would recognize as a *democracy*, one in which choices are made through mutual adjustments among players wielding relatively equal amounts of power. However, this seldom occurs in practice. Although the elite control of natural resource decision making is being challenged by an increasing number of organized interest groups, inequality in power is still rampant, as is discussed shortly. Thus, polyarchy as a route to environmental problem solving is beset by a number of problems. First, the political ascendancy of a small number of powerful interests skews the decision process to their liking. Dryzek[102] refers to this as *corporatism*, a situation in which access to and control of decision making is concentrated in the hands of political, corporate, and producer elites, who may talk rhetorically of the common good but, in reality, pursue self-interest. Thus, polyarchies have a tendency to become vehicles for entrenching the political status quo. Second, the increased number of participants demanding a voice in decision making can be a problem. It is true that a variety of viewpoints helps to make decision making more sensitive to environmental events and can foster a more open experimental approach to policy making, but, on the downside, polyarchies are disjointed and rife with conflict among the numerous players. The resulting slowness in making choices means that they are relatively ineffective in dealing with rapid environmental change. Finally, polyarchies require economic growth for the same reason as do market systems: to produce enough wealth to smooth social relations so that potential conflict is kept within reasonable bounds and collective decision making can proceed. The effects of continued economic growth on the environment have already been mentioned.[102]

Autarchy

Autarchy means self-government and is implied by Dryzek[102] in his discussion of the radical decentralization of society that, he believes, would lead to a practical form of anarchy. The latter is evident in the small-scale self-sufficiency that is promoted by some environmental thinkers as an ecologically rational response to the excesses of modern industrial society. In this vision of sustainability, small communities would be mapped on to local bioregions in such a way that inhabitants would be able to maintain a close watch on the health of the natural environment on which they depend for their livelihood. The assumption is that such dependence would generate an ongoing concern with sustainable practices and with mutual, benign control of one another's behavior. Although some indigenous people have developed such self-sufficient communities, to all extents and purposes the latter have disappeared from the western world. Dryzek[102] believes that this is unfortunate for tragedies of the commons are rare under such social arrangements, whereas ecological problems such as tropical deforestation occur at their most destructive when indigenous systems of cooperation and communality have broken down. Examples of autarchy in the West are to be found in such isolated gestures as community forestry, but these are a far cry from genuine autarchy. There are so few examples of autarchy in the West that it is not clear whether it would provide a credible way of dealing with environmental problems. A number of critics point out that small-scale devolution of industrial society is not practicable anyway,[104,105] whereas Dryzek wonders how such small communities would survive when surrounded by large and aggressive societies.[102]

Environmental disputes usually involve disagreements over the type of social choice mechanism that should be used in environmental decision making. As one might expect, individuals have distinct preferences that reflect their underlying ideological assumptions. In trying to understand the relationship between ideology and preferences for these different kinds of social choice mechanisms, it is useful to think of the latter arranged along two ideological dimensions: central versus decentralized decision making and economic growth versus restraint (Figure 3.1). In doing so, it is possible to arrange types of social choice mechanism according to their central assumptions. Thus, centralized control and economic growth are, as we have seen, cornerstones of the Imperial tradition, whereas decentralization and restraint reflect the more Arcadian vision.

The Dominant Mode of Social Choice

There are conflicting perspectives on the relative balance of social choice mechanisms in North America. One view is that prior to 1970 and the upsurge of modern environmentalism, decisions about natural resource and environmental matters rested in the hands of the *iron triangles*, the political, economic, and technical elites who assume a dominant role in environmental (and social) decision making in industrial societies. Depending on the issue at hand and the political sector involved, the composition of these tight-knit, enduring groups will vary. In his analysis of pesticide policy making in the United States, for instance, Bosso[107] speaks of enduring links between executive bureaus, congressional committees, and client groups with a stake in particular programs. In a similar vein, Dryzek[102] comments: "Once all nuclear issues were the preserve of a tight and exclusive 'iron triangle,' which, as part of its pro-nuclear ways, managed to keep the waste question out of the limelight. This triangle was composed of the Joint Committee on Atomic Energy of the US Congress, the federal Atomic Energy Commission, and the nuclear indus-

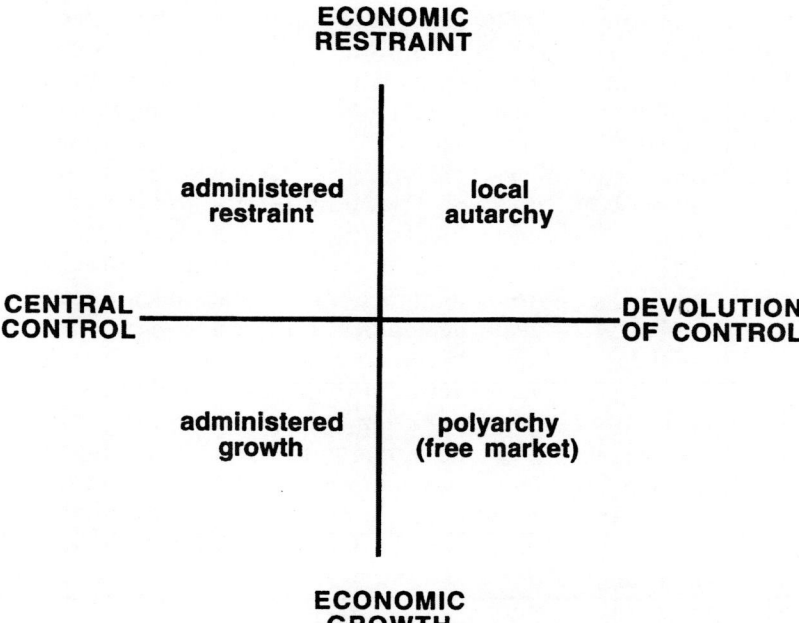

FIGURE 3.1. Social choice mechanisms.

try" (p. 112). Because iron triangles do not exist in a vacuum, their members do not have unlimited power to do as they may but are better seen as extremely influential players in the policy-making arena. The power of these networks has waned in recent decades as participation in the policy process has diversified. However, this is not to say that they have faded from the scene but, rather, that the expression of power follows different, more subtle channels.

In contrast to the elitist (corporatist) view of decision making is the perception of a more genuinely pluralistic society (polyarchy), one in which a challenge to the hegemony of the political-corporate elites arose in the 1970s. This period of interest-group politics took on the appearance of a pluralist competition between vested interests with a net increase in both the complexity of and the conflict surrounding decision making.[107] On the face of it, one might have anticipated some political benefits from this growing diversity of opinion within the environmental policy forum. Whether this is so is a matter subsequently discussed. Dowie[72] suggests that we have moved into a more conservative problem-solving era

in which national environmental organizations have begun to reach a political accommodation with the ruling elites in an attempt to develop a more cooperative approach to environmental problems. At the same time, "a rejuvenated, angry, multicultural, and decidedly impolite movement for environmental justice" has emerged as the seat of radical protest.[72] Hence, environmental problem solving is no longer the exclusive domain of elites but takes place against a backdrop of diverse, often radical, opinion.

Both pluralist (polyarchy) and corporatist interpretations of social choice can be discerned in individuals' perceptions of natural resource decision making. For example, Salazar and Alper,[108] in their review of forest politics in British Columbia, observed that some participants recognize elements of pluralism (interest group politics) in land-use negotiations in the province but that, more generally, "pluralism . . . is seldom applied to the B.C. political economy; political analysts are almost uniform in their view that political power in the province is concentrated" (p. 385). It follows that many would see an "exploitation axis," comprising a "policy-making linkage between the state (provincial government) and economic producer groups" in which the "forest industry, forest industry unions, and the Ministry of Forests may be seen as elements of corporatist dominance of forestry policy . . ." (p. 385). My sympathy with corporatist interpretations of natural resource politics stems in part from the history of forestry in North America, especially during the eighteenth and nineteenth centuries, which is one of domination and exploitation of the resource by a small number of powerful capitalists.[109,110] Whether this has changed significantly in the twentieth century is discussed in chapters 4 and 5.

Although western democracies exhibit a mixture of social choice mechanisms, a persuasive argument can be made for the idea that we live in a predominantly corporatist society, one in which power is concentrated in the hands of political-economic elites. For instance, Ophuls and Boyan[111] (p. 314) argue:

Our current political system is statist, not democratic. The federal government is a bureaucratic and electoral behemoth dedicated to one primary end: the satisfaction of human appetite at the expense of nature. Popular participation, such as it is, is token, minimal, symbolic; and the behemoth is largely beholden to organized and monied interests.

Fischer[15] concurs, referring to this kind of social organization as that of a "technocorporate state" arranged hierarchically into three levels. At the top is a small number of elites who make the basic governing decisions. They include the executive branch of government, senior legislators and administrators, together with the more powerful members of the capitalist class. Fischer believes that although these elites compete over the direction of policy, they cooperate to shape policy agendas in favor of their own (Imperial) interests and ends. The second level in this system is composed of the scientific and managerial experts who provide the technical information and skills upon which the elites depend for the implementation of policy. In return for a so-called comfortable position in society, they offer (on the whole) their allegiance to the elites whom they serve. These experts may on occasion fashion policy; however, their primary function is to provide the technical support upon which the power of the political-economic elites depends. At the bottom of this hierarchy is a depoliticized mass public, cut off from meaningful involvement in decision making by the state's attempt to structure the dialogue over environmental problems as technical, rather than as matters of value. Emerging from the silent majority, however, are organized and vocal interest groups who are seen by elites as a worrisome threat to their agendas. In Fischer's view, then, modern western societies are corporatist states in which there is a strong movement toward the use of technical information to suppress or avoid normative discourse, that is, democratic discussion of purposes and ends.

Clearly, this is but one view of how society works, not one with which everyone would agree.[108,112] However, it happens to be consistent with my own understanding of forestry politics, which is explained more fully in chapter 4. I've found it convenient to modify Fischer's views slightly and consider the natural resource decision-making process to involve two main groups: a policy-making elite and the public with the mass media acting as a conduit between them (Figure 3.2). The policy-making community is composed of an iron triangle of Fischer's political-economic elites together with their technical advisors. In contrast, the public is more diverse, with organized interest groups playing the most significant role in their persistent attempts to become involved in decision making. Most notable among the latter are environmental groups, unions, professional associations, producer groups, small businesses, cooperatives, and so on. Occupying a niche between these two often-opposing groups are the media, which inform, or scandalize, the public about the machinations of the ruling elites while providing the latter with an insight into grassroots reactions.

The expression of power within this technocorporate state is not "necessarily repressive, coercive and violent. Rather, it forms a continuous, productive network in which both the weak and strong participate. This tight web of

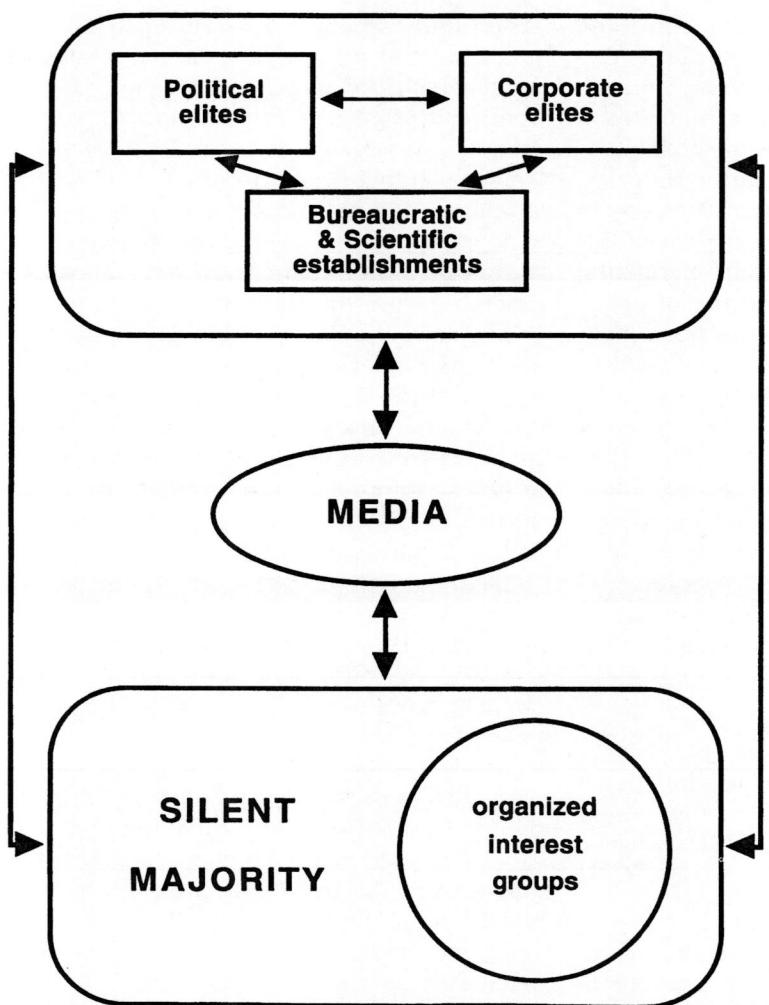

FIGURE 3.2. The technocorporate state.

power relations pervades families (and) societies ... it structures people's behavior and perceptions of what is normal or abnormal, true or false. In this respect, power is closely connected with the production of knowledge and with what is given the status of truth"[113] (p. 84). With this sentiment, human choices are shaped by powerful elites who have access to privileged information and the means to manipulate public agendas and perceptions. One of the most ardent advocates of this interpretation is Noam Chomsky who has long maintained that, in what passes for a democracy, U.S. public opinion is manipulated by a mass media that offers a " biased view of the world filtered through the goals and requirements of concentrated economic power."[114] However, this shaping of the public's perception of environmental problems is not limited to elites, as Gould found in his study of several contaminated communities in the Great Lakes region[115] (pp. 173–175):

Industry, environmental organizations and the various levels of government attempt to manipulate public perception of local environmental conditions to promote their political and/or economic interests. To the extent that the specific type of local pollution problem lends itself to concealment, the invisibility of pollutants may be manipulated to promote more favorable local perceptions of the state of the local environment. Private capital and government attempt to minimize *primary social visibility* of local pollutants through various methods of concealment, diffusion and transport . . . the political strategy of environmental organizations involves efforts at environmental consciousness-making, largely to counter industrial and governmental unconsciousness-making efforts.

Promises of political devolution and participatory decision making (autarchy) implicit in recent discussions of environmental management seem hollow in the face of the reality of the technocorporate state. Environmental and grassroots organizations are well aware of these structural impediments to change and are unlikely to stop their challenge to the political status quo. As a result, the prospect for endless conflict remains with us.

Internecine Conflict in the Technocorporate State

Destructive conflict occurs between, as well as within, all levels of the technocorporate state. Although we may live in a corporatist state (at least in my view), this does not mean that the public suffer this situation gladly. One would expect to find perpetual conflict between the public's organized groups and the ruling elites, between the rulers and the ruled. Just as influential are the internecine struggles within each of these levels of society. Conflict, of course, isn't necessarily evil per se. It can lead to adaptive behavior if approached constructively. For instance, both Dryzek[14] and Smil[104] argue that pluralistic discourse, despite its endless haggling, is the only effective route to adaptive behavior. The more open the debate on ecosocial issues, the more likely that monumental blunders on the part of elites will be avoided.[106] However, open debate between equals is not what actually happens in practice, as I have expounded on previously. How this conflict unfolds in practice and its consequences for environmental problem solving are matters to which we now turn.

Conflict between Decision-Making Elites

Although the accommodation between political and economic elites can be mutually ben-

eficial, there is persistent conflict between them.[15] As one might expect, political scientists have developed differing models of the interaction between governments and corporate interests.[116] One view is that the state, in essence, serves the interests of, and is dependent on, the capitalist class. In contrast, there is the perception that bureaucracies and states can become powerful enough to resist domination by external economic pressures. A more middle-of-road view combines both perspectives, assuming that although states may be dependent on one occasion, under different circumstances they may be able to exert some degree of autonomy. Take the history of forestry in New Brunswick as an example. The province was from the earliest days of settlement dependent on income from exports of lumber, particularly to Britain.[117] When the privileged access to the British market was curtailed in the middle of the last century, the provincial government was compelled to search elsewhere for revenue to sustain the provincial economy. This search involved attempts to attract external capital to the province so that its rich timber resource could be exploited to the full. The result was that "there existed in New Brunswick in the period between 1860 and 1940 almost a client state perilously dependent on its forest industry"[110] (p. 163). Little has changed in the last half-century, as the province continues to be dependent on income from the forest industries to sustain its economic well-being.[118] This record implies a dependent relationship, with the state being subservient to industry; however, it does not convey the complexity of actual events. Even while successive New Brunswick governments encouraged and supported the development of its forest industry, they also made persistent efforts to rest greater returns from their lease of Crown land (publicly owned forests) to the major corporations. Their limited success, in this regard, is attributed by Cashore[119] to the amount of public land that had been alienated (sold) to private interests over the course of the province's history. That is, the major corporations could always turn to private supplies of timber to fulfill their needs and so were less dependent on access to Crown land, which was the provincial government's only trump card in bargaining. The relationship between government and industry, therefore, while mutually advantageous can be rather brittle.

Turning to conflict within the bureaucracy, it seems self-evident that adaptive management would be enhanced if there were to be more constructive cooperation between agencies. However, on closer inspection, this appears to be a naive view. Students of government, it seems, take two quite different positions on the matter.[120] There are those who believe that the promotion of competition between agencies would lead to innovative management proposals that, in turn, would provide decision makers with a broader array of policy options. The contrasting view is that functional rivalry between agencies would simply lead to wasteful duplication and greater preoccupation with self-protection. However, these scholarly debates may be moot for, in practice, agencies appear "to worry far more about organizational maintenance than extending the agency's reach or maximizing its budget. Consequently, the cost of competing with functional rivals, in terms of threats to external political support and ultimately the bureau's existence, will generally be perceived to be too great"[120] (p. 701).

A case in point is the political maneuvering surrounding California's Mono Lake Basin, the source of a substantial portion of Los Angeles' water supply.[120] Beginning in 1893, the U.S. Forest Service held jurisdiction over the Basin, managing it according to multiple-use criteria. However, over time, the increasing population in Los Angeles set in motion a series of interconnected events that led to a decline in the region's ecological integrity. Chief among these problems was an increasing demand for water, which resulted in a 50 percent decline in the lake's volume. In turn, this change sparked a shift in the balance of fish and wildlife with the result that both began to decline. Concurrent mining operations in the area also threatened to mar the aesthetic appeal of the Basin for recreational users. Here, then, was an opportunity for the National Park Service (NPS) to assert its authority and seek to gain

jurisdiction over the area. However, events did not turn out this way[120] (p. 710):

Given the lack of substantial timber production, the apparent environmental havoc that the U.S. Forest Service appeared to be unwilling or unable to prevent, the ire of environmentally sensitive individuals and groups, and the aesthetic beauty involved, conditions appeared ripe for an NPS gambit to gain control. The Park Service could claim to be the agent of environmental purity and organize those dissatisfied with the status quo around the idea of transferring administrative control . . . (As things turned out) the NPS consciously turned its back on the situation at Mono Lake. Given every conceivable opportunity to get involved in the debate over the lake's future, and despite beliefs at the bureau that the area rightfully belonged in its portfolio, agency management stopped any positive response dead in its tracks. . . . Agency leaders believed that the possibilities . . . for aggrandizement were offset by the political risks. . . . The Forest Service was no more aggressive in defending its turf than the NPS was in acquiring control. Rather than responding creatively, the Forest Service wished to stay out of the political crossfire and watch passively as the gradual deterioration of Mono Lake spurred others to decide the area's administrative fate.

One conclusion from Mono Lake is that, at least in this case, the two main agencies involved declined to become involved in competitive struggle even though the opportunity afforded itself. Kunioka and Rothenberg[120] observe that this conclusion applies more generally to federal land-management agencies where, typically, there is an absence of competitiveness and a preoccupation with autonomy. While agencies do little to promote themselves, they may also offer few innovative solutions to environmental problems. In light of this, an argument might be made for some effort by the political elites to foster constructive competition between agencies. How this might be achieved is another matter.

For instance, even when agencies do seek to expand their jurisdiction, they meet with resistance from other bureaucracies. Thus, in the late 1980s, the Canadian government sought to respond to the public's environmental concerns by developing a Green Plan for sustainable development. This afforded the Department of Environment, a previously poor relation in the federal cabinet, with the opportunity to expand its authority and influence.[121] However, because this agency was identified with a fervent desire for environmental regulation, other federal agencies became nervous about the possibility of interference in their affairs. In a time of fiscal austerity, increased budgetary allocations to environment would also mean less for themselves and more difficulty in carrying out their traditional mandates. In addition, the domestic political climate within Canada is very much in favor of economic growth. Few politicians and bureaucrats are inclined to promote the radical changes needed to achieve sustainability in the use of natural resources. Hoberg and Harrison[121] conclude, therefore, that for the Green Plan to succeed, the Department of Environment would have to surmount the obstacle of opposition to growth in its influence from within the federal government itself. Considerable time and energy is consumed, therefore, in agencies' attempts to protect their domains from incursion by others. This fosters neither cooperation nor the innovative thinking that one would hope to encourage in pursuit of adaptive management.

Conflict between Experts and Policymakers

The relationship between policy-making elites and their technical-scientific advisors has been, and remains, uncomfortable. Part of the problem lies with the differing conceptions of the role of science in policy making that are held by the two groups. Many in the scientific community, as well as in society at large, subscribe to a positivist model of the relationship between knowledge and power, which tends to "presume that: (1) The overriding purpose of science is prediction with precision, scope and accuracy, including prediction of the consequences of policy alternatives. (2) Such science-based predictions are prerequisites to major policy decisions intended to ameliorate or solve the problems of society. (3) Scientists are different from others who participate in

these decisions because their scientific input is objective and value-free"[122] (p. 296).

One consequence of this view is that science is advanced as a crucial resource in rational planning and policy making, one that deserves greater attention and funding.[123–130] Sufficient numbers of politicians and the general public believe in this view of science as impartial discoverers of the truth[131] to help sustain substantial funding for these exercises in positivist research. For instance, American ecologists were able to obtain generous support for the International Biological Program (1968–1974) from Congress on the assumption that the new systems ecology would "provide the basic science approach to problems of environmental management"[132] (p. 422). This same tradition underlies the recent initiative of the American Ecological Society,[126] the current concern with ecosystems management (see chapter 6), and the belief that rational action on such matters as global warming must await definitive scientific evidence.[133]

Associated with the positivist view of science and policy is the desire by scientists to retain their autonomy from what they see as political interference in research.[134] In part, this resistance to control is based on their belief that political pressure jeopardizes their credibility as sources of objective information. Indeed, many scientists actually believe that their work is or can be "objective" (i.e., impartial), a persistent belief that is part of the positivist myth.[122,135] It is understandable, therefore, that they should become frustrated at what they perceive to be the misuse of their research in policy-making contexts[79] and have "aggressively fought to reaffirm and protect the image of a neutral science"[131] (p. 222).

Scientists may have grounds for complaint, for there is considerable evidence to support the conclusion that scientific information is used in policy making more as a weapon in support of particular decisions than the basis on which that decision is made.[136–139] Most frequently, scientific information is acknowledged to the extent that it is consistent with the ideology and self-interest of the group of policymakers in question.[140] When it is not, it may be ignored and discredited, along with the scientists who produced it. Thus, Ozawa[131] argues that science is often used as both a shield behind which policymakers can hide and a persuasive tool, for when they have to make unpopular decisions.

The positivist model of science and policy has been rejected by what I referred to in chapter 1 as the technoskeptics, among whose ranks one finds both scientists and policymakers. More data, they believe, will not resolve our ecosocial problems, which have more to do with values and human behavior than anything else.[141] Indeed, one of their arguments is that positivist scientists do a grave disservice by implying that scientific ingenuity may well resolve the problems that face us.[142-144] It follows that there is a considerable body of opinion that rejects the belief that science can be the neutral arbiter of environmental disputes. For example[106] (p. 145):

Whenever science attempts to influence policy, three necessary conditions for efficient scientific research and analysis—autonomy, disciplinarity and a low level of criticism—are immediately broken, leading to endless technical debate rather than the hoped-for consensus which can limit arguments about policy. The technical debate concerns the interpretation to be given to the existing body of evidence, but no matter how large this body may be, widely divergent interpretations may be maintained, making argument practically endless. As debate continues, many long-settled technical issues are reopened for investigation, and attempts definitely to resolve one issue often succeed only in opening up many more technical issues for consideration; technical uncertainties grow rather than diminish as more research is done.

However, this is not to say that such dispute is useless and should be avoided. In fact, Collingridge and Reeve[106] argue in favor of the salutary role of criticism and dispute in policy making. In their view, policy blunders of considerable magnitude can be made when scientific information is not subject to adequate scrutiny or when one particular scientific interpretation becomes dominant. Thus, scientific dispute can have the ironic role of ensuring that no single school of thought influences policy making without being challenged. In

the face of such dispute, policy making would remain a primarily political process.[121] Critics of the positivist conception of the role of science in policy making, therefore, reject both the idea of a neutral science and the possibility that traditional science can resolve environmental conflicts. Instead, they argue for a more facilitating role of science, one that encourages greater discourse and participation between stakeholders at all levels of the policy process.[131,135] This is discussed further in chapter 7.

Conflict among Public Groups

Few members of the public are involved in any significant way in natural resource decision making, particularly at the policy level. Those who do get involved are usually members of an organized group such as an environmental interest group, worker's union, or producer's association. In a broad sense, all of these organizations share a common ambition in trying to protect the interests of their members in face of the political and economic power of the ruling elites. In a society dominated by Imperial beliefs, however, one in which adversarial modes are built into the political and legal systems, it is not unexpected to see more conflict than cooperation between interest groups as they scramble to influence policy. This conflict is further exacerbated by internal strife within groups, as different ideological positions and subgroup interests reveal themselves. To illustrate these problems and the deleterious effect they have on adaptive problem solving, I refer briefly to the problems experienced by two groups: environmentalists and labor unions.

In the late 1980s, Milbrath[145] estimated that while many Americans (70–80 percent) were concerned about some aspect of the environment, relatively few (5–10 percent) were involved in environmental organizations, with even less (fewer than 1 percent) engaged in environmental activism. However, the environmental movement in the United States has been in steady decline during the 1980s and 1990s, the national organizations such as the Sierra Club, Audubon, and Greenpeace losing "20 to 30% of their membership since 1990 and stand accused of mismanagement, faulty strategies and priorities, cozying up to corporations and, overall, selling out the concerns of their members"[146] (p. 150). At the same time, many organizations are in a state of disarray as factions within them maneuver for control of organizational agendas. According to Sessions,[74] this turmoil has resulted from pressure from both the political right and left. For instance, the environmentalist agenda was dealt a severe blow by the rise of political conservatism in the United States following the election of Ronald Reagan to the Presidency. With him came an increased emphasis on the Imperial mode and the virtues of economic growth, environmental deregulation, and free trade, all of which are anathema to Arcadian environmentalists. At the same time, environmentalists of the left (ecosocialists/eco-Marxists) attempted to redirect the priorities of environmentalism away from traditional concern with the ecological integrity of the natural world toward matters of "environmental justice."[147] The latter is a form of environmentalism that combines concerns about urban pollution with an attempt to draw attention to the fact that the most common victims of environmental hazards are minorities and the poor.[30,148]

The consequences of these pressures on the environmental movement are seen as unfortunate, at least by those who are sympathetic to the traditional environmentalist agendas. Thus, the growth of social justice concerns within the movement has, according to Sessions,[74] broadened and diluted the mandate of environmentalism to a point where it has lost itself in a confusion of aims with the result that it has become less politically effective. For instance, the split in Earth First! between its biocentric (wilderness preservation) and anthropocentric (social justice) wings, which resulted in some of its founders resigning, is a reflection of a broader schism that plagues the environmental movement.[73] Organizations and movements that are embroiled in such ideological disputes can hardly present an effective and unified front in pursuit of their aims. How this affects problem solving will be dealt with in a moment.

Labor unions have similar fraternal disputes. For instance, the intense controversy over logging practices in British Columbia has considerable implications for the economic well-being of the province's forestry workers. It is a familiar story. Major forestry multinationals need access to pristine valleys much prized by environmentalists who wish to see logging operations stopped in these spectacularly beautiful rain forests. The ensuing conflict has implications for employment among woodcutters and millworkers. If the environmentalists win their battles, then more land will be set aside for preservation, and fewer traditional jobs will be available as companies downsize their operations. One would have thought that, under these circumstances, the labor unions would have aligned themselves firmly behind company interests and against environmental concerns. However, this does not seem to be happening in this case.[149] For instance, the forest policy of the Pulp, Paper, and Woodworkers of Canada (PPWC), a union representing workers in pulp and sawmills as well as chemical manufacturing, has taken on a green tone with the advocacy of value-added production, limited corporate control of public lands, increased silviculture, and a reduction in the use of pesticides. This agenda seems not to sit well with some officials of the International Woodworkers of America (IWA), a union representing loggers, who are reported by Tatroff[149] (p. 28) as pointing out "that it's easy for others like the PPWC—which doesn't represent loggers—to demand changes that could kill IWA jobs." Because of these underlying tensions, communication between the two unions, and their supporters, on forest policy issues can become difficult to say the least. Thus, a recent public debate involving representatives of the two unions was dismissed by one of the union participants as a "circus."[149] It seems that one union sees the interests of its members better served by supporting current corporate plans while the other takes a longer-term view that involves reducing the influence of the major corporations through greater diversification in the use of the forest. Needless to say, these are substantially different positions that are unlikely to be resolved soon.

Conflicts between Rulers and the Ruled

The way in which society makes decisions about natural resources and other environmental matters is, as we have seen, riddled with conflict and confrontation. Though we may like to convince ourselves that we live in a pluralistic democracy, in reality we see a continuing struggle for control of environmental agendas between powerful elites and a variety of relatively weak public interest groups. For instance, control of forest policy in British Columbia has pitted two advocacy coalitions against one another.[150] Although such groupings are not without their internal conflict, they share sufficient common ground to coalesce around development and environmental agendas. Thus, the development coalition (composed of major figures from the forest industry, provincial bureaucracy and forestry profession working within the Imperial mode) promotes the liquidation of the province's old growth forest and its conversion into managed second-growth plantations.[150] The capital needed to do so is attracted to the province by ceding long-term control over the resource to outside industrial interests.[150] Thus, preference is for strategic policy making to remain under the control of political-economic elites, whereas the more tactical implementation of such policy is left to technical managers. Traditional planning, therefore, is "dominated by professional foresters directing a team of experts (biologists, land-use planners, etc.) to develop a forest management plan in an environment largely isolated from the public"[151] (p. 1). However, it is this central control, and the policies that go with it, that has aroused the ire of environmentalist and other public groups in the postwar years. The environmental coalition in British Columbia, as elsewhere, seeks to rest some decision-making authority from this iron triangle and, in doing so, introduce more Arcadian, ecocentric values into forest policy. Lertzman et al.[150] point out that, as the years have gone by, the environmental coalition has learned a great deal about environmental management and have also become skillful in questioning the basic legitimating beliefs of the development coalition. Behind these conflicts over public participation in decision making, therefore, "lie different interpretations of democracy. Officials stress the importance of representative democracy and the power of elected officials to make decisions based on public mandate. Concerned citizens, on the other hand, see grass roots action as the essence of democracy. They see citizen participation as a check against expert elitism as well as a way to ensure their needs are met"[152] (p. 360). The problem for government and industry, therefore, is how to respond to pressure from public interest groups.

Governments, in particular, are faced with a thorny dilemma. In a liberal democracy, the function of the state is one of encouraging economic growth and the success of the private sector (capital accumulation) while, at the same time, safeguarding public interests.[153] Thus, resource-dependent states are locked into economic arrangements with resource industries that they jeopardize at their peril. Under such circumstances, the state may feel compelled to quell any public discontent that threatens the well-being of business and industry.[153] The upsurge of environmental discontent is one such threat because the environmentalist agenda calls for regulation of, and costly changes in, industrial production. However, many states and industries have responded by espousing, at least publicly, the need for more openness in decision making, something to be achieved by greater public participation.

For instance, the U.S. Congress has mandated a role for public participation in governmental decisions through the National Environmental Policy Act (1969) and the National Forest Management Act (1976), which require agencies to allow the public access to natural resource decision making.[154] Despite these official sanctions for the principle of public participation, the actual level of public involvement achieved has been modest. If one recognizes that public participation in decisions can range from a passive (being informed about events; the right to raise objections to plans) to a more active (determining agendas; sharing final decisions) role,[152] then the latter forms of participation are conspicuous by their absence. Indeed, one critical view is that the response of government agencies to the demand for participation has been laggardly, a token gesture characterized by a mixture of concessions and resistance.[150,155,156] Thus[157] (p. 15):

For the most part, the old involvement and participation techniques did not change. They continued to be bureaucratic techniques to exchange information, to request comments on issues or proposals that had already been formed, or to hold public meetings or consultations about restricted alternatives. Participation had been narrowed into a set of techniques designed to secure administrative compliance with statutory and regulatory requirements.

This dilution of public involvement is understandable in light of agencies' tendency toward self-protection. One would expect to find that, while agencies may listen, they would only hear whatever fits their policy agendas. Mohai[158] has explored this at some length, concluding that the "assertion that a change occurred in the 1970's causing the Forest Service to become more responsive to environmental groups out of its desire to avoid conflict appears to be correct. However, this change did not go so far as to change the professional ideology (utilitarian) of the agency itself, which continues to exert a very important influence" (p. 155). At the same time, the traditional control of management decisions by professionals is not something that they are keen to give up. Many natural resource professionals, for instance, are adamant about maintaining their control of management decisions,[159] one reason for this being their unwillingness to assume responsibility for decisions that they themselves have not made. Why, they ask, should public interest groups have the privilege of making decisions, the consequences of which fall onto the shoulders of professional managers?[160] It is not surprising, therefore, to find that recent articles in professional journals are still trying to persuade their audiences that public participation is both necessary and important.[161] This is not to say that professional apprehension about public participation is unwarranted. All public participation processes bring with them a variety of headaches,[162] the nature of which are discussed more fully in chapter 7.

Disillusionment with the public participation process has pushed many environmentalists beyond mere talk. They have concluded, for example, that participatory exercises simply reproduce the inequalities of power inherent in society,[152] with themselves at the bottom of the heap. Subsequent attempts by radical environmentalists to redress the balance of power infuriate more traditional professionals, who have spent their working lives within the political status quo trying to avoid political involvement and who see environmentalists as special interests trying to usurp the officially sanctioned decision process for their own selfish ends. Nevertheless, despite this resistance, environmental groups persist in attempting to mobilize support among the general public,[163] seeking windows of opportunity within which to push their cause. Such opportunities occur when there is a confluence of political circumstances that favor environmental action. It is at these times that environmental groups have the most influence on events. Almeida and Stearns[164] suggest that a combination of elite instability and external support for the environmentalist cause serves to tip the balance in their favor, no matter how temporary this might be. This confluence of events seems to explain the success of victims of Minamata disease in obtaining redress for their suffering.[164] Thus, a combination of interagency conflict, symbolic gestures of sympathy toward the victims by government, and the growth of a national sentiment against pollution all served to afford the victims an opportunity to press their case. Subsequently, however, this window closed as the government took a more conservative turn and developed institutionalized ways of defusing the antipollution social movements.[164]

Although there continues to be resistance to environmental activism, both government and industry have moved in recent years to improve the dialogue over environmental management. For instance, the 1996 American Forest Congress was a major attempt to draw in a wider array of stakeholders into the forest policy debate. Despite these constructive steps, there is a great deal of distrust among the environmentally concerned. Some, for instance, dismiss practices like environmental mediation as a ploy to quell environmental protest.[153] Others feel a deep sense of injustice at what they consider to be the corporate crimes they see at every turn, the consequences of which we must all suffer.[165] It is citizens who feel this way and, presumably, see no other way to express their grievances that turn to civil disobedience and violence, the incidence of which is so numerous that I hardly need give examples. The fact that violence does happen tells us something about the state of the decision process and the level of frustration felt among dissidents. One would hope that the sociopolitical causes of this violence might be recognized, but the message of both this and the preceding chapter is that the majority of people prefer not to delve into either the psychological or the sociopolitical roots of the problem. This is unfortunate for it simply provides a fertile breeding ground for continued misunderstanding. Facing up to both the psychological and the sociopolitical obstacles to change is, therefore, a crucial first step in achieving adaptive problem solving. Unfortunately, it is not easy to encourage what can be, at times, a traumatic experience. The prospects for doing so are matters to which we turn in the remainder of the book.

References

1 Cotgrove, S. 1982. *Catastrophe or cornucopia: The environment, politics, and the future.* Chichester: Wiley.

2 Orr, D. 1992. *Ecological literacy: Education and the transition to a postmodern world.* Albany: State University of New York Press.

3 Vallentyne, J. 1974. Limnology and education in the next decade. *Journal of the Fisheries Research Board of Canada* 31:513–519.

4 Hackman, J. and C. Morris. 1975. Group tasks, group interaction process, and group performance effectiveness: A review and proposed integration. In *Advances in experimental social psychology*, Vol. 8, ed. L. Berkowitz, 45–99. New York: Academic Press.

5 Luszki, M. 1958. *Interdisciplinary team research: Methods and problems.* Washington, D.C.: National Training Laboratories.

6 Van Dyne, G. 1972. Organization and management of an integrated ecological research program. In *Mathematical models in ecology* , ed. J. Jeffers, 111–172. Oxford: Blackwell.

7 Leckie, G. 1975. Interdisciplinary research in the university setting. Center for Settlement Studies, Occasional Paper No. 9, University of Manitoba, Winnipeg, Canada.

8 Barmark, J., and G. Wallen. 1981. The development of an interdisciplinary project. In *The social process of scientific investigations, Sociology of the sciences yearbook, volume IV 1980*, ed. K. Knorr, R. Krohn, and R. Whitley, 221–235. Dordrecht, Holland: D. Reidel Publishing Company.

9 Cuff, W. 1983. An evaluation of the Port Hacking Estuary Project from the viewpoint of Applied Science. In *Synthesis and modelling of intermittent estuaries*, ed. W. Cuff and M. Tomczak, 273–292. Berlin: Springer.

10 Savory, A. 1988. *Holistic resource management.* Washington, D.C.: Island Press.

11 Neck, C., and C. Manz. 1994. From groupthink to teamthink: Toward the creation of constructive thought patterns in self-managing work teams. *Human Relations* 47:929–951.

12 Guzzo, R., and M. Dickson. 1996. Teams in organizations: Recent research on performance and effectiveness. *Annual Review of Psychology* 47:307–338.

13 Nutt, P. 1990. *Making tough decisions: Tactics for improving managerial decision making.* San Francisco: Jossey-Bass.

14 Dryzek, J. 1990. *Discursive democracy: Politics, policy, and political science.* Cambridge: Cambridge University Press.

15 Fischer, F. 1990. *Technocracy and the politics of expertise.* Newbury Park, Calif.: Sage.

16 Merrifield, J. 1989. *Putting the scientists in their place: Participatory research in environmental and occupational health.* New Market, Tenn.: Highlander Research and Education Center.

17 Jackson, S. 1992. Team composition in organizational settings: Issues in managing an increasingly diverse work force. In *Group processes and productivity*, ed. S. Worchel, W. Wood, and J. Simpson, 138–173. Newbury Park: Sage.

18 Garcia, M. 1989. Forest Service experience with interdisciplinary teams developing integrated resource management plans. *Environmental Management* 13:583–592.

19 Janssen, W., and P. Goldsworthy. 1996. Multidisciplinary research for natural resource management: Conceptual and practical implications. *Agricultural Systems* 51:259–279.

20 Meridith, D., N. Hatfield, and P. Harvey. 1996. A nontraditional team approach: Making it

work in traditional bureaucracy. *Journal of Forestry* 94:17–20.

21 Kennedy, J. 1991. Integrating gender diverse and interdisciplinary professionals into traditional U.S. Department of Agriculture-Forest Service culture. *Society and Natural Resources* 4:165–176.

22 Magill, A. 1988. Natural resource professionals: The reluctant public servants. *The Environmental Professional* 10:295–303.

23 Dorney, R. 1989. *The professional practice of environmental management.* New York: Springer-Verlag.

24 Petulla, J. 1987. *Environmental protection in the United States.* San Francisco: San Francisco Study Center.

25 Nesmith, C., and P. Wright. 1995. Gender, resources, and environmental management. In *Resource and environmental management in Canada: Addressing conflict and uncertainty,* ed. B. Mitchell, 80–98. Toronto: Oxford University Press.

26 Tarnapol, P. 1991. Women and the Society of American Foresters. *Women and Natural Resources* 12:24–27.

27 Otero, R., and N. Brown. 1996. Increasing minority participation in forestry and natural resources: The MINFORS Conference. *Journal of Forestry* 94:4–7.

28 Dearden, P., and L. Berg. 1993. Canada's National Parks: A model of administrative penetration. *The Canadian Geographer* 37:194–211.

29 Porter, T. 1997. Crown land now a battleground: N.B. natives may try to stop forestry operations. *Daily Gleaner,* 1, 7 November, Fredericton, N.B., Canada.

30 Sexton, K., K. Olden, and B. Johnson. 1993. "Environmental justice": The central role of research in establishing a credible scientific foundation for informed decision making. *Toxicology and Industrial Health* 9:685–727.

31 Szasz, A. 1994. *Ecopopulism: toxic waste and the movement for environmental justice.* Minneapolis: University of Minnesota Press.

32 Luloff, A. 1995. Regaining vitality in the forestry profession: A sociologists perspective. *Journal of Forestry* 93:6–9.

33 Schnaiberg, A., and K. Gould. 1994. *Environment and society: The enduring conflict.* New York: St. Martin's Press.

34 Good, T. 1986. Budworm and Bafflegab. *New Maritimes* June: 14.

35 Klein, J. 1990. *Interdisciplinarity: History, theory, and practice.* Detroit: Wayne State University Press.

36 Hollingshead, A. 1996. Information suppression and status persistence in group decision making: The effects of communication media. *Human Communication Research* 23:193–220.

37 Miller, A. 1984. Professional collaboration in environmental management: The effectiveness of expert groups. *Journal of Environmental Management* 18:365–388.

38 Belbin, R. 1981. *Management teams: Why they succeed or fail.* London: Heinemann.

39 Janis, I. 1989. *Crucial decisions: Leadership and policymaking in crisis management.* New York: The Free Press.

40 Kowitz, A., and T. Knutson. 1980. *Decision making in small groups: The search for alternatives.* Boston: Allyn and Bacon.

41 Jehn, K. 1995. A multimethod examination of the benefits and detriments of intragroup conflict. *Administrative Science Quarterly* 40:256–282.

42 Alexander, E. 1994. The non-Euclidean mode of planning: What is it to be? *Journal of the American Planning Association* 60:372–376.

43 Friedman, J. 1994. The utility of non-Euclidean planning. *Journal of the American Planning Association* 60:377–379.

44 Barmark, J., and G. Wallen. 1986. The interaction of cognitive and social factors in steering a large scale interdisciplinary project. In *Interdisciplinary analysis and research: Theory and practice of problem-focused research and development,* ed. D. Chubin et al., 229–239. Mt. Airy, Md.: Lomond Publications.

45 Scheidel, T. 1996. Divergent and convergent thinking in group decision-making. In *Communication and group decision-making,* ed. R. Hirokawa and M. Poole, 113–130. Beverly Hills: Sage.

46 Taylor, J. 1986. Building an interdisciplinary team. In *Interdisciplinary analysis and research: Theory and practice of problem-focused research and development,* ed. D. Chubin et al., 141–154. Mt. Airy, Md.: Lomond Publications.

47 Narayanan, V., and L. Fahey. 1982. The micropolitics of strategy formulation. *Academy of Management Review* 7:25–34.

48 Entwistle, N. 1981. *Styles of learning and teaching.* Chichester, U.K.: Wiley.

49 Hyman, I., et al. 1973. Patterns of interprofessional conflict resolution on school child study teams. *Journal of School Psychology* 11:187–195.

50 Barker, M. 1974. Information and complexity: The conceptualization of air pollution by specialist groups. *Environment and Behavior* 6:346–377.

51 Perkins, J. 1982. *Insects, experts, and the insecticide crisis: The quest for new pest management strategies.* New York: Plenum.

52 Brown, J. 1984. Professional hegemony and analytic possibility: The interaction of engineers and anthropologists in project development. In *Applied social science for environmental planning,* ed. W. Millsap, 37–59. Boulder, Colo.: Westview Press.

53 Erickson, P. 1979. *Environmental impact assessment: Principles and applications.* New York: Academic Press.

54 Hirschhorn, L. 1988. *The workplace within: The psychodynamics of organizational life.* Cambridge, Mass.: MIT Press.

55 Putnam, L. 1996 . Conflict in group decision-making. In *Communication in group decision-making,* ed. R. Hirokawa and M. Poole, 175–196. Beverly Hills: Sage.

56 DeBiasio, A. 1986. Problem solving in triads composed of varying numbers of field-dependent and field-independent subjects. *Journal of Personality and Social Psychology* 51:749–754.

57 Davies, M. 1985. Cognitive-style differences in belief persistence after evidential discrediting. *Personality and Individual Differences* 6:341–346.

58 Greene, L. 1976. Effects of field dependence on affective reactions and compliance in dyadic interactions. *Journal of Personality and Social Psychology* 34:569–577.

59 Gruenfeld, L., and T. Lin. 1984. Social behavior of field independents and dependents in an organic group. *Human Relations* 37:721–741.

60 Kennedy, J. 1988. Legislative confrontation of groupthink in U.S. natural resource agencies. *Environmental Conservation* 15:123–128.

61 Hirschhorn, L., and C. Barnett. 1993 . *The psychodynamics of organizations.* Philadelphia: Temple University Press.

62 Clark, T. 1993. Creating and using knowledge for species and ecosystem conservation: Science, organizations, and policy. *Perspectives in Biology and Medicine* 36:497–525.

63 Dunbar, R., J. Dutton, and W. Torbert. 1982. Crossing mother: Ideological constraints on organizational improvements. *Journal of Management Studies* 19:91–108.

64 Taylor, S., and R. Bogdan. 1980. Defending illusions: The institutions struggle for survival. *Human Organization* 39:209–218.

65 Holling, C., and G. Meffe. 1996. Command and control and the pathology of natural resource management. *Conservation Biology* 10:328–337.

66 Kessler, W., and H. Salwasser. 1995. Natural resource agencies: Transforming from within. In *A new century for natural resources management,* ed. R. Knight and S. Bates, 171–187. Washington, D.C.: Island Press.

67 Clarke, J., and D. McCool. 1996. *Staking out the terrain: Power and performance among natural resource agencies.* Albany: State University of New York Press.

68 Brunson, M., and J. Kennedy. 1995. Redefining "multiple use": Agency responses to changing social values. In *A new century for natural resources management,* ed. R. Knight and S. Bates, 143–158. Washington, D.C.: Island Press.

69 Westley, F. 1995. Governing design: The management of social systems and ecosystem management. In *Barriers and bridges to renewal of ecosystems and institutions,* ed. L. Gunderson, C. Holling, and S. Light, 391–427. New York: Columbia University Press.

70 Dietz, T., and R. Rycroft. 1987. *The risk professionals.* New York: Russell Sage Foundation.

71 Harshbarger, D. 1973. The individual and the social order: Notes on the management of heresy and deviance in complex organizations. *Human Relations* 26:251–269.

72 Dowie, M. 1995. *Losing ground: American environmentalism at the close of the twentieth century.* Cambridge, Mass.: MIT Press.

73 Sessions, G. 1996. Critical notice of Earth First! Environmental Apocalypse. *Trumpeter* 13:197–200.

74 ———. 1995. Political correctness, ecological realities, and the future of the ecology movement. *The Trumpeter* 12:191–196.

75 Shaiko, R. 1993. Greenpeace U.S.A.: Something old, new, borrowed. *Annals, AAPSS* 528:88–100.

76 Brown, G., and C. Harris. 1992. The U.S. Forest Service: Toward the new resource management paradigm? *Society and Natural Resources* 5:231–245.

77 DeBonis, J. 1995. Natural resource agencies: Questioning the paradigm. In *A new century for natural resources management,* ed. R. Knight and S. Bates, 159–170. Washington, D.C.: Island Press.

78 Mattson, D. 1996. Ethics and science in natural resource agencies. *BioScience* 46:767–771.

79 Villanueva, A. 1996. Conflict and cooperation in environmental administration. *Social Science Journal* 33:421–435.

80 Anderson, W. 1992. The reasoning of the strongest: The polemics of skill and science in medical diagnosis. *Social Studies of Science* 22:653–684.

81 Parmerlee, M., J. Near, and T. Jensen. 1982. Correlates of whistle-blowers' perceptions of organizational retaliation. *Aministrative Sciences Quarterly* 27:17–34.

82 Williams, N., G. Sjoberg, and A. Sjoberg. 1980. The bureaucratic personality: An alternative view. *Journal of Applied Behavioral Science* 16:389–405.

83 O'Day, R. 1974. Intimidation rituals: Reactions to reform. *Journal of Applied Behavioral Science* 10:373–386.

84 Monthey, R. 1994. Organizational adaptations to ecosystem management in the Pacific northwest. In *Managing forests to meet peope's needs: Proceedings of the 1994 Society of American Foresters/ Canadian Institute of Forestry convention.* 18–22 September. Anchorage, Alaska: Society of American Foresters.

85 Hilborn, R. 1992. Can fisheries agencies learn from experience? *Fisheries* 17:6–14.

86 Springer, J. 1985. Policy analysis and organizational decisions: Toward a conceptual revision. *Administration and Society* 16:475–508.

87 Bullis, C. 1991. Communication practices as unobtrusive control: An observational study. *Communication Studies* 42:254–271.

88 Miller, A. 1984. Professional dissent and environmental management. *The Environmentalist* 4:143–152.

89 Perrucci, R., et al. 1980. Whistle-blowing: Professionals' resistance to organizational authority. *Social Problems* 28:149–164.

90 Swift, J. 1983. *Cut and run: The assault on Canada's forests.* Toronto: Between the Lines.

91 Norgaard, R. 1992. Coordinating disciplinary and organizational ways of knowing. *Agriculture, Ecosystems, and Environment* 42:205–216.

92 Lee, R. 1992. Ecologically effective social organization as a requirement for sustaining watershed ecosystems. In *Watershed management: Balancing sustainability and environmental change,* ed. R. Naiman, 73–89. New York: Springer-Verlag.

93 Staw, B., L. Sandelands, and J. Dutton. 1981. Threat-rigidity effects in organizational behavior: A multilevel analysis. *Administrative Science Quarterly* 26:501–524.

94 Bella, D. 1987. Organizations and systematic distortion of information. *Journal of Professional Issues in Engineering* 113:360–370.

95 Deshpande, R., and A. Kohli. 1989. Knowledge Disavowal—Structural determinants of information-processing breakdown in organizations. *Knowledge: Creation, Diffusion, Utilization* 11:155–169.

96 Trist, E. 1993. The assumptions of ordinariness as a denial mechanism: Innovation and conflict in a coal mine. In *The psychodynamics of organizations,* ed. L. Hirschhorn and C. Barnett, 165–175. Philadelphia: Temple University Press.

97 Knuth, B., et al. 1995. Fishery and environmental managers' attitudes about and support for Lake Trout rehabilitation in the Great Lakes. *Journal of Great Lakes Research* 21:185–197.

98 Clarke, L. 1989. *Acceptable risk: Making decisions in a toxic environment.* Berkeley: University of California Press.

99 Collingridge, D. 1992. *The management of scale: Big organizations, big decisions, big mistakes.* London: Routledge.

100 Lyles, M., and I. Mitroff. 1980. Organizational problem formulation: An empirical study. *Administrative Sciences Quarterly* 25:102–119.

101 Caldwell, L. 1994. Disharmony in the Great Lakes Basin: Institutional jurisdictions frustrate the ecosystem approach. *Alternatives* 20:26–31.

102 Dryzek, J. 1987. *Rational ecology: Environment and political economy.* New York: Blackwell.

103 Harrison, K. 1990. *Between science and politics: Assessing the risks of dioxins in Canada and the United States.* University of British Columbia. Prepared for delivery at the annual Meeting of the Canadian Political Science Association, Victoria, B.C., 27–29 May 1990.

104 Smil, V. 1993. *Global Ecology: Environmental change and social flexibility.* London: Routledge.

105 Beckerman, W. 1995. *Small is stupid: Blowing the whistle on the greens.* London: Duckworth.

106 Collingridge, D., and C. Reeve 1986. *Science speaks to power: The role of experts in policy making.* London: Francis Pinter.

107 Bosso, C. 1987. *Pesticides and politics: The life cycle of a public issue.* Pittsburgh: University of Pittsburgh Press.

108 Salazar, D., and D. Alper. 1996. Perceptions of power and the management of environmental conflict: Forest politics in British Columbia. *Social Science Journal* 33:381–399.

109 Dumont, C. 1996. The demise of community and ecology in the Pacific northwest: Historical roots of the ancient forest conflict. *Sociological Perspectives* 39:277–300.

110 Gillis, R., and T. Roach 1986. *Lost initiatives: Canada's forest industries, forest policy, and forest conservation.* Westport, Conn.: Greenwood Press.

111 Ophuls, W., and A. Boyan. 1992. *Ecology and*

the politics of scarcity revisited: The unraveling of the American dream. New York: W. H. Freeman.

112 Dobuzinskis, L. 1992. Modernist and postmodernist metaphors of the policy process: Control and stability vs. chaos and reflexive understanding. *Policy Sciences* 25:355–380.

113 Hausler, S. 1993. Community forestry: A critical assessment. The case of Nepal. *The Ecologist* 23:84–90.

114 Edwards, D. 1996. Sci-Fi or Chomsky? *Ecologist* 26:76–77.

115 Gould, K. 1993. Pollution and perception: Social visibility and local environmental mobilization. *Qualitative Sociology* 16:157–178.

116 West, P. 1994. Natural resources and the persistence of rural poverty in America: A Weberian perspective on the role of power, domination, and natural resource bureaucracy. *Society and Natural Resources* 7:415–427.

117 MacNutt, W. 1963. *New Brunswick: A history: 1784–1867.* Toronto: Macmillan.

118 Baskerville, G. 1995. The forestry problem: Adaptive lurches of renewal. In *Barriers and bridges to renewal of ecosystems and institutions,* ed. L. Gunderson, C. Holling, and S. Light, 37–102. New York: Columbia University Press.

119 Cashore, B. 1988. *The role of the provincial state in forest policy: A comparative study of British Columbia and New Brunswick.* Master's thesis, Carleton University, Ottawa, Canada.

120 Kunioka, T., and L. Rothenburg. 1993. The politics of bureaucratic competition: The case of natural resource policy. *Journal of Policy Analysis and Management* 12:700–725.

121 Hoberg, G., and K. Harrison. 1994. It's not easy being green: The politics of Canada's green plan. *Canadian Public Policy* 20:119–137.

122 Brunner, R., and W. Ascher. 1992. Science and social responsibility. *Policy Sciences* 25:295–331.

123 Adams, P., and A. Hairston. 1996. Calling all experts: Using science to direct policy. *Journal of Forestry* 94:27–30.

124 Bolin, B. 1994. Science and policy making. *Ambio* 23:25–29.

125 Ewart, A., ed. 1996. *Natural resource management: The human dimension.* Boulder, Colo.: Westview Press.

126 Lubchenko, J., et al. 1991. The sustainable biosphere intiative: An ecological research agenda. *Ecology* 72:371–412.

127 Malone, T., and R. Corell. 1989. Mission to planet earth revisited: An update on studies of global change. *Environment* 31:6–11, 31–35.

128 Marcin, T. 1995. Integrating social sciences into forest ecosystem management research. *Journal of Forestry* 93:29–33.

129 Orians, G. 1986. The place of science in environmental problem solving. *Environment* 28:12–17, 38–41.

130 Stern, P., O. Young, and D. Druckman, eds. 1992. *Global environmental change: Understanding the human dimensions.* Washington, D.C.: National Academy Press.

131 Ozawa, C. 1996. Science in environmental conflicts. *Sociological Perspectives* 39:219–230.

132 Kwa, C. 1987. Representations of nature mediating between ecology and science policy: The case of the international biological programme. *Social Studies of Science* 17:413–442.

133 Wildavsky, A. 1995. *But is it true? A citizen's guide to environmental health and safety issues.* Cambridge, Mass.: Harvard University Press.

134 Parlour, J. 1978. The roles of specialists and institutions in the development of environmental information. *Journal of Environmental Management* 7:219–234.

135 Robinson, J. 1992. Risks, predictions, and other optical illusions: Rethinking the use of science in social decision-making. *Policy Sciences* 25:237–254.

136 Florio, E., and J. Demartini. 1993. The use of information by policymakers at the local community level. *Knowledge: Creation, Diffusion, Utilization* 15:106–123.

137 Greenberg, M. 1992. Impediments to basing government health policies on science in the United States. *Social Science and Medicine* 35:531–540.

138 Hoberg, G. 1990. Risk, science, and politics: Alachlor regulations in Canada and the United States. *Canadian Journal of Political Science* 23:257–277.

139 Martin, B. 1988. Scientific controversies: Nuclear winter: Science and politics. *Science and Public Policy* 15:321–334.

140 Stockdale, J. 1988. Facts or ideology? Science and environmental policy. *Wisconsin Sociologist* 25:144–152.

141 Carver, J. 1976. Energy, information, and public policy. *American Behavioral Scientist* 19:279–285.

142 Brown, G. 1993. Science + advice science policy advice: The role of scientific expertise in policy-making. *Research Technology Management* 36:9.

143 Ludwig, D., R. Hilborn, and C. Walters. 1993. Uncertainty, resource exploitation, and con-

servation: Lessons from history. *Ecological Applications* 3:547–549.

144 Ludwig, D. 1993. Environmental sustainability: magic, science, and religion in natural resource management. *Ecological Applications* 3:555–558.

145 Milbrath, L. 1989. *Envisioning a sustainable society.* Albany, N.Y.: State University of New York Press.

146 Sessions, G. 1995. Postmodernism and environmental justice: The demise of the ecology movement? *The Trumpeter* 12:150–154.

147 Perrolle, J. 1993. Comments from the special issue editor: The emerging dialogue on environmental justice. *Social Problems* 40:1–4.

148 Pinderhughes, R. 1996. The impact of race on environmental quality: An empirical and theoretical discussion. *Sociological Perspectives* 39:231–248.

149 Tatroff, D. 1993. Clear-cut thinking. *Our Times* 12:26–34.

150 Lertzman, K., J. Rayner, and J. Wilson. 1996. Learning and change in the British Columbia forest policy sectior: A consideration of Sabatier's advocacy coalition framework. *Canadian Journal of Political Science* 29:111–133.

151 Higgelke, P., and P. Duinker. 1993. *Open doors: Public participation in forest management in Canada.* Report to the Canadian Pulp and Paper Association, Forestry Canada, Lakehead University, Thunder Bay, Ontario, Canada.

152 Wiedemann, P., and S. Femers. 1993. Public participation in waste management decision making: Analysis and management of conflicts. *Journal of Hazardous Materials* 33:355–368.

153 Modavi, N. 1996. Mediation of environmental conflicts in Hawaii: Win-win or co-optation? *Sociological Perspectives* 39:310–316.

154 Gericke, K., and J. Sullivan. 1994. Public participation and appeals of Forest Service plans— An empirical examination. *Society and Natural Resources* 7:125–135.

155 Nelkin, D., and M. Pollak. 1979. Consensus and conflict resolution: The politics of assessing risk. *Science and Public Policy* 6:307–318.

156 Wynne, B. 1982. *Rationality and ritual: The Windscale inquiry and nuclear decisions in Britain.* BSHS Monograph 3. Chalfont St. Giles, U.K.: British Society for the History of Science.

157 Cortner, H., and M. Shannon. 1993. Embedding public participation. *Journal of Forestry* 91:14–16.

158 Mohai, P. 1987. Public participation and natural resource decision-making: The case of the RARE II decisions. *Natural Resources Journal* 27:123–155.

159 Miller, A., and W. Cuff. 1986. The Delphi approach to the mediation of environmental disputes. *Environmental Management* 10:321–330.

160 Baskerville, G., and P. Duinker. 1986. Pest management in plantations: An institutional analysis. In *Pest management in plantations: A consultative approach,* ed. N. Sonntag et al. Vancouver, B.C.: ESSA Ltd.

161 Diemer, J., and R. Alvarez. 1995. Sustainable community, sustainable forestry: A participatory model. *Journal of Forestry* 93:10–14.

162 deLeon, P. 1990. Participatory policy analysis: Prescriptions and precautions. *Asian Journal of Public Administration* 12:29–54.

163 Steel, B., J. Pierce, and N. Lovrich. 1996. Resources and strategies of interest groups and industry representatives involved in federal forest policy. *Social Science Journal* 33:401–419.

164 Almeida, P., and L. Stearns. 1998. Political opportunities and local grassroots environmental movements: The case of Minamata. *Social Problems* 45:37–60.

165 Cable, S., and M. Benson. 1993. Acting locally: Environmental injustice and the emergence of grass-roots environmental organizations. *Social Problems* 40:464–477.

4

Conventional Problem Solving

The term *problem solving* is applied to a variety of activities ranging from puzzle solving, such as cleaning up a toxic spill, to the more complex and uncertain tasks of planning and policy making. In all cases, the typical problem-solving process is less rational than some would like to see.[1] Consequently, practical problem solving strays far from the rational-comprehensive ideal so revered by technical professionals. According to this latter way of thinking, problem solving should pass through a number of discrete stages arranged in a logical sequence, with each step building on the previous in a coherent manner (Figure 4.1). It is assumed, for instance, that it is possible to begin the problem-solving sequence by achieving consensus on the basic nature of the problem in question, which allows subsequent development of a comprehensive picture of it during the problem definition stage. Once this exhaustive understanding has been achieved, a variety of possible solutions are generated in some unspecified, but rational, way. A logical, often computer-assisted, decision process allows the most promising solution to be selected for implementation. Once choices have been made, the policy is implemented as planned by managers and other operational experts, closely following the policy guidelines laid down. After a suitable interval, the impact on the problem of the chosen alternative is carefully evaluated against some established criteria of effectiveness. The results of this evaluation are then used to modify the original conception of the problem, and new routes to its solution may be explored, thereby starting the problem-solving cycle rotating once again. On the whole, however, given the rational and comprehensive nature of the earlier phases, it is assumed that redirection of efforts at such a late stage would most likely be unnecessary.[2]

In practice, however, problem solving is much more disorderly and nonlinear. Problems may be denied, suppressed, or formulated in ways that divert attention away from those responsible. Often there is a mismatch between the seriousness of the problem and the attention afforded it.[3] Commonly, actions are taken without a full understanding of the problem, and, even worse, politically acceptable solutions are imposed on problem situations contrary to available wisdom. In other words, problem solving in the real world is a confused mess in which problem solvers are usually at the mercy of events rather than in control of them. This makes discussion of the problem-solving process rather difficult unless, for the sake of convenience, one assumes a more rational sequence of events than is found in the real world. For descriptive purposes, therefore, the next two chapters are organized along the lines indicated in Figure 4.1. Chapter 4 outlines the conventional problem solving that dominated natural resource (and other

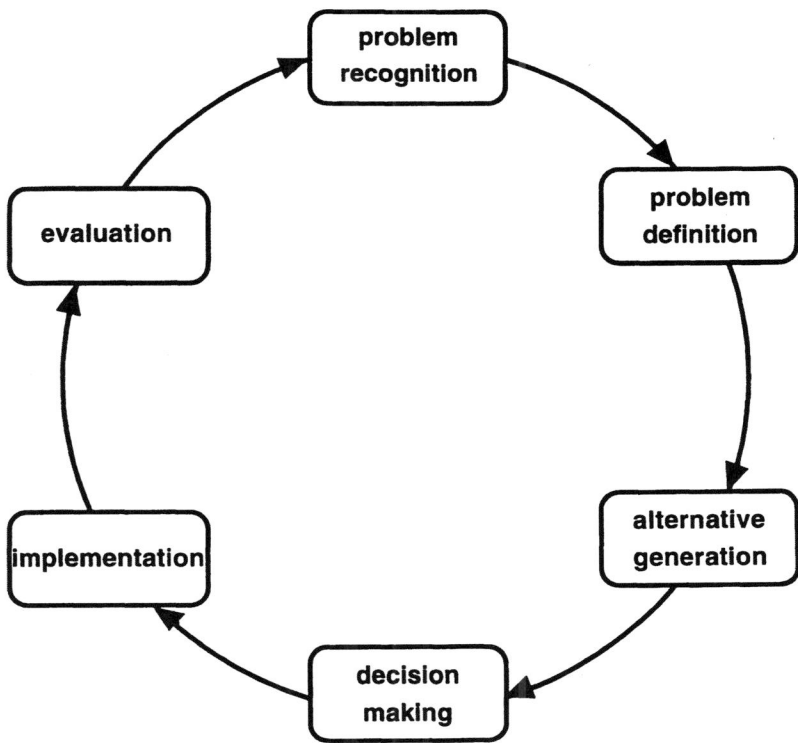

FIGURE 4.1. Rational problem-solving stages.

forms of) management prior to the 1970s. The impact on environmental problem solving of the subsequent upsurge of environmentalism in the 1970s and 1980s, with its politics of confrontation, is discussed in chapter 5.

The essential argument set out in the present chapter is that the conventional problem solving exhibited by the iron triangles of yesteryear was, and is, ineffective in dealing with the environmental problems that confront us. The uneasy alliance between political necessity, corporate vested interest, and bureaucratic self-interest is hardly a recipe for innovative approaches to complex environmental problems. Rather, I argue that conventional problem solving is prone to *type III* errors, the tendency to address the wrong problem or to obscure "a more profound problem by preoccupation with a lesser issue."[4] Thus, cleaning up toxic dumps may be a laudable activity and wonderful news for local residents, but it leaves in place the sociopolitical system that created the problem in the first place. The

lesser issue (dump site) provides an opportunity for decisive action by governments who accrue political dividends, while the more fundamental questions of political economy are ignored by all but the most persistent environmentalists.

In talking about conventional problem solving, therefore, I'm referring to the kind of policy-planning process that has dominated natural resource and environmental management since the First World War. As mentioned previously, policy making has been dominated by iron triangles operating within a corporate state in pursuit of traditional Imperial goals.[5,6] In Canadian forestry, for instance, the rampant exploitation of the eighteenth and nineteenth centuries gave way to an attempt by provincial governments to control the worst excesses of the lumber barons through regulation of harvesting.[7] This administrative phase of forestry[8] resulted in close cooperation between industry, government, and forestry experts, an iron triangle that has dominated policy making for much of this century. It was only in the 1970s and onward that this arrangement was questioned by other sectors of society. The question of interest here is how effective this conventional problem-solving system has been, and is, in solving the environmental problems of the day. I shall explore this matter further in the context of a detailed case study, that of the spruce budworm (SBW) problem in New Brunswick, Canada.

Introduction to the Spruce Budworm Case

The New Brunswick Environment (E)

New Brunswick (NB), one of Canada's smallest provinces, is located on the eastern seaboard, where it is bounded to the west by Maine and to the north and east by the provinces of Quebec and Nova Scotia, respectively. Despite being small by Canadian standards, it covers roughly 28,000 square miles (72,500 square kilometers), approximately half the size of England in area but not population. While close to 50 million souls are crowded within England's shores, only 724,000 are spread thinly over the NB landscape. Some 85 percent of NB's land surface is forested, primarily with a mixed Acadian forest representing a transition zone between the deciduous hardwoods to the south and the coniferous softwoods of the boreal forest to the north. Birch, maple, and beech are the typical hardwoods, and the most frequent conifers are balsam fir; white, red, and black spruce; and white pine.[9] Because of its many rivers, which were used as routes to the interior by settlers and a means of transporting lumber to the coast, most of the province has been easily accessible to industrial forestry. As a result, virtually the whole of the forested area has been harvested several times. There is no significant old growth or wilderness forest left in NB, unlike the more recently settled provinces in western Canada. The few pockets of old growth that are left are of dubious significance, although they have

become bones of contention between environmentalists and the forest industry.[10]

As a result of 200 years of continuous cropping, the forest has changed profoundly from that first encountered by the first European settlers. The majestic pine and spruce that abounded in the presettlement forest, like pillars in a natural cathedral, have long since been replaced by what some refer to as toothpicks, small trees in dense thickets.[11] Although significant change has taken place, there is considerable disagreement over its nature and extent. Professional foresters are inclined to argue that, while it has changed in terms of age structure and timber quality, the species composition of the forest remains much the same as in presettlement times.[9] While acknowledging that this is probably the case, those of a more Arcadian persuasion tend to emphasize the decline in forest quality that has accompanied settlement.[12] European exploitation began in the eighteenth century with the harvesting of the massive white pine, much sought after by the Royal Navy as masts for their warships. As the oldest and best trees were removed, or became less accessible, attention turned to smaller specimens as new markets were developed for a lumber industry. With the subsequent decline of white pine in the middle of the nineteenth century, industry turned to white spruce, which sustained the lumber mills until the early years of the present century. At this point, the preceding decades of unrestrained exploitation finally started to make themselves evident as white spruce, in turn, declined in quality, leaving the forest industry in dire straights. The response of the provincial government, in the 1920s and 1930s, was to encourage pulp and paper production so that use could be made of the remaining spruce-fir forest.[7,13] We see, in this sequence of events, a pattern so typical of resource exploitation. As one species is worked to commercial extinction, technological innovations allow industry to begin the cycle again with a new species.[14] In the meanwhile, future decision-making options are reduced as more capital is invested in new industrial infrastructure, and the economic well-being of the province becomes increasingly dependent on a single industry.

With the shift to pulp and paper production in NB, and the new interest in the spruce-fir forest, industry came face to face with the spruce budworm (SBW) (*Choristoneura fumiferana*), an indigenous pest whose depredations had, until this point, been of little commercial significance. The SBW is endemic throughout eastern Canada and the northeastern United States, breaking out into epidemics at regular intervals. Conventional wisdom has it that epidemics occur when large areas of unbroken mature and overmature balsam fir are present, together with several consecutive years of warm, dry spring weather. The SBW caterpillars feed on new foliage and, during epidemics, may virtually defoliate the host tree. Successive years of defoliation results in extensive stand mortality over a wide area.[15] Balsam fir is thought to be the most vulnerable species, followed by white spruce and some of the other spruce species. This cycle of forest growth followed by insect outbreak, tree die-back, and the subsequent collapse of the epidemic is a natural cycle that has maintained forest composition and insect life in harmony over the centuries. More recently, it has been suggested that the frequency and intensity of SBW outbreaks has been increasing due to past harvesting practices that favored the regeneration of balsam fir. Thus, historical studies have shown that, in previous centuries, the interval between outbreaks appears to have been from 60 to 100 years, whereas in the twentieth century the interval has been reduced to 19–37 years.[16] In New Brunswick, for instance, the major outbreak of 1913–1920 was followed by a prolonged outbreak that started to build up in 1949, barely 30 years later.

The response of the provincial government to the 1950s outbreak was, at the prompting of the major forest companies, to mount a chemical control program to combat the epidemic.[17] As a result, there commenced in 1952 what was to become the world's most extensive forest-spraying program. At its height, in 1976, approximately 4 million hectares (57 percent of New Brunswick's area) were sprayed with chemical pesticides. This annual attack on the SBW continues to the present day, albeit in a much-reduced form, making it the most sustained forest-spraying operation the world has seen.[17]

New Brunswick Sociopolitics (S)

The decision to implement a SBW spraying program in 1952 was made by a small group of politicians and corporate executives against the advice of a substantial body of scientific opinion. Forest entomologists of the day advised great caution in embarking on large-scale use of DDT, as we shall see. In addition, there was considerable disquiet within the professional forestry community and among the general public about this course of action. However, it was more than 20 years before the public became sufficiently aroused to act on their doubts. This elite control of decision making had its roots in the political origins of NB as a British colony. New Brunswick began to emerge as a political entity at the end of the bitter struggle between Britain and France for domination of North America. In 1763, NB was ceded to Britain along with the rest of Nova Scotia, of which it was still a part. A decade later, the rebellion of the 13 American colonies against the British Crown drove a flood of Loyalist refugees into Canada. During 1782–1784, for instance, 35,000 Loyalists sailed for Nova Scotia, many of whom settled along the St. John River in what was to become the province of New Brunswick. It is understandable that the new settlers should bring with them an antirepublican stance, a belief in British Imperial rule and a certain predilection for an aristocratic society. These sentiments were not shared by many of the established settlers who had arrived earlier from, and retained strong ties with, the New England states. Indeed, for a brief period, it was uncertain whether NB would remain loyal to the Crown or join forces with the rebellious colonies to the south. However, the proximity of the military and naval bastion at Halifax appears to have ensured that NB remained within the Empire.[13]

The form of government that developed under colonial rule was similar to that of other British colonies of the time: an appointed governor and council with direct responsibility to London, together with a relatively weak legislative assembly. Membership of both the council and the assembly was restricted to the privileged classes who attempted to ensure their standing by personal accumulation of landholdings and, in the case of members of the legislature, catering to their constituencies through dispersing public largesse in the form of money and jobs, a form of patronage vestiges of which remain to this day.[18] At the same time, the granting and sale of land remained under the control of the executive council until 1837, when the legislative assembly assumed control of this politically sensitive task.[13] Over the course of time, large areas of Crown land were granted or sold to the wealthy, while smaller grants of 100-acre lots went to new settlers. As a consequence, only 54 percent of NB remains as Crown land, a far smaller proportion than found in most other Canadian provinces. Nevertheless, much of the political maneuvering in early NB was over gaining ownership of land and control over revenues from exploitation of the Crown forests.[13] Indeed, the struggle for control of Crown land became, and remains, a central political theme in NB politics. All of these struggles were between elites, however. The common people were not involved in any significant way. They had neither the wherewithal nor the political connections to become major players in carving up the new territory. Some achieved a modicum of independence on their 100-acre land grants, but many remained dependent on the whims of those who controlled the production and marketing systems that were taking shape.[19] Thus, acquiescence to decision making by political elites started early and, despite a history of regular but ineffective dissent, remains to the present day.[19,20]

The image of NB portrayed by regional historians and political commentators, therefore, is that of a "timber colony" that evolved into a "paper plantation" and a "client state" dependent on foreign capital.[7,19,21,22] That is, the political struggles that took place over ownership of land would have been fruitless in the absence of merchant capital to exploit the forest resource. Initially, the financing of industry came from local entrepreneurs, but, with the inevitable changes in technology and the size of industrial enterprises, capital investment became increasingly dependent on foreign sources. According to Sandberg[23] (p. 2–3), the consequence is that NB became a client state:

> over-dependent on a few key , often externally based companies . . . (remaining) financially dependent on, and ideologically committed to a few large monopoly capitalist firms. In the same way that the state became a client to large mercantile ventures, so too did individual households. Merchant capital employed debt bondage when engaging workers in seasonal spells of work.

In other words, in the last century, merchants provided woods workers with the means to spend a winter working in the woods, in the expectation that they would pay off their debts in the spring. Often, however, they would work for up to seven months in the woods, sawing and cording logs, only to find that at the end of the day their total returns were paltry. In this way, "merchants dominated woods workers, small woodlot owners and smaller sawmillers through patron-client relations based on credit"[23] (p. 4). As time went by, land ownership became concentrated in the hands of large sawmillers, which, in turn, set the stage for the arrival of multinational pulp and paper companies. The first large pulp and paper mill in 1915 was promoted by "Americans and Upper Canadians (from what is now Ontario and Quebec) with good local political connections, who were part of a continental network of financiers and industrial promoters"[23] (p. 5). At the present time, there are 12 pulp and paper mills in NB, their parent companies either owning or leasing about 68 percent of NB's forests and accounting for about 60 percent of forestry revenues. The 140 or so small sawmills and other processors account for the remaining 40 percent.[24] Thus, the forest industry in NB has not developed much beyond its eighteenth- and nineteenth-century roots in primary production.[20] It remains a source of fiber rather than an industrial center for finished wood products. The result, according to Sandberg[23] (pp. 11 and 21), is that:

The pulp and paper mills hold incredible power in (some NB) communities . . . This is perhaps to be expected given the lack of alternatives and local high unemployment. For the same reason, the large, mostly foreign, pulp and paper mills are . . . instrumental in the electoral success and survival of the provincial governments. . . . Corporations now dominate the forest. With the sanction and active enthusiasm of the client states, they determine the strategies for development . . . Alternative approaches become almost 'unthinkable'. Critics are few and quickly marginalized. A large part of the electorate is dependent on this system or has been lulled into the belief that such developments are inevitable.

However, there has always been an undercurrent of resentment among the rural population against their dependent state. Parenteau[25] (p. 21) believes that rural woods workers "have placed the responsibility for widespread poverty, unemployment and insecurity in the province squarely on the shoulders of the big corporations that now dominate the industry" and have engaged in a 60-year battle over the low prices paid for wood pulp. Sometimes these battles have moved beyond words, as woodcutters blockade pulpmills[26] and burn mechanical harvesters[27] in an attempt to draw attention to their grievances. Unfortunately, these protests often pit one set of workers against another, as each tries to protect its livelihood, and there is little progress toward constructive change.

Thus, the SBW problem arose in the context of a forest that had been degraded after almost two centuries of exploitation and that was located in a resource-dependent province reliant on a few major corporations for its economic well-being. The power to make decisions about forest management has traditionally been held by political and economic elites, with the latter playing an increasingly significant role as the province became dependent on the external capital investments needed for the construction and operation of industrial infrastructure.

New Brunswick Mind-sets (P)

The Europeans who settled NB brought with them the dominant Imperial mode of thinking. In many ways, the forest was a problem for them, something that stood in the way of agriculture. Indeed, in those early days, lumber barons believed they were doing others a service by clearing the forest.[7] However, this Imperial attitude was not shared by everyone. There is a long history of dissenting opinion in the region, starting with the resistance of the Micmac Indians to the despoliation of their land. The protest continued with a variety of settlers, often in positions of authority, who tried to restrain the worst excesses of their compatriots, often to their own disadvantage.[28] Much of this protest was in favor of better stewardship rather than promoting a more radical Arcadian vision. However, the latter was a constant, albeit marginal, feature of the political landscape seen in such developments as the conservation movement that arose at the end of the nineteenth century, an uneasy alliance between the practical concerns of scientific managers and lumber workers with the more Arcadian views of environmental preservationists.[7] How the SBW was dealt with within the context of this complex psychological, social, and ecological stew is a matter to which we now turn.

Problem Recognition

In the preparatory phase of problem solving, the dim outlines of a problem are first recognized and then fleshed-out as more information becomes available. Thus, problem recognition (or identification), the initial realization that a problem may exist, is the crucial first step in the problem-solving process. While it precedes problem definition, it begins to merge with this latter stage as more information is accumulated and the problem becomes more clearly evident. The distinction between the two stages, therefore, is arbitrary but useful.

Although environmental problems are changes in the biosphere that are potentially harmful, someone has to decide that these changes are important enough to warrant attention. Because environmental problems are embedded in a great deal of informational noise, they do not appear on the scene clearly defined in some easily recognizable form, despite the belief among many scientists and managers that problems are "objective entities that present themselves" to the observer.[29] In practice, all environmental problems are socially constructed. This is not to say that they are figments of a feverish imagination but that everyone's perception of problems is subjective, a matter of personal interpretation. Thus, there are real events happening in the environment, but their nature and significance depend on how they are interpreted. Any claim about the existence of a problem, therefore, is a marketing exercise in which someone becomes convinced that a problem exists and then tries to persuade others about the importance of his or her perception of reality. Inevitably, the individual is drawn into the public arena in which competing interests jostle for position in their attempt to convince others that their particular problem, or version of a problem, is more worthy of attention than anyone else's. It is perhaps a sad reflection of the frenetic nature of mass society that issues "must vie for attention against a backdrop of the limited processing capacity of government"[30] (p. 8). As a result, the attention afforded any single environmental problem is often fleeting, as windows of opportunity open and, just as quickly, close.

One of the more persistent features of human personality is that we feel compelled to protect our personal ways of seeing things.[31] As a result, we are inclined to engage in selective attention, paying heed to events that confirm our worldviews and ignoring those that do not. This is just as true of objective scientists as it is of the lay public.[32,33] Because all of us are programmed from birth to scan our surroundings for potential danger, problem recognition can be viewed, in large part, as the perception of risk. As we grow more sophisticated at this, the scanning process becomes automatic, receding into the background as we use our personal radars at a subconscious level

to monitor the world around us. Nevertheless, scanning for threats is a ubiquitous perceptual phenomenon that occupies much of our daily routine, although we are usually unaware of it. Because of our limited ability to manage the anxiety generated by threats to our well-being, we also engage in selective attention, carefully blocking out whatever threatens to overwhelm us. Some dangers cannot be avoided and must be dealt with, but the rest we ignore as best we can. The process of risk perception (i.e., problem recognition), therefore, is a very subjective endeavor, one that is strongly influenced by personal and social factors.

As one might expect, there are considerable differences between the way in which the public and the governing elites perceive and estimate risk. In environmental controversies, for instance, antagonists are likely to recognize problems that they perceive to be the result of an opponent's behavior or worldview, while ignoring any problems with their own perspective. As a consequence, most disputes assume a predictable form. Those who abide by the Imperialist worldview try to ignore the problems that result from industrial production and economic growth, whereas Arcadians go to great lengths to bring these problems to light. Imperialists respond by drawing attention to the utopian and irresponsible nature of Arcadian behavior and ideals. In modern industrial society, therefore, environmental problem recognition has become a battleground in which competing groups seek priority for their worldview or version of particular problems.[34] They do so by concocting stories or myths about the problem in question and attempt to garner support for their version of reality.[35] These stories are mixtures of fact and fiction that serve to make events meaningful to those involved while providing a compelling argument in support of their position. In previous decades such debates were muted, but they have increasingly assumed the form of a drama played out on a public stage where all sides seek control of the public discourse and policy making becomes a struggle over alternative realities.[30] Typically, those in powerful positions portray themselves as both reasonable and competent while more marginalized groups often take the moral high ground, arguing that their cause is morally just. In part, both of these postures are ways of seeking political advantage. By asserting their competence and limiting issues to narrow technical matters, dominant elites can restrict the number of potentially disruptive voices in a policy dispute. In contrast, when environmentalists recognize ethical and social justice issues, they are attempting to widen their opportunity for participation in debates by extending the discourse beyond mere technical expertise.[30] Thus, the extent to which powerful establishments make provision for public involvement in decision making, and their tolerance of dissent, will determine the nature and intensity of conflict over problem recognition. For many individuals, conformity to established practice and attention to officially sanctioned "problems" is their route of choice through this political and moral minefield.[36] Others are driven by their indignation over perceived wrongs to speak out.[37] This is not to say that such whistle-blowers and environmental activists are always correct in their perceptions of a problem. They, too, are subject to psychological biases and sociopolitical pressures that may distort their vision of reality. Thus, no version of a problem holds a monopoly on the truth.

The implication of all this political maneuvering is that problem recognition evolves over a period of time. Seldom does any single group of responsible people sit down one day and decide unequivocally that a problem exists. Instead, the many different players in the drama enter the picture at different stages, bringing with them their personal biases and vested interests, and adopting different postures about the seriousness of events. As the situation unfolds, therefore, there is seldom consensus on the nature of the problem. Indeed, there may be those who remain convinced that the whole affair is a storm in a teacup and that there was never any significant problem in the first place. Problem recognition during the days of exclusive iron triangles was relatively uneventful compared to the contentiousness of the more diverse interest-group politics that followed. How this drama unfolded in the SBW case is something to which we now turn.

SBW Problem Recognition

In 1945, natural resource bureaucrats in both the federal and provincial governments, as well as managers in the forest industry in eastern Canada, were aware that a major SBW epidemic was building. Federal entomologists, whose task it was to monitor the state of forest insect pests, had been watching an outbreak of SBW in Ontario since 1942. Three years later, as the pest spread inexorably eastward, their warnings were becoming more urgent.[38] Similar vigilance was evident in the northeastern United States where Department of Agriculture scientists were tracking the apparent movement of the epidemic into the New England states.[39] The implications for New Brunswick were that SBW would reach epidemic levels, possibly in 1949, threatening the spruce-fir forests of northern NB with potentially dire consequences for the pulp and paper industry and the NB economy. Many of the older men in the forestry community had personal recollections of the last major outbreak of 1913–1920 that had devastated large tracts of the softwood forest and, according to some, had contributed to the collapse of the lumbering industry in the 1920s, from which it had not recovered[40] (p. 128). Accordingly, 1945 saw the formation of the Forest Insects Control Board (FICB), a Canadian federal committee that was given the mandate of cooperating with the provinces and forest industries in controlling outbreaks of insect pests, especially that of SBW. Thus, the scientific community recognized a biological problem that they brought to the attention of the political-corporate elites who immediately perceived the political and economic implications of the impending epidemic. There was little disagreement over the nature and gravity of the *threat* posed by the epidemic, but less agreement on what should be done about it or who should pay. There was also a sense of frustration, even anger, among the scientific community that the problem had been allowed to develop in this way.

Forest entomologists had known since 1924 many of the basic features of SBW ecology.[41] They were aware that it was a cyclical endemic pest, one that would once again reach epidemic proportions 30 or so years into the future (1950s). Indeed, pulp and paper companies had been leery of situating new plants in NB for this very reason; there seemed to be little sense in investing money in a region that was periodically devastated by an endemic pest. Thus, one reason for the limited action to forestall a new outbreak of SBW during the intervening years seems to have been the absence of commercially attractive management methods. As early as 1911, entomologists had been arguing that the only practicable way of reducing damage in any future epidemics was to commence a program of forest management (selection harvesting and silviculture) that would reduce the SBW's favorite food supply (balsam fir). By reducing and breaking up the large stretches of mature and overmature stands of balsam fir, it was argued that growth of less vulnerable species, such as spruce, would also be promoted. This silvicultural hypothesis had been offered to the authorities regularly by government and academic entomologists for the 30 years preceding the 1940's buildup.[42] However, with the collapse of the 1913–1920 SBW outbreak, these admonitions appear to have taken second place to other, more pressing concerns. So we see, in 1945, representatives of industry accepting the need for increased forest management but acknowledging that progress toward this goal had been slow.[43] There were legitimate reasons for this tardiness. Implementation of an effective forest management program would have required driving new roads into areas most susceptible and vulnerable to attack. Not only was this extremely costly but it was not clear where to construct these roads because the areas of high risk were not adequately mapped. At the same time, since the pulp and paper industry was just beginning to be established in NB in the 1920s and 1930s, its tenure was uncertain and its future problematic. Indeed, future wood supply was the least of its problems, a much more pressing issue being the availability of adequate energy supplies to power the mills. In light of these uncertainties, and given the long time-horizon of future SBW threats, one can understand why there may have been hesitation in making the massive investments required for intensive forest management.

Thus, the initial phase of "problem recognition" was confined to a small group of politicians, bureaucrats, industrialists, and scientists, the so-called iron triangle of policy making. Within this policy community, there was little argument about the nature of the threat, couched as it was in terms appealing to the prevailing Imperial mode of thinking. Here was a biological threat that had to be controlled in the interests of the economic well-being of both industry and province. Any disagreements that arose within the elite groups were not about the SBW threat as such but, rather, over technical and economic matters.

Problem Definition

Bounding

An important early step in problem definition involves deciding on the scope of the problem to be attacked. Scientific definitions of the SBW problem, as that of an insect pest that was out of control, is consistent with the technical rationality adopted by objective-analytic personality types. Attention is restricted to the technical aspects of the problem, leaving extraneous political and economic matters to others. The latter, of course, were the major preoccupations of decision-making elites, who, in approaching the problem from the perspective of political rationality, saw the SBW problem primarily in terms of its political and economic consequences. In this scheme of things, the role of science was to monitor threats to the elites' political and economic ends, as well as providing technical fixes to remedy the looming problems.

By restricting the problem to that of an insect pest stepping out of bounds, other more controversial problem definitions were avoided. For instance, given the 30 years of relative inaction over forest management,[43] the problem could have been defined more broadly as a matter of inadequate forest policy. This problem definition might have raised questions about the political status quo, one in which the provincial government had limited power to control forest policy.[7,21] Or the problem could have been defined in more socioeconomic terms, raising questions about the wisdom of continued dependence on the pulp and paper industry that, in turn, was dependent on a vulnerable resource base. Such a problem definition could have led to a search for alternative value-added uses for the forest resource and the diversification of the forest industry in its reliance on fiber and lumber production. The need for alternative resource-use policies was a pressing issue at the time because of the social catastrophe that was occurring in rural NB. The era of small family farms, which had often depended on sales of pulpwood to supplement their incomes, was coming to an end. Between 1941 and 1961, the number of farms in NB fell by almost one-half.[19] Recognition of this broader, social problem, along with the imminent SBW epidemic, might have offered the opportunity for radical sociotechnical approaches to both problems, but such innovative thinking did not arise until the early 1970s, as we shall see. Instead, the government of the day appears to have squandered money on infrastructure projects that did little to move the forest industry away from reliance on primary production.[18]

One could reasonably conclude that a form of groupthink pervaded the policy-making community at the time. Although there was a lively debate within the scientific community over control methods, there does not appear to have been any such ferment over problem definition. Speaking of the state of affairs in Nova Scotia, Webster[28] (p. 151) observes that:

While a few voices . . . were strongly critical of the prevailing forest regime before the Second World War, evidence of dissent amongst foresters and officials in the Department of Lands and Forests in the post-war era is elusive. It may be that this sort of evidence is merely hidden, cloaked by mechanisms used by the Department to keep evidence of intra-departmental dissent concealed. Even if this in the case, there still remains no public evidence of such dissent, some of which would surely have emerged had any significant degree of dissent existed within the bureaucracy.

Limiting the problem definition to that of a technical issue was convenient, however, for it allowed the implementation of a technical solution made possible by the advent of DDT and the availability of war-surplus aircraft. The possibility of expanding the problem definition to include more social issues seems not to have arisen, as the need to take immediate protective measures became paramount. Thus, the problem was seen as a biological crisis that needed prompt attention, and further social planning was lost in the mists of some future time. Running throughout the deliberations from 1945 onward was the evident alarm about the possible destruction of valuable wood supplies and the paramount need to protect the resource. The forest industry, notably NB International Paper, wanted its pulpwood reserves protected, while the provincial government was preoccupied with protecting its economic mainstay. The goal of subsequent problem-solving efforts, therefore, was to protect the softwood forest of NB.

Information Search

Given the technical definition of the problem, it was assumed by the majority of those involved that if a solution was to be found it would be technical in nature. On the basis of available information, there was scientific consensus that there were only three possible routes to SBW control: (1) chemical control by aerial spraying with DDT, (2) biological control using parasites and diseases, and (3) forest management (silviculture) aimed at changing forest composition.[44] In the 1940s, therefore, scientific efforts were made to explore these various options, although all involved were aware that there was little time to develop biological and silvicultural control methods. It might also be pointed out that all of this activity was set within the Imperial mode of thinking. The purpose of research was to find ways of manipulating either the insect pest or its host species, not to explore ways in which human beings might change their lifestyles to live with, rather than seeking to control, SBW epidemics. Such Arcadian perspectives were not prevalent at the time, nor are they now.

Experiments with the aerial spraying of DDT commenced in 1945 with the spraying of test sites in the Nicolet National Forest in Wisconsin[45] and in the Suffield Experimental Station in Alberta.[46] These were to continue in different locations over the next few years, allowing scientists and managers from both the United States and Canada to cooperate in determining effective dosages and delivery systems. There was no doubt in anyone's mind that DDT could kill the pest. The main questions that needed to be resolved were primarily to do with logistical and operational detail, together with an estimation of the impact of DDT on the forest ecosystem. Even in 1949, many of the forest entomologists involved were very reluctant to give their approval to the large-scale deployment of DDT. One academic member of the Forest Insect Control Board, for instance, argued that, in principle, he thought spraying was wrong and that, in their present state of ignorance, DDT spraying might do more harm than good.[47] This scientific caution among traditional entomologists was not shared by economic entomologists who were more eager to try DDT.[48]

Meanwhile, the prospects for biological control of SBW were considered to be rather slim. In 1945, it was known that the biological control of indigenous species was less effective than previous successes in controlling exotic pests. Indigenous species have had millenia to adapt to the complex of parasites and disease in which they are embedded, so that manipulation of this complex is difficult, if not impossible. However, research on potential viral diseases and other pathogens was conducted by both Canadian and American scientists who scoured western North America and Europe for potential candidates. In one case, thought was given to importing gypsy moth and European budworm eggs to help cultivate their viral pathogens.[49]

Forest management using silvicultural practices was the strategy preferred by most of the forest entomologists involved, based on their scientific knowledge of SBW biology and ecology that had been amassed over the years. In particular, its life cycle, feeding habits, and dispersal mechanisms were reasonably familiar, and something was known about the parasites and diseases that kept it in check.[41] Less well known were the factors that instigated an outbreak and, most importantly, what caused epidemics to collapse. It was also apparent to many foresters that the spruce-fir forest on which postwar industrial expansion was based had been significantly shaped by previous SBW outbreaks. For instance, the major portion of the forest was in only two age classes, the oldest dating back to an earlier SBW outbreak around 1880 and stands of younger trees that owed their prominence to the more recent outbreak that ended in the early 1920s.[9] The power of the SBW to alter the forest profoundly was, therefore, quite evident. Thus, most of the senior scientists involved wanted to respond to this power with an ecologically based program of management using harvesting and silvicultural practices to deal with the outbreak. In meetings of Forest Insects Control Board, and other venues, representatives of industry were reminded frequently about the limited progress being made in this regard.[50] Other scientific observers of the scene, while skeptical about the activities of the forest industry, remained optimistic about the long-term prospects for silviculture[51] (p. 396):

operators generally have not arrived at a realization of either the necessity or the advantages of rational silvicultural practices. In other words, they do not consider them profitable under the present conditions. This is no doubt due, primarily, to the fact that virgin stands are still available. As time goes on and as the depletion of the forest progresses, the present attitude towards silviculture in commercial forestry will gradually become altered; in fact, some companies are already looking towards scientific management of their forests with a view to ensuring continuous production.

Despite conventional wisdom being in its favor, the silvicultural route to SBW control was, and remains, unproven.[42,52] Many of its advocates have displayed appropriate scientific caution in recognizing the need for large-scale experiments to see if it would work.[53] In any case, the onrush of events swept away these cautious deliberations as officials scrambled to find an immediate solution to the SBW problem.

Information Integration and Modeling

The problem faced by scientists in advising government, in this crucial period between 1945 and 1950, was to make sense of the available evidence about SBW control. Because there was a clear preference among scientists for silvicultural solutions, and disagreements about the wisdom of starting on a regime of DDT spraying, it might be worthwhile to look at this scientific debate in more detail. In what follows, I draw heavily on Miller and Rusnock.[42]

The function of the spruce budworm in the forest ecosystem of northeastern North America is a complicated matter. Given the vastness of the affected areas and the wide variety of forest, topographical, and climatic conditions found across this range, as well as the different histories of human intervention in these forests, it should come as no surprise that a comprehensive view of spruce budworm and forest ecology has been difficult to achieve. In the 1940s, a particular conception of ecological stability underscored scientists' understanding of the susceptibility of forests to SBW attack. Stability was viewed in terms of an ecological climax in which plant and animal communities persist in a relatively unchanged form unless interrupted by some severe disturbance. Such associations are thought to have an inherent resistance to disease, insect outbreaks, and various other natural dangers. On the other hand, when such associations are broken up, these natural controls are disrupted, and the forest is rendered susceptible to insects and disease.[41,53-55] Westveld et al.[56] (pp.145–146) describe this notion of stability:

In studying the biology of forest insects, entomologists have found that in mixed forests there are environmental conditions ideal for maintaining natural control agents such as parasites, predators, and birds that attack a variety of forest insects. Consequently these agents are already present in considerable numbers whenever any noxious insect species suddenly increases to epidemic proportions. These natural factors may multiply enough to help check an infestation . . . But when the original forest cover is radically disturbed, these natural controls are partially lost. Fire, poor logging practices, cultivation or other disrupting agents may bring about a stand that is wholly unrelated in composition to the original forest. Because of the changed site conditions, species that were unable to maintain themselves in the original associations become established. In the absence of vigorous competitors, these intruders take possession of extensive areas, often in relatively pure stands outside their natural range, and on sites to which they are only temporarily well adapted. Such stands are especially susceptible to forest pests and other damaging agents. . . . Dangerous situations arise when cutting practices, continued persistently and over a long period, favor species other than those characteristic of the site. Under incorrect cutting methods, stand composition and quality deteriorate and the forests become easy prey to a host of enemies.

The presettlement forests of the northeast were thought to possess this kind of stability and, with it, the inherent resistance to insects and disease, including budworm. Losses may have occurred due to budworm attack, but they would be small, on the whole, and relatively nondisruptive.[41] The budworm's new status as a forest pest was due, on the whole, to human activity in the forest (logging, fire, insecticide spraying, etc.), which has disrupted the original composition and produced unstable conditions. Early speculation about susceptibility to attack stressed the importance of *food supply* as the main cause of the development of spruce budworm outbreaks. A decline in food supply through mortality of mature fir and spruce was correspondingly thought to explain the collapse of outbreaks.[57] It was also thought that favorable weather conditions helped in the development of spruce budworm outbreaks, which are associated with relatively dry, warm springs recurring over several years.[58] Analysis of patterns of observed damage have led some to speculate that spruce budworm epidemics migrate outward from relatively small areas of infestation, the "epicenters" that were assumed to arise in central Canada and from which the epidemic spread eastward on the prevailing winds.[59,60] Most practical recommendations for minimizing SBW damage are directly related to factors that affect the vulnerability of forests, the most important of these being the presence of large and contiguous areas with a high proportion of mature balsam fir. Silvicultural "solutions," therefore, involve breaking up these large tracts in the hope that feeding conditions will be less than favorable for the larvae.[42] Although these measures aimed at reducing the vulnerability of forests to budworm attended to a few of the more important factors involved, a more holistic approach was recommended by some scientists who sought to restore the stability and balance of the climax forests thought to exist in presettlement times.[42] Generally, this strategy would require intensive forest management with selection timber harvests at relatively short intervals aimed at maintaining a continuous, all-aged forest of appropriate (climax) species composition.

Although silviculture was seen as the most effective long-term means of managing the SBW problem in the late 1940s and early 1950s, a radically different conception of forest dynamics was gathering momentum that rejected the notion of silvicultural control of SBW as misguided. It was only at a much later date that this alternative view of forest dynamics gained ascendency, one in which it was believed that periodic devastation of these forests by SBW is an intrinsic part of their ecology and is not a sign of instability. In this way of thinking, SBW outbreaks do not significantly alter species composition; rather, they ensure that "overmature" stands of spruce-fir are replaced by younger stands of similar composition. These forests do not enjoy the long-term stability, with its inherent insect and disease resistance, suggested by the older climax models of forest dynamics. Indeed, in this view, the idea of a budworm-proof forest is chimerical.[61,62] On the contrary, such forests are thought to be periodically recycled through the action of budworm, the result being the production of another budworm-susceptible forest at a later date and, inevitably, another spruce budworm outbreak. Human activity has not had a significant role in creating these dynamic forest conditions that are thought to

be naturally occurring and of considerable antiquity.[61,63] These skeptical views about the silvicultural hypothesis were developed further by Baskerville,[61,64,65] who concluded that not only were its underlying assumptions unsubstantiated but remedies based on them would produce effects opposite to those desired. In his view, any attempt to change the forest composition, through silviculture, to one less vulnerable to budworm would run counter to normal ecological trends and, thus, introduce instability into the system. The critique went further, though, questioning both the techniques and the goals of proposed forest management and silvicultural solutions. Baskerville argued that although selective cutting had considerably increased the proportion of spruce relative to fir in the overstory, the advance regeneration remained predominantly fir. He was therefore skeptical about the efficacy of selective cutting in changing species composition.[61] Furthermore, he contended that even if silvicultural methods attained their goal (i.e., an all-aged, selection management forest), this would render the forests more rather than less susceptible, as it would mean the preservation of large areas of mature forest, giving ideal conditions for the survival of budworm. In short, he argues that the SBW problem is an inherent, not a man-made, feature of the forest and that the silvicultural suggestions have, therefore, little hope of success. However, this newer notion of forest dynamics did not play any significant role in scientific thinking at the time.[42]

The adoption of a climax model of forest dynamics and its silvicultural implications by many of the forest ecologists involved in the SBW debate of the 1950s is an interesting example of the way in which underlying assumptions influence professional judgment even among "objective" personality types. In the absence of definitive, unambiguous data, which is always the case in dealing with complex problems, selective attention is paid to those ideas that conform to one's basic assumptions. Forest entomologists and other forestry professionals were inclined to select data in support of their intuitive sense of forest dynamics, a viewpoint that was informed by the dominant climax model, as we have seen. It was only with a shift in theoretical ecology away from climax models that data were interpreted differently. However, this has not led to greater levels of "truth" but only to another set of interpretations based on a cyclical model of forest dynamics. The implication of all this is that the interpretation of scientific data, and its integration into models, is a subjective process that is always influenced by the observer's assumptions. Thus, environmental disputes can seldom be "resolved" by new data but recur as different sets of assumptions come into play.[42]

Alternative Generation

In rational-comprehensive planning, the generation of solutions follows an intensive effort to identify and formulate the problem. In reality, however, "solutions" are often decided before this takes place. Prior choice of a plan of action may be made on political grounds or, in the case of SBW, forced on decision makers by the perceived absence of any feasible alternative. As the epidemic continued to build in the late 1940s, policymakers in NB must have felt that they had few, if any, options. Previous policy decisions in the 1920s onward had locked the province into a developmental path that was heavily dependent on the pulp and paper industry. The threat posed by the SBW to pulpwood stands was clear, and, as we have seen, the best available scientific advice was that silvicultural methods offered the best long-term remedy but that, in the short term, there was little option but to use chemical control methods, however grudgingly this conclusion was reached by some of the more ecologically minded scientists.

Experiments with the dusting of forests with lead arsenate had been tried as early as 1921 in an attempt to control such pests as the gypsy moth and hemlock looper. However, technical difficulties and the shear danger of low-level flying over broken terrain in inadequate aircraft limited its further development.[45] The advent of DDT, and its apparent success in both public health and agriculture, provided fresh impetus for those entomologists who were interested in chemical control methods. The profession of entomology had been polarized from the 1880s onward between "economic" and "ecological" entomologists. The former allied themselves with agricultural producers and held the utilitarian view that entomology should be "concerned primarily with the habits of pests that were relevant to the development of effective control measures"[48] (p. 35). In contrast, ecological entomologists showed more interest in basic science and the development of a broader understanding of pest dynamics, something that inclined them toward biological control. Thus, the availability of DDT and powerful, war-surplus aircraft offered an opportunity for economic entomologists to explore avenues previously denied them. For the first time in their professional history, they could offer decision makers a viable and relatively inexpensive control method for forest pests. All that the ecological entomologists could offer was more of the same—the silvicultural remedy that they had been promoting, without success, for 30 years.

This convergence on aerial spraying as the only viable option underscores the crucial importance of the early stages of problem solving on subsequent deliberations. Given that the "problem" was defined in the Imperial mode as a biological epidemic that threatened dire economic consequences, it follows that the basic solution sought would be to control the pest by technical means. The reasoning involved here is typically analytic-reductive. That is, the problem is reduced to its narrowest technical elements, and solutions are sought in refinements of method, rather than in re-evaluating forest policy goals and assumptions. However, a more innovative approach was within the grasp of those involved in decision making, if they had attended to the concerns of some elements of the forestry community. Among the latter, there was a feeling that the forest industries had acquired more wood supplies than they could possibly use, given the limited mill capacity in the area.[66] With the aid of hindsight, it is evident that, rather than being seen as a "pest," the SBW could have been put to use as a silvicultural tool.[64] That is, while valuable stands were protected by spraying, the SBW could have been allowed to consume, and hence regenerate, portions of the softwood forest of no immediate interest to industry. These regenerated areas would have required no further protection but would have grown into useable pulpwood in time for industrial use in later decades. It is a clever way of thinking that makes use of imaginativeness and breadth of vision, but the possibility of such a strategy was not considered in the 1950s and did not surface until 20 years later when Baskerville suggested starving the SBW into submission.[67] Similarly, because the problem was not defined in more psychosocial terms, the possibility of systemic changes in the forest industry away from reliance on pulp and paper production (an OH perspective) or the need for changes in lifestyles (an SH perspective) and attitudes (an SA perspective) to accommodate the budworm was not contemplated. Nor have systemic changes been given serious attention in subsequent years. Thus, a combination of technical and economic imperatives limited the options considered by decision makers to that of chemical spraying with DDT.

Decision Making

The decision to proceed with aerial spraying of DDT was, therefore, a foregone conclusion. Given its technical feasibility, the political-economic pressures, and the absence of any viable alternative, the program quickly became a fait accompli. In 1952, an agreement was reached between Premier McNair, acting on behalf of the Province of New Brunswick, and N. B. International Paper to spray an area of 200 sq. mi. (75,000 ha) in their Upsalquitch holdings. This was despite the qualms of the ecological entomologists, local professional foresters, and a growing realization that DDT was not the miracle cure it was originally thought to be. An understanding of the notions of pest resistance, and the indiscriminate destruction of beneficial insects, was becoming more widely known outside scientific circles.[68] In other words, the decision to spray was taken by a handful of political and economic leaders contrary to the advice of some of their scientific

advisors and with the grudging acquiescence of the rest.

Although this first trial was "successful" in the sense that SBW larvae were killed in sufficient quantities to protect the threatened wood supply, it was also found that the sprayed area was reinfected immediately from surrounding areas of infestation. Thus, if protection was to be afforded the NB forests, it was clear that they would have to be sprayed regularly until the epidemic collapsed or the trees were harvested. In the fall of 1952, therefore, the newly elected Premier Flemming was faced with the onerous decision of whether to proceed with what was clearly going to be a long-term, expensive spray program on an even grander scale than previously contemplated. There was no doubt in the mind of the major forestry corporations that such a program was imperative. Their representatives had held a meeting at the Upsalquitch airbase with provincial scientists and bureaucrats at which time it was decided to send a telegram to the Premier recommending that because spraying was the only hope of averting catastrophe, plans should be set afoot to spray up to 1 million acres in 1953.[69] This decision had, in Webb and Irving's[70] view, drawn "exhaustively upon the expert knowledge, opinion and experience then existing in Canada and the United States" (p. 118). As befitting his standing as a professional forester in his previous life, however, Premier Flemming paused before committing himself and sought further advice from the local forestry community that, as one might expect, was ambivalent about extending the program even further. For instance, two colleagues at the Maritime Forest Ranger School took opposing stands. One argued that, after working on the project, he felt it was successful, and as long as the area of spraying could be extended to deal with the problem of reinfestation, then it was a costly but necessary step.[71] In contrast, a more tentative opinion was offered by his colleague who pointed out that there were too many unknowns for him to be happy with extending the program.[72] His list of unknowns is an excellent summary of the void into which the decision to spray would take the province.

Economic and political imperatives won the day, however, for, despite any misgivings he might have had, Premier Flemming gave his stamp of approval to the program. Subsequently, as the program took on a life of its own, the area covered increased far beyond the 1 million acres initially approved, as we shall see.

Implementation

The 1952 spray operation may yet stand as the most carefully deliberated, most meticulously planned and most carefully observed forest spraying operation yet to occur in Canada . . . that action kicked off one of the boldest ventures yet attempted in eastern Canadian forest management as well as one of the most controversial, poorly understood, misrepresented and universally unloved ones[70] (p. 118).

These observations by Webb and Irving, who were intimately involved in the program, echo the feelings of other professionals who take pride in the program's technical accomplishments while expressing frustration at the political reaction to their efforts. Following the decision to proceed with a more extensive spraying program, representatives of the major pulp and paper companies agreed to form, in cooperation with the Province, a Crown corporation (Forest Protection Limited, FPL) that would implement the spray program.[73] Although the province is a majority shareholder, "during the early years of FPL's operation, the company was, for all practical purposes controlled by the industrial sponsors and particularly Canadian International Paper. During this period, industry managed Crown Lands as if they were their own . . . Industry control of the company was so pronounced that on at least one occasion, when the directors were deliberating whether to approve the spray program, the two directors representing the Province were asked to withdraw and did so"[73] (pp. 22–23).

The FPL corporation began a spraying operation in 1953, the logistics of which were daunting. Spraying is aimed at killing sufficient numbers of emerging larvae to protect the tree from excessive defoliation. Larvae are susceptible only during a three-week period between late May and early June because, after this, the larvae burrow into buds and are protected from contact with insecticide. At the same time, the areas most affected by infestation continually shift due to the migratory powers of the insect. While the epidemic began in the north, it soon shifted to the south of the province and then back again. Within this broader pattern, and especially after 1960, forest stands at high risk were scattered throughout the affected region in a patchwork quilt of threatened areas.[70] Spraying this kind of complex, dynamic target is difficult to say the least, especially in light of the limited timeframe available. All of this was made even more difficult from 1973 onward by the imposition of buffer zones around habitation and bodies of water that could not be sprayed. In total this amounted to about one-third of the provincial forest. Because this included 90 percent of small-woodlot owners, their land was disproportionately affected by the SBW depredations, as well as being a reservoir for SBW breeding.[70]

Despite those operational difficulties, many in the forestry community were impressed by what was accomplished: "As a feat of logistics, the program has been outstanding. To track the annual distribution of budworm populations and resultant damage over millions of hectares of forest with such precision and effectiveness of crop protection was exceptional"[9] (p. 51). In contrast, the antispray groups that arose in subsequent years were less enthralled by this technical virtuosity, feeling that the technological effort was misplaced, directed by a misguided forest policy.

To begin with, the media presented the spray program in stirring military terms, a tendency that was perhaps understandable so recently after World War II and with the rampant Imperial mode so blatant at that time. One editorial in a leading local newspaper, under the heading "Operation Budworm," almost leapt off the page in its enthusiasm:[74]

There is something exciting about this project. For those who thought all the horizons had been reached, this is the opening of a new frontier in the forests. The small band of people taking part in the work must have something of the pioneering spirit about them, and for the pilots (many of whom piloted bombers in the war) this war against a destructive insect must have a satisfying significance. These people all understand that mankind must never relax his fight against the weeds and the parasites of nature, and they understand perhaps more than the rest of us that we have fewer forest lands to be destroyed than we once had.

At the same time, residual doubts within the professional forestry community were sufficiently strong to warrant an invitation to Premier Flemming from the Canadian Institute of Forestry to address their annual meeting where he characterized the battle with SBW, in biblical terms, as Man's continuing struggle against insect plagues. While New Brunswick's storehouse was being threatened by the SBW, Operation Budworm was "an epic of what science and skill can do when they are combined with courage and resolution. . . . Paul Bunyon . . . could (not) have handled the situation much better . . . as all well informed people clearly recognize, Operation Budworm represented only one battle in a campaign which unfortunately must go on for some time."[75] In 1952, it seems, there was no embarrassment in using symbols such as Paul Bunyon, a mythic figure who epitomizes wanton forest destruction to modern environmentalists. Although the Premier also made it abundantly clear that sensible people would be in favor of the program, his admonitions failed to quell the undercurrent of distaste for it. As a consequence, there developed a recurring cycle that was to repeat itself throughout the controversy, in which public disquiet was met by official attempts to calm the situation. For instance, in 1956, two letters from interested citizens appeared in the provincial capital's daily newspaper claiming widespread apprehension that the spray program might upset the balance of nature[76] and wondering why scientific advice about forest management had not been followed.[77] In response to this unsettled state of affairs, and the imminent announcement of a 3 million acre spray program for 1957, the president of FPL recommended to Premier Flemming that he go on radio to further clarify the situation,[78] because:

The public which we believe to be in general supporters of forest spraying, have been feeding on a fare in the press and radio which must be confusing to them, to say the least. Our own attempts to keep the public informed about the overriding importance of the insect outbreak and of our actions to minimize its effects on trees have, I think, fallen a bit short of the mark because we cannot speak directly to the people of the Province and so perhaps our story lacks authority or may seem so to the reader.

As the 1960s unfolded, the undercurrent of concern about the spray program remained muted, apparently unaffected, it seems, by the furor created elsewhere by Rachel Carson's *Silent Spring*. Although the book and its message received front-page treatment in local media,[79] there was little political reaction among the general public. In fact, public expressions of opinion remained remarkably polite. In one instance, a gentleman writing in a local newspaper was almost apologetic in explaining how he, and his property, had been inadvertently sprayed with DDT. "During a night-watch brought about by a respiratory disorder resulting from this air attack, it occurred to me that the instigators of this program would perhaps be a little resentful if their own homes were thus polluted in someone else's interests."[80] Even the antispray stance of Michael Wardell, the swashbuckling publisher of the capital's daily newspaper, failed to extend the level of public concern.[81]

However, as the decade drew to a close, the broader climate of opinion began to change. Almeida and Stearns[82] draw attention to this phenomenon as one in which local environmental action is sometimes triggered by a confluence of events, not the least of which is a general shift in the national climate of opinion. This seems to have happened at the end of the 1960s with, for instance, a move by the federal governments in both Canada and the United States to ban the use of DDT in aerial spraying. The NB provincial government had apparently been anticipating something of the kind, for experiments had been conducted in 1967 with Fenitrothion as a possible replacement for

DDT. Subsequently, NB stopped using DDT altogether in 1969 and switched to Fenitrothion as the major insecticide used in the spray program. The official reason for this was concern about the damage to fish and wildlife caused by DDT.[83] Indeed, traces of DDT where starting to be found in Atlantic lobsters off the coast of Maine and NB, an unpleasant shock that began to link forestry practices with a threat to the lucrative shellfish industry. Fenitrothion remained the primary insecticide used in the spray program, with smaller amounts of aminocarb, malathion, and phosphamidon also being used. In response to public pressure, experiments were also conducted with the bacterial insecticide Bt *(Bacillus thuringiensis)*, and, despite misgivings among many of those involved in the program, it is now a major component of the current, much diminished, spraying regime.[84] These efforts by the provincial government to adjust to public concerns, however, failed to diminish the gathering environmental storm that was to break over the province in the 1970s.

Evaluation

Although the period 1945–1970 was not without its difficulties and controversy, in terms of natural resource decision making it was a relatively uncomplicated era, at least in NB. A decision-making elite was firmly in control of forest management, while the mass of the urban public seemed indifferent to both the environmental and social problems in the rural areas. As a result, evaluation of the SBW control program remained a prerogative of government and industry, who were inclined to use technical and economic arguments in support of their policies. From 1970 onward, however, a growing clamor of dissent emerged from those unwilling to accept official evaluations of the spray program and the forest policies within which it was embedded. This period of interest-group politics is discussed more fully in the next chapter but is introduced here to round off the SBW case.

The political climate that had developed in North America and Europe in the late 1960s and early 1970s was one of libertarian experimentation and rebellion against authority. This had been evident in Europe and the United States for some time where environmental and antiwar groups had been mobilizing to make their voices heard in policy-making domains that had previously been the exclusive reserve of decision-making elites. In the early 1970s, this wave of protest and interest-group politics eventually arrived in NB, albeit in a rather attenuated form, with a concerted effort by environmentally concerned citizens to take collective action against what they saw as the depredations of industrial forestry. At the same time, rural forestry workers and landowners intensified their efforts to form woodlot owners associations to act as collective bargaining units in their ongoing struggles to rest a better return on their labors. To add to the turmoil, a variety of government commissions painted a sobering picture of the state of the forest and forest industry, recommending immediate government action to implement a more rational scheme of forest management. In 1971, further embarrassment came when Bridges Brothers, owners of a blueberry farm, brought suit against FPL, the Crown corporation responsible for spraying operations, claiming that the spray program had damaged their crops by destroying the pollinating insects that were crucial to the setting of fruit. Not only did they seek compensation for losses but they wanted a permanent injunction against further spraying of their property. Meanwhile, the forest industry was arguing that the intensely competitive nature of international markets in general, and the pulp and paper market in particular, made it essential that they be given security of fiber supply at a competitive price. At the same time, the media were beginning to pay more attention to environmental issues that, in turn, tended to alarm the urban populations who had previously been indifferent to many of these issues. It goes without saying, therefore, that the period from 1970 to 1985 was a time of intense turmoil, one that can be characterized more by its pervasive conflict than by cooperation.

Sitting in the midst of this mayhem, the provincial government was surrounded on all sides by interest groups pressing their various causes at a time when the forest itself was in a perilous state. Navigating through this political minefield was, to put it mildly, a difficult task. The political challenge faced by legislators during this period was to keep revenues from the forest industry flowing into provincial coffers while trying to mediate between, and placate, conflicting interests. Unfortunately, in a client state dependent on its forestry revenues, NB politicians simply had little room to maneuver, as it was not possible for them to recognize any fundamental problem with chemical control methods. All that the legislators could do in response to the growing demands from environmentalists was to make incremental changes in the program. For instance, at the time, "there were virtually no laws regulating the aerial application of pesticides in New Brunswick"[17] (p. 86). In 1973, therefore, apparently in response to public pressure, the provincial government responded with a Pest Control Act that tightened operational procedures by, among other things, introducing buffer zones around habitation and bodies of water that were not to be sprayed. As mentioned earlier, not only did this make the practice of spraying much more complex but it provided widely dispersed breeding grounds for the SBW. In the eyes of the prospray forestry community, this was to create endless headaches and immense damage to small-woodlot owners whose land fell disproportionately within these no-spray areas.[9,85] In an attempt to justify its policies, the provincial government offered economic and technical justifications for continuation of the spray program. The limited nature of these arguments is worth further attention.

Economic Evaluation of the Spray Program

For the past 40 years, the economic value of the SBW spray program has been assumed as a fundamental given by prospray advocates primarily on the commonsense grounds that the province's wood supplies have been protected in amounts sufficient to maintain the economic viability of the pulp mills. Environ-

mentalist have taken quite a different perspective, however[17] (p. 114):

Proponents of the spray programme are virtually unanimous in their assertions that its benefits far outweigh all cost factors. Industry representatives continuously extol the virtues of the use of chemical insecticides, juxtaposed with anecdotal evidence of mass unemployment and a crippled New Brunswick economy if chemicals are not used. These spokespeople tend to reduce the complexities of the spray regime into a simplistic, erroneous and sensational statement: use chemical insecticides to protect the forests or do not use insecticides and destroy the province.

According to Versteeg,[17] the "disaster scenario" presented by government and industry was based on a dearth of credible economic analyses. Speaking in 1984, he concluded that no adequate analyses had been conducted to that date and that assertions of economic viability were simply based on unsupported assumptions.

On the other hand, a thorough cost-benefit analysis of something as complex as the SBW spray program presents government with a formidable challenge. A comprehensive evaluation would have required that comparisons be made between the chemical spray regime and alternatives based on either biological or silvicultural control methods, neither of which had been used to any significant extent in the 1970s and 1980s and about the economics of which little was known. Further, both direct and indirect costs of each regime would have to be assessed. It is the latter that are the most difficult to incorporate into any quantitative evaluation. For instance, Versteeg[17] (p. 118) argues that these indirect costs, which are omitted from earlier economic analyses, include: "dollars spent by federal and provincial departments on planning for and monitoring of the spray operations, legal fees incurred in defending expensive lawsuits, and costs of consultants' reports and Task Force Reports on various adverse impacts of the programme, the unmeasurable costs of health care resulting from spray-induced illness, and the costs of damage to the ecosystem resulting from the use of toxic insecticides."

A recent economic analysis conducted under the auspices of the Forest Pest Management Caucus, a national body of pest management experts, tries to address some of these deficiencies.[86] Working within a set of assumptions about the degree of foliage protection needed, probable amounts of damage, and projected costs of operations, a comparison was made between chemical (Fenitrothion) and biological *(Bacillus thuringiensis, Bt.)* control methods, together with one silvicultural option (planting). The report concluded that, strictly in terms of timber protection, it pays to control spruce budworm. "For each dollar spent in controlling spruce budworm using fenitrothion, between $4.20 and $6.00 is returned in the form of additional timber benefit" (p. 92). The benefits of *Bt* used alone would be proportionately less because of its greater initial cost and vulnerability to the vagaries of weather, compared to that of chemical methods. On the other hand, planting to replace damaged stock was found to be uneconomic in the absence of any attempts to control the pest.

Needless to say, these conclusions would be comforting to prospray advocates but more ammunition for environmentalists who would point to the continuing discounting of economic externalities and intangibles. The controversy over economic efficacy of the spray program is unlikely to be resolved easily because of the radically different evaluation criteria adopted by the pro- and antispray advocates. The former, arguing within the Imperial framework, are concerned primarily with what are referred to in forestry as "timber values," while the latter couch their criticisms in Arcadian terms, seeking to extend the debate to include a variety of ecological and social values.

Technical Evaluation

Technical evaluation typically involves determining whether a program achieves its goals, its effectiveness compared to alternatives, and whether there are unexpected, unwelcome outcomes. (These three categories were suggested by a source I can no longer find.) Perhaps we can look at these in turn.

Did the Spray Program Achieve Its Goals?

The primary index used to answer this question has been the amount of protection afforded susceptible stands of trees. On the face of it, the scientific community has abundant evidence that the use of insecticides has achieved this objective.[9,84,85] There seems to be little doubt that large areas of susceptible forest, which would otherwise have been consumed by the SBW, remain intact. Whatever doubts may exist about this conclusion stem from the realization, which started to take shape between 1970 and 1980, that the SBW has more diverse feeding habits than previously recognized and, under certain conditions, may extend its depredations beyond mature balsam fir to younger trees of this species as well as to red and black spruce that were thought to be less vulnerable to attack.[85] Any evaluation of the effectiveness of spraying, therefore, would have to take this into account. Not only would the degree of protection provided mature balsam fir have to be assessed but also that afforded these other species and age classes. Evaluations of spraying efficacy based solely on the extent to which vulnerable trees are protected may, however, miss a more important point: why so many trees were protected in the first place. For instance, a government official, commenting on the SBW situation in Ontario, observed that "the companies were demanding the spraying of low value balsam fir and white spruce while they actually intended to continue harvesting the remaining stocks of higher value black spruce"[87] (p. 222). The point being made is that while the protection of large amounts of mature balsam fir and white spruce maintains a strategic reserve of timber, it also maintains a large food supply for the SBW, requiring constant spraying to deny the insect access to the feast. A variety of forestry scientists justify such a strategy by arguing that it affords managers the time and opportunity to implement a reasoned management plan.[85,88] However, as mentioned earlier, it may also preclude the possibility of using the SBW in its natural role as a recycler of aging forest. The scientific community seems to stumble when it is faced with these broader questions about the purpose of the spray program. Such questions take the evaluation of the spray program beyond technical matters to that of policy goals and, hence, beyond the scope of technical research.[87]

Is It More Effective than Alternative Means?

A great deal of research has been conducted on the SBW and the chemical spray program over the last 40 years.[88] While some of this has led to technical innovation in operational delivery of spray, there was no significant shift to alternative approaches prior to nor during the 1970s. Indeed, Irving[85] (p. 36), speaking as managing director of the spraying agency, commented: " What this comes down to is that the considerable science and technology directed at this problem over the last 30 or 40 years has disappointingly little to show for the effort. To the resource managers and their political masters it means that options for coping with the insect in the here and now are very little changed from those of the 1950's—use insecticides and/or salvage, or let nature take its course." There are a number of psychosocial reasons for this state of affairs.

The pursuit of practical alternatives to aerial spraying of insecticides is, of course, constrained by ecological realities. Given the vast scale of the problem, and the endemic nature of the pest, it is not possible to summon strictly technical fixes even with the most creative thinking. However, innovations of a more sociotechnical nature are possible, even though they would require a much broader conception of the problem than typically adopted by the agencies involved in pest control. Why these alternatives (such as using SBW as a silvicultural tool) have not been pursued is open to debate, but one contributing factor could well be organizational rigidity. The widespread occurrence of this latter behavior in North America was noted in chapter 3 and is underscored by Boisjoly et al.[89] (p. 228), who point

out that the pathologies of bureaucratic behavior are well documented and include "lack of communication, distortion of information as it passes up the hierarchy, jealously of existing lines of authority, bias in favor of the status quo, bureaucratic turf protection, power games," and so on. Given the ubiquitous nature of these organizational problems, it would be surprising if some or all of them did not occur in New Brunswick. Obtaining information on this is, as Webster[28] has pointed out, very difficult. Some limited evidence of psychosocial problems within and between the agencies involved in the SBW spray program came to light in a Delphi study conducted some years ago.[90] The Delphi process is one in which the opinions of experts and others are sought by mail, their replies being summarized by the research team and then circulated to the participants anonymously in a series of so-called rounds. An exchange of views is achieved but at a distance and without the various conflicts that accompany face-to-face confrontation. While the Delphi process is subjective and anecdotal, it provides a forum in which participants can raise issues and identify problems that they believe to be important in regard to the issue at hand. The SBW Delphi study included natural resource professionals and environmentalists, the views of the former being of interest here. Of the various organizational hindrances to innovation seen by members of this professional group, two are relevant here: coordination of management and research and resistance to change at all levels in the system.

Coordination of the large number of organizations and personnel involved in forest management is difficult. In particular, mention was made by some of the Delphi professionals of conflict between the major organizations over various aspects of SBW management. Although it is difficult to find additional evidence bearing on this contention, there are independent sources that add credence to the perceptions of the Delphi respondents. For instance, Baskerville argues that a "most pernicious form of institutional resistance to adaptive management is bureaucratic territoriality"[91] (pp. 35–36). In one case, when new policy tools were due to be transferred to the provincial management agency from the federal research agency:

senior management in the research agency became concerned that sufficient recognition be given to the "contribution" of the service, and determined that transfer could occur only within their terms. The proponents of this notion stated an explicit proprietary interest in the model and the policy design tools, and expressed great concern that this ownership be recognized. Their motives were to insure that, in the total picture, a major scientific contribution of their unit was given adequate recognition. The result of their intervention in this manner, was a breakdown in communications between the research agency and the management agency with respect to the budworm/forest policy design tool, and no progress toward implementation for almost two years, a period in the process when rapid communication and response was most essential.

Similar problems with coordination and communication also arose in a major research initiative on SBW control methods undertaken jointly between Canada and the United States in the early 1980s.[92] The organizational problems encountered by this massive research project have been described at length elsewhere,[93,94] but one example might be useful for illustrative purposes[94] (p. 7):

The immensity of the SBW problem introduced many jurisdictional/organizational 'turf' difficulties in managing the research program. Commitment to certain common goals or procedures by administrators and investigators is often necessary for both efficiency and effectiveness. The CANUSA Program did not particularly succeed in overcoming many of these problems . . . (one result was that) there were essentially four programs; E-Canada, E-U.S., W-Canada, W-U.S.

Some professionals in the Delphi study also mentioned that these problems of research integration were not helped by the climate within organizations that, in their view, provided little opportunity for genuine discussion of assumptions and fostered a general insularity that resulted in little interest in other people's work. Thus, reluctance to adopt new methods stemmed from what respondents referred to as a form of defensiveness in which any critical evaluation or suggestion of change

was seen as a personal attack and a condemnation of past practices. Constructive dialogue was limited by an excessive sensitivity about admitting that current practices may no longer be appropriate. Again, independent verification of these views is difficult to obtain, but a case in point is provided by Baskerville[91] (pp. 28–29):

The fear of evaluation stems from the basic fear that such an evaluation might show the programs are not working. Yet this information is precisely what is needed . . . When the presentation of alternative policy includes an evaluation of the existing policy, it is incredible how quickly subsequent discussion centers on a defense of the status quo. For example, there has been no real evaluation, by the management agency, of the spray rule used for the past twenty six years . . . There have been many defenses of the rule, but no systematic evaluation, which would lead to an improved approach to the future. . . . It seems incredible that, after twenty six years, there are still arguments about whether or not the policy of crop protection with insecticides prolongs the outbreak! A seriously debilitating feature of these arguments is . . . the continued use of a disjointed incrementalism approach to the problem, and the continual defense that this approach requires . . ."

Though none of these examples provide definitive evidence of widespread organizational rigidity in the 1970s and early 1980s, they do offer some support for the perceptions of the Delphi respondents, and they are in line with problems commonly found in other jurisdictions. Yet, the fact remains that the silvicultural alternative to pesticide spraying remained unexplored on a large scale, dismissed by some, but favored by others. For instance, Irving[85] argues that, based on his experience with the 1970–1980 outbreak, forest composition cannot be changed in any significant way by silviculture and concludes that:

We have learned . . . that a great number of our earlier assumptions were in error. The fir component in a forest does not make it more susceptible to budworm attacks, only more vulnerable to damage. Pure red and black spruce stands are quite capable of supporting budworm populations . . . Another fallacy has been the assumption that breaking up fir stands into smaller stands interspersed with hardwoods greatly reduce susceptibility and vulnerability. There has been especially little evidence to support this on intensively-managed private lands in New Brunswick" (p. 41).

Irving is probably correct to say that the forest management option is not credible, assuming that there is no fundamental change in the way in which the forest industry operates. However, there are others who believe that forest management remains a viable alternative under certain conditions,[12,95-97] one of the most important of which is systemic change in the structure of forestry operations in NB. For instance, Kettela[84] mentions that experience with the spray program resulted in the conclusion that, in some circumstances, infestations should be allowed to kill specific forest stands as part of an overall forest management system. This is a version of the earlier suggestion by Baskerville that the SBW epidemic might be broken by allowing the insect to destroy selected areas, thereby depriving the insect of its food supply.[98] Such a strategy, however, would have required a province-wide collaboration in mounting a massive harvesting and salvaging operation, something that the forestry community appears to have perceived and rejected as a form of "socialized forestry." However, these changes are not necessary, according to Webb and Irving,[70] because they see clear evidence that the spray program was justified and has achieved its intended objective without any serious environmental perturbations and health risks. Opinions to the contrary are due in large part, they say, to misconceptions on the part of the general public:

Distorted public images in such emotionally-sensitive areas as the environment and public health have, of course, become commonplace. This is readily enough understood in that predominant part of the lay public that depends for its information on media that are too often inexacting, sensation seeking, or as with the CBC, promoting an anti-establishment environmental prejudice of its own . . . it is a less-understandable and often more frustrating fact that serious misunderstandings, misconceptions and committed mind-sets critical of the New Brunswick approach are prevalent in more technically-educated and purportedly scientifically-oriented circles as well . . . What is truly wanted is the most informed debate possible on the subject *within those elements of society best equipped to arrive at commonsense judgments* of the matter and to help find a more acceptable solution (p. 119) (emphases added).

As one might expect, in this controversial arena, for every prospray viewpoint there is a counter argument offering radically different conclusions. For instance, Howard[99] (pp. 58–59) takes a more skeptical view of events:

Armed as they were with inadequate scientific information, the New Brunswick government and the forest industries were able to sketch a wholly unrealistic accounting of the risks and benefits of the spray programme. Far from having to deny responsibility for poisoning, big business and big government argued that no poisoning existed, at the same time attesting to the value of the programme. Such one-sided accounting was made all the more possible and powerful by the government's unquestioning acceptance of the industrial threat of devastating financial costs if the programme were challenged. It was economic pressure tactics again, this time based on blind faith in science.

However, given the economic and political realities, one wonders what the provincial government might have done to extricate itself from this morass. The SBW problem had been handed down from one set of politicians to another over a 40-year period. None had done much more than make incremental change. Anything more radical would have required a restructuring of the forest industry, retraining of forestry workers on a massive scale, and a potential flight of capital from the province. Thus, politicians in a client state who try to keep the economic system working must necessarily maintain a brave face and hope that the situation does not disintegrate further. Unfortunately, while this kind of political rationality might achieve the compromises needed to maintain the status quo, it does not address the problem at hand.

Unexpected Impacts

The externalities of the spray regime, its unexpected consequences, became a source of considerable headache for the iron triangle governing forestry in NB. The controversy that surrounded the alleged environmental and public health impacts of the SBW control program became increasingly vituperative as the 1970s unfolded. Yet, even though the basic disagreements were essentially matters of conflicting ideologies, the debate revolved around technical issues, to the detriment of the aims pursued by local environmental interest groups. In addition, the intensity of the controversy exacerbated the tendency of all concerned to commit type III errors, as we shall see. Both environmental and public health issues became a matter for heated debate.

Environmental Impacts.

Questions about environmental impacts are usually answered by ongoing scientific monitoring of target and nontarget organisms (including people). A variety of federal and provincial agencies, such as the Canadian Forestry Service and Environment Canada, have been involved, their research being coordinated by a monitoring committee.[73] Conflicts have arisen in this process over appropriate standards of scientific rigor in the conduct of impact research. Agencies that adopt strict positivist criteria for evaluative work often find the quality of research carried out by field scientists falls below what they consider to be an acceptable level. In contrast, field scientists, who have to make do with the limited facilities available to them in particular situations, point out that they do the best they can under the circumstances. These disputes appear to be yet another example of the conflict between *pure* scientists, who prize positivist rigor, and *applied* scientists, who are more willing to use and see the value in professional judgment. The resulting disputes reduce the credibility of any advice that scientists may have to offer policymakers who are faced with what seems to be a fractious scientific community unable to agree on policy-relevant information.

A case in point is the vigorous interchange, during the 1970s, between National Research Council of Canada (NRCC) and Forestry Canada scientists, which is discussed at length by Miller and Rusnock.[100] Since the introduction of Fenitrothion in 1967, evidence had been

accumulating about its effects on various components of the ecosystem. Generally speaking, Fenitrothion is a nonpersistent, broad spectrum insecticide that degrades relatively quickly in forest environments and is thought to exhibit low mammalian toxicity. The insecticide is not very toxic to fish, one of the reasons it was chosen to replace DDT, but has a particularly severe effect on pollinating insects, an impact of considerable importance for blueberry growers, as we have seen. Significant reductions in populations of forest songbirds after Fenitrothion use have also been reported.[100] In response to concerns about potentially detrimental impacts, the NRCC convened a symposium that concluded that areas of ignorance remained unaddressed and that research had not been designed to answer key questions concerning the environmental safety of Fenitrothion.[101] It is clear that the NRCC felt that caution should be the order of the day in spray operations, that variables (i.e., spray formulations) should be strictly controlled, and that research should be integrated into ongoing operations rather than left to shift for itself. In a blunt statement, the NRCC chairman contended[102] (p. 576):

Operationally, the use of pesticides in Canada for the control of spruce budworm must be described as involving brute force and educated ignorance . . . The actual operational control program is so complex and chaotic that, except in the crudest sense, we have no knowledge of how much spray will actually reach the spruce budworm or where the remainder will lodge in any given operation. With some outstanding exceptions, most of the scientific work done to date lacks rigor, planning, and control, and one wonders whether there has been more concern with appearing busy than with shedding light on the risks and benefits. On the basis of both published information and the evidence presented at the symposium, the mammalian toxicology of the commercial insecticide fenitrothion per se has been studied in relative depth. Ironically, it appears that the chemical itself could be one of the safer pesticides when used correctly. However, in order to spray it on pests (be they spruce budworm or others), fenitrothion is normally mixed with solvents, emulsifiers, and diluents. The toxicity of these mixtures is not well understood and needs urgent attention. . . . Subtle and yet poorly understood risks may be associated with these complex and variable mixtures of chemicals. This unsatisfactory situation will not improve until the knowledge, controls, and regulations of operational programs apply to total mixtures and to the precise method and conduct of the application, nor until ad hoc changes are eliminated.

These observations resulted in muted protest from Forestry Canada that responded with a defense of its monitoring practices.[103] Although conceding many of the criticisms of the NRCC report, it interpreted them in a different light, suggesting that the NRCC had an insufficient grasp of reality, especially as applied to forest management practice. The author conceded that monitors of spray practice had little say in the operational practice of spraying, but this is not to be wondered at:

Those operations take place within a demanding time frame; they are at the mercy of the weather and are bounded by the requirement that they kill budworm to save foliage at high efficacy. Each unit of work, a 5000 Ha block, has a price tag of around $25,000, and has to pay its way. The forest cannot be rented like a farmer's potato field and modified as experimental strips with a range of treatments. Forest spray operations do not lend themselves readily to research manipulation, and it cannot be expected that operators can subordinate their prime responsibility to serving research needs. The spray airstrip has a pressing schedule in May-June, and it is costly to set aside spray planes, pilots, mechanics, mixing tanks and materials to await the design of the experimenter and the whim of the weather. These are the facts of life (pp. 2–3).

In other words, the situation in the field does not lend itself to laboratory precision and experimentation based on positivist thinking. All that monitors can do is make use of professional judgment, intuition, and observation and try to get along using this practical intelligence. More refined, positivist experimentation is impracticable.

Our interest in this debate lies not with the correctness of either argument but rather with the apparent insolubility of these interpretational disputes due to the adoption of contradictory modes of reasoning by protagonists. When positivist modes of reasoning are applied to the evaluation of complex, messy problems, standards of rigor more appropriate

for laboratory science are demanded that field scientists cannot meet. This was particularly true in the early days of the SBW program when research was a by-product of operations. In more recent years, monitors have become "a little less dependent on operational spraying because some scientists (have) access to aircraft and study areas that (are) truly experimental and under their control"[88] (p. 6). Despite these research innovations, epistemiological differences between scientists still result in disputes analogous to that described earlier. For instance, a similar debate over the impact of Fenitrothion between Environment Canada and Forestry Canada occurred in the late 1980s, with the same unsatisfactory outcome.[104,105]

One might conclude that science can play an important role in program evaluation, but it suffers from the same ideological conflicts typical of nonscientific forms of judgment. When research has policy implications, as in the SBW case, research findings receive greater scrutiny, and yet more scientific controversy is generated. The resulting arguments may serve to alert authorities to a potential problem, but whether this advances the policy review process is not clear. As Collingridge and Reeve[106] have pointed out, scientific conjecture may not help the policy process other than in ensuring that one scientific viewpoint does not come to dominate decision making. At the very least, technical evaluation and scientific controversy are potentially useful as an early warning system. In the spruce budworm case, however, innovative policy has not resulted from scientific conjecture.

Public Health Impacts.

When the tide of environmental concern that was making itself felt elsewhere finally reached New Brunswick in the late 1960s, there was a feeling of powerlessness and mistrust among some members of the general public over the way in which the forests were being managed.[107] Initially, concern seems to have focused on environmental rather than human health, but this was to take a more serious turn when discontent with spraying came to a head in an unexpected and unfortunate way.[108] The inchoate nervousness about the aerial spray program finally became more focused when it was learned that local children were dying of *Reye's syndrome*, a rare condition that recent research had linked to the spraying of pesticides. "First diagnosed in Australia in 1963 . . . It has since been reported in children from infancy to 18 years all over the world. The victim is usually recovering from a viral infection such as chicken pox or flu. Suddenly the child develops a startling relapse, commencing with intractable vomiting, lethargy, convulsion, delerium and, often within hours, coma. Early diagnosis and vigorous treatment lead to survival rates from 34 to 75%. Untreated, the mortality rate may reach 100%. The cause of Reye's syndrome is not known. Most cases begin with a viral infection, but the virus infection alone is not significant to trigger the syndrome. It appears that an additional cofactor is required. It is not clear whether the cofactor impairs the body's defence against the virus or whether the virus hinders the body's ability to withstand an assault by the cofactor"[17] (p. 91).

Because of the lack of facilities and expertise in dealing with such a rare disease in New Brunswick, several local children had been flown down to a Children's Hospital in neighboring Nova Scotia where a group of Dalhousie University researchers were becoming puzzled by the arrival of so many cases from the same area. As a consequence, a research program was initiated that, over the course of the next two years, began to point to the pesticides used in forestry as the possible unknown agent that triggered Reye's syndrome.

The Dalhousie research group, led by Dr. John Crocker, published its initial findings in 1974.[109] In this study, the hypothesis was offered that the pesticides used in the forest protection program in New Brunswick could have the effect of enhancing the virulence of common, but otherwise relatively benign, viruses. Because these were preliminary findings, derived from animal studies, the research team remained suitably cautious in interpreting its findings. Perhaps as a result, this first study received little media attention in New Bruns-

wick. Matters took a decidedly different turn, however, with the Crocker team's second study in which they attempted to determine which components of the spray led to viral enhancement.[110] "The testing results were dramatic. A statistically significant higher incidence of mice who were exposed to the emulsifiers as opposed to the active ingredients in the pesticide formulation died with symptoms similar to those of children with Reye's syndrome. This work was due to be published in the June 1976 issue of *Science*. However, a newspaper journalist in Cape Breton (Nova Scotia) broke the story in March 1976. Just weeks before, the Nova Scotia government had approved a plan to use synthetic chemicals to control the budworm infestation in the Cape Breton Highlands. Reaction to the story was swift and dramatic. A special session of Cabinet was convened in which Dr. Crocker explained his research. That afternoon, the Minister of Health announced that the spray programme would be cancelled"[17] (pp. 91–92). Such decisive governmental action was not forthcoming in New Brunswick where, following an interview with Crocker aired on CBC Radio on April 1,1976, the provincial government convened a committee of experts to review the Dalhousie research.

The panel of six medical and agricultural experts, chaired by W. G. Schneider, President of the National Research Council of Canada, met in late April 1976 to review the available evidence on the association between forest spraying and Reye's syndrome. They concluded that, apart from the prevailing consensus on the multifactorial nature of Reye's syndrome, little else was known with any degree of certainty.[111] As far as they could tell, the incidence of Reye's syndrome in New Brunswick was no greater than other jurisdictions for which statistics were available (about 1 per 200,000 per year in the United States).In addition, the occurrence of the New Brunswick cases seemed to cluster in the winter months rather than during or shortly after the spraying season in May and June. Significantly, winter is a peak period for viral infections such as influenza. The panel conceded that the Dalhousie studies offered some support for the idea that environmental contaminants might contribute to the development of Reye's syndrome but added that these studies required replication with some refinements in the design of control groups. Given the paucity of available information on the composition of the spray formulations, the incomplete provincial medical record keeping, and the rather tangential relevance of the laboratory research of the Crocker group, the panel was unable to come to any definitive conclusions about possible links between exposure to spraying and Reye's syndrome. In their final comments they called for more research and more adequate monitoring of the situation.

In formulating a response to the Schneider Report, the government of New Brunswick pointed to the inconclusive nature of its findings and the absence of any proven association between spraying and Reye's syndrome. This allowed the 1976 spray program to go ahead as planned in the absence of any scientific proof to suggest otherwise. In making this judgment, those involved adopted a conventional approach to risk management, which requires unequivocal proof of harm before action is taken to change a particular policy. This stringent criterion dominates natural resource decision making, particularly when the courts are involved.[17] Unfortunately, it also runs the risk of accepting type II errors (false negatives), thereby tolerating harmful policies.[112] Before these findings became public, however, a mother of a young victim of Reye's syndrome told the CBC radio audience that she believed her child had died as a result of exposure to the spray chemicals. The next day, a number of local citizens, moved by the interview, formed the Concerned Parents Group.[17] This group carried the brunt of public protest against government policy in the coming decade.

From the beginning, the Concerned Parents demanded the complete and immediate cessation of pesticide spraying until such a time as the government could demonstrate the safety of the chemicals being used. The initial position taken by the Concerned Parents was based on very little empirical data. Indeed, the problem of obtaining adequate support for their claims was to handicap the group

throughout its existence. Unlike the more circumscribed and well-known cases like Love Canal[113] and Woburn, Massachusetts,[112] the potential health effects of the SBW spraying program were spread over an enormous geographical area and involved thousands of people. Under these conditions, data collection was difficult, to say the least. Despite this, the Concerned Parents did make a concerted effort to build on their flimsy beginnings with both epidemiological and toxicological evidence in support of their position.

Formal, scientifically acceptable epidemiological studies require considerable expertise and a great deal of money, neither of which were available to the Concerned Parents. To make matters worse, Reye's syndrome was not, at that time, a reportable illness in New Brunswick, which meant that the provincial Department of Health did not have complete records on this and other relevant illnesses. In the face of this inadequate reporting system, and without financial resources, the Concerned Parents were reduced to informal and unsystematic data collection. Compared to other childhood illnesses, the incidence of Reye's syndrome in New Brunswick was very low, although this was of little comfort to those parents who had lost a child. Nevertheless, during the period in question (1972–1982), the Concerned Parents claimed 21 confirmed cases in Atlantic Canada of which 17 were in New Brunswick. Thus the annual incidence, which was based primarily on information from the patients treated in Halifax, was something less than 2 cases per year. Though later studies threw these data into doubt, the Concerned Parents used them to claim that the incidence of Reye's syndrome in New Brunswick was six times that found in neighboring Nova Scotia, a province that had curtailed spraying operations some years earlier.[114]

A similar fate befell the Concerned Parents' use of statistics in relation to other illnesses, such as cancer and birth defects, which they also thought to be influenced by exposure to pesticides. Once again, they were faced with a paucity of useful information. Their foray into the byzantine world of interprovincial statistics led them to the conclusion that the age-adjusted rates for brain, liver, skin, and blood cancer were considerably higher in New Brunswick than in Nova Scotia.[114] This contention was supported, in part, by information revealed in a piece of investigative journalism broadcast in a CBC radio documentary.[115] Unfortunately for the Concerned Parents, an expert panel set up expressly to evaluate this claim of a high incidence of certain kinds of cancer in the province was unable to confirm any link to the spray program.

In the area of epidemiology, therefore, the Concerned Parents were never able to establish a firm basis either for their claim that there was a higher incidence of selected illnesses in New Brunswick or that these illnesses were related to spraying. The closest they did come to the latter was their involvement in a study of seasonal variations in levels of blood cholinesterase in children living in sprayed areas of New Brunswick.[116] A consistent seasonal pattern was demonstrated in which levels of cholinesterase were found to be substantially lower during the spraying season compared to samples taken before and after this period. Although such findings are only suggestive of a possible link between enzyme metabolism and the spray program, the authors did raise this possibility at the end of their paper, thereby offering the Concerned Parents further support for their argument. Such a speculative leap, however, was not readily accepted by local scientists.[70]

The Concerned Parents relied primarily on the research produced at Dalhousie University in forming their toxicological arguments. Although concern was focused on Fenitrothion in the early studies done by Crocker et al.,[109] attention quickly turned to the emulsifiers used in the spray formulation when it became apparent that the insecticide itself did not appear to have significant viral-enhancing properties. In a series of studies on laboratory animals, the Crocker group was able to demonstrate a significant viral-enhancing effect of Toximol MP8 and Atlox 3409, two of the more commonly used emulsifiers.[117] Fenitrothion remained in the picture as a cause for concern as the result of the accidental exposure to the pure insecticide of one of the technicians

working on the Dalhousie research. This provided the research team, albeit in an unfortunate way, with the opportunity to study acute Fenitrothion poisoning.[118] Of particular interest was the finding that the insecticide appears to have been retained in the victim's tissues and to have had continuing physiological effects, for up to 8 months or more. This was contrary to conventional wisdom that maintained that Fenitrothion was quickly metabolized and, therefore, not stored in the body.

At this point in the proceedings, the Concerned Parents must have felt that they had developed a strong toxicological case. Not only were they able to use the research produced by the Crocker group in their lobbying efforts but assiduous searching of the scientific literature had provided them with additional ammunition on pesticide poisoning.[119] Unfortunately for their cause, a pattern of argument developed in their confrontation with government that proved to be both intractable and fruitless. The disagreement, as in so many environmental disputes, revolved around differences in opinion over the correct use of data. In response to the Concerned Parents' toxicological claims, the provincial government offered two unwavering, and traditional, rebuttals: one should use great caution in extrapolating from laboratory findings on animals to humans, and, more particularly, the extremely high doses used in such research tell you little about the effects of spraying under normal conditions. Thus, as mentioned earlier, in estimating risk the government was inclined to type II errors, the willingness to accept the possibility that it was ignoring real effects (false negatives). The Concerned Parents were never able to deal effectively with this line of argument, stemming as it does from a value position rather than a scientific "objective" judgment. As one might have anticipated, the Concerned Parents were inclined to type I errors, a strategy in which one accepts false positives, the possibility that one is acting in response to a link between pesticides and ill-health that, in reality, does not exist.[112]

From 1977 to 1982 there ensued a period of bad-tempered skirmishing in which both sides became increasingly frustrated over their apparent inability to agree on the significance of evidence. The Concerned Parents would find and bring to the attention of the media some new study that in their eyes demonstrated cause for concern, only to have the government respond that no single study could shift the weight of evidence toward proof of harm.[120] During this period, the Concerned Parents had requested of the government that the whole question of spraying and public health should be examined by a Royal Commission. Little came of the request until January 1982 when CBC Radio broadcast the documentary "Poison Mist" in which new evidence on the incidence of illness in New Brunswick was presented, along with some of the latest research from Dalhousie University on the role of interferon in the etiology of Reye's syndrome.[121] The ensuing furor finally moved the government to action, and an expert panel was convened the following month to study, once more, the relationship between spraying and Reye's syndrome,[122] which was quickly followed by two additional panels on cancer and reproductive anomalies.[123,124]

The Reye's syndrome panel reached similar conclusions to those of the previous study conducted in 1976. When compared to other jurisdictions that employed similar case-finding techniques, the incidence of Reye's syndrome in New Brunswick was comparable to, if not lower than, such locations (such as Colorado, Michigan, and Ohio). The more relevant comparison with Nova Scotia was not possible because of the voluntary reporting system in that province. In addition to this low incidence of Reye's syndrome, the panel was unable to find any geographical (proximity to sprayed areas) or temporal (proximity to the spraying season) associations between exposure to the spray program and Reye's syndrome. These findings deflated the Concerned Parents' arguments but left intact their belief that spraying could have a deleterious effect on human health that was not being detected by scientific studies (personal communication, Cathy Richards, past president of the Concerned Parents).

The Spitzer[124] panel on forest spraying and cancer concluded that although there appeared to be a greater incidence of certain kinds of cancer in New Brunswick than in neighboring provinces, the total incidence of cancer was comparable with that of other jurisdictions. However, no firm associations were apparent between forest spraying and these cases of cancer, and, in the absence of more extensive research, more definite conclusions were unlikely. Once again, one sees here the positivist faith in the virtue of more analytical research as the route to clarification. Finally, Hatcher and White[123] offered similar observations in regard to reproductive anomalies. There was some indication of a higher incidence of certain kinds of birth defects (neural tube anomalies) in some New Brunswick counties, but their possible association with aerial spraying remains unproven and would remain so in the absence of appropriate (analytic) research.

The net outcome of the panels' deliberations, therefore, was a modest clarification of the situation and a great deal of residual uncertainty. This is by no means an unusual state of affairs, according to Brown,[125] who suggests that environmental controversies seem to follow a predictable sequence with similar outcomes. In response to public pressure, government agencies conduct official studies that usually find no association between contaminants and health. Sometimes community activists will react to these findings with a more intensive campaign of litigation and confrontation, but, in New Brunswick, the panels' findings appeared to have demoralized local antispray groups. Coupled with the decline in the epidemic and a subsequent reduction in spraying, a certain ennui descended on the province, at least as far as the spraying controversy was concerned. Whether the matter has been put to rest depends, of course, on one's opinion about the veracity of the panels' methods and findings, something to which we now turn.

All the expert panels adopted the positivist mode of thinking characteristic of scientifically trained professionals (chapter 2). The way in which this limits the effectiveness of evaluation has important implications for the credibility of their findings and inferences about the impact of SBW spraying on public health in NB and elsewhere. Two fundamental aspects of positivist thinking, reductionism and objectivity, are particularly troublesome in the present context. I shall discuss each, in turn, with some relevant examples.

Reductionism enters into analytical epidemiology because of the need for a precisely defined starting point, some category of defining instances that allows one to focus attention on a specific disease.[126] This is such an obvious first step that it is hardly ever given a second thought, even though it is logically problematic. If a disease is defined in a narrow, exclusionary way by discarding instances of other unrelated diseases, then what one is doing is presupposing that these unrelated illnesses do not share a common etiology with the entity in question. However, in many cases one cannot know this because the proposed epidemiological study seeks to establish what these causal relationships are in the first place. Thus, you have a paradox. In order to study the causes of a disease using analytical epidemiology, one has to define the disease entity as if one knew already what these causal factors were and, therefore, can exclude other diseases that do not share a common etiology. Consider the following example.

The traditional epidemiology practiced by the expert panels in New Brunswick was one in which evidence was sought on the incidence of severe health problems such as Reye's syndrome, cancer, and debilitating birth defects. There is good reason to believe, however, that pesticides may have a more pervasive, less obvious effect on human health than this strategy implies. For instance, a great deal of animal research concludes that pesticides may compromise the immune system as well as having a mutagenic effect.[127] If this is also the case in humans, then, in addition to the more severe effects sought by the New Brunswick panels, there may exist a more subtle form of ill health. A compromised immune system could well lead to a greater susceptibility to a variety of infectious diseases that, together with an increase in mutations, might result in more sickly children exhibiting general ill health. Bertell,[128] in her study of an analogous health problem, the effects of low-level radiation, suggests that such a subtle degradation of public health would be hard to detect. Indeed, the health changes that *could* be detected would represent, in her opinion, only a minute proportion of the whole. For instance, it is only in recent years that the more subtle effects of pesticides are being recognized in such problems as the impact of "estrogen mimics" on the reproductive development of animals, including human children.[129,130] Thus, cancer mortalities would be the most gross but least likely outcome of exposure to radiation (and pesticides?), the more likely outcome being benign tumors and nonmalignant health problems. It follows that an analytical focus on a limited number of severe disease entities misses the point, failing as it does to address the prospect of a more pervasive, but more subtle, effect of pesticides in promoting general ill health.

Consider, for example, the adoption by the Spitzer Task Force[122] (p. 27) of the standard definitional criteria for Reye's syndrome proposed by the Centers for Disease Control (CDC):

1. **Acute noninflammatory encephalopathy, that is, less than eight white blood cells in the cerebrospinal fluid or no evidence of brain or meningeal inflammation on autopsy.**
2. **Microvesicular fatty metamorphosis of the liver at biopsy or autopsy, or a threefold rise in at least one of the following blood tests of liver function: serum glutamic oxaloacetic transaminase, serum glutamic pyruvic transaminase, or ammonia.**
3. **No other reasonable explanation for the neurologic liver abnormalities.**

The application of these strict criteria led to some interesting consequences for the study. Because Reye's syndrome was not, at the time, a reportable disease in New Brunswick and local physicians were largely unfamiliar with the illness, the province did not have a reliable record of its incidence. The Spitzer panel was obliged, therefore, to conduct an intensive case-finding search for likely cases in all provincial hospital records (together with some out-of-province hospitals to which patients may have been transferred). This initial survey, which used much broader criteria than the CDC standard definition, identified 3234 cases of illness between 1972 and 1981 that were similar in appearance to Reye's syndrome and some of which could have been misdiagnosed by local physicians (e.g., encephalitis, hepatitis, influenza, mononucleosis, and so on). A careful screening of these records, using the CDC criteria, reduced the number of cases to 12 confirmed cases of Reye's syndrome, 4 possible, and 7 doubtful[122] (pp. 28–31). Now, from the perspective of analytical epidemiology, this is a perfectly logical procedure, identifying, as it does, a precisely defined disease entity and excluding "unrelated" illnesses. From a more holistic, commonsense perspective, this approach excludes from consideration the possibility that pesticides may have been a factor in the etiology of some, or all, of the excluded 3211 cases. Given the *viral enhancement* hypothesis being examined by the Spitzer panel, it does seem odd that only one form of viral disease should receive attention, especially one that is so rare. These definitional strictures were very confusing to the general public who saw what seemed to be obvious cases of pesticide-related illness dismissed from contention. For instance, some years prior to the panel's deliberations, a rural school had been, according to the local children, accidentally sprayed. Shortly thereafter, three children in the same grade contracted encephalitis-like illnesses, two of whom died. In the furor that erupted, parents were assured by medical authorities that these illnesses were not Reye's syndrome and were unrelated to the spray program.[131] From the definitional perspective adopted by the provincial government (the CDC criteria), the former statement is reasonable. However, neither they nor the subsequent Spitzer panel could exclude some causal connection for no one knew what the relationship might be between the aerial spray program and encephalitis.

One can argue, therefore, that a narrow focus on a precisely defined disease entity, while being in accord with the tenets of analytic epidemiology and positivist science, actually misses the point. It does not address the problem of the more pervasive effects of pesticides on a broader spectrum of illness through the viral enhancement mechanism. By focusing on a rare illness, therefore, the Spitzer panel ran the risk of conducting a trivial study, one that involved a type III error. Sadly, the same criticism can be leveled at the other major studies on cancer[124] and reproductive anomalies.[123] In restricting their search to a limited number of severe kinds of cancer and reproductive anomalies, the panels, once again, did not address the issue of more pervasive ill health. With regard to pesticides and cancer, for instance, one should bear in mind Bertell's[128] point, noted earlier, that excess cancer fatalities and malignant tumors are the health outcomes least likely to occur in response to exposure to low-level environmental toxins. In addition, the Hatcher and White[123] panel seems to recognize some of the limitations of its approach in noting that "the number of reproductive parameters monitored were few, and the available information was very limited" (p. iv). Thus, analytical methods appear to afford one the opportunity to look for clearly defined needles in a poorly documented haystack. The conclusions one can draw from such an approach are strictly limited.

Reductive thinking also results in "isolation," in which analytical thinkers seek to understand complex systems by isolating a few variables from their context so that they can be studied under "controlled" conditions. This is problematic even when dealing with simple systems, for the complex interactions between variables, which determine the way in which each individual variable expresses itself, are lost in the process. Although such reductionist strategies have been successful in the physical sciences, they are of dubious utility in biological and human contexts. Perhaps I can demonstrate this point in the present case.

From the onset of the spruce budworm spraying program in 1952, the forests and people of New Brunswick have been exposed to DDT (up to 1968), Fenitrothion, Aminocarb (Matacil), Phosphamidon, Malathion, and recently, the biological agent Bt. Fenitrothion and Aminocarb have been the primary insecticides in use since 1969 in both water and oil-based formulations. Adequate suspension of the active ingredient also requires the use of organic emulsifiers and solvents such as Atlox 3409F, Toximul MP8, Dowanol TPM, and Cyclosol 63. Thus, the composition of spray formulations varied over the years and from batch to batch, depending on the specific mixtures being used at the time. Determining the precise nature of these formulations was made more difficult by the proprietary nature of information on insecticide composition and the simple fact that some of the solvents contained fuel oils, the composition of which was both complex and unknown (to the expert panels).[17]

One can only conclude that the province had been exposed to a chemical soup between 1952 and 1982, to which one might add agricultural pesticides, medications (e.g., aspirin in the case of Reye's syndrome), emulsifiers in household chemicals, food additives, toxins produced by cooking food, natural pesticides in foods, cigarette smoke, radon, and so on.[132,133] A truly exhaustive epidemiological study would need to isolate the variable in question from the effects of these confounding factors in the hope of estimating its separate effect. Even if this were possible, the synergistic effect of interaction between variables would be lost. Thus, one cannot help but agree with the National Research Council of Canada[134] (p. 18):

Humans . . . are seldom, if ever, exposed to a single chemical, and the toxicological consequences of exposure to multiple agents are unknown. Available data show that simultaneous exposure to more than one agent can result in synergistic, additive, inhibitory, or protective effects that are dependent on the agents and the circumstances involved. It is therefore difficult to develop unequivocal cause-effect relationships with epidemiologic methods unless the effect produced by the agent is unique or unusual . . . Epidemiological studies, which are generally retrospective rather than prospective, are of little use in forecasting the health risks associated with a particular chemical because the level of exposure to and effects of the chemical must be known before predictions can be made.

Thus, analytical toxicology is faced with a paradox in isolating variables for study. The studies reviewed by the expert panels appear to

have followed a common toxicological pattern[134] (p. 66): "In general, long-term animal assay procedures are conducted to assess potential carcinogenicity of a single chemical and only a few studies have been conducted to assess the effects of interactions among chemicals." The problem with this strategy is that chemicals are thought to act both directly and indirectly on cells. In the case of carcinogenesis, direct action on the genetic material may initiate cancerous growth. More indirectly, however, some chemicals may act to either promote or inhibit the action of other chemicals that influence the growth of cancer cells. This indirect action can be either additive or synergistic. Chemical interactions are, therefore, of the utmost importance. When one turns to the toxicological studies reviewed by the expert panels, only in the case of Reye's syndrome[111,122] was interaction considered. All of the studies reviewed by the other panels[123,124] assessed the effect of technical-grade Fenitrothion (virtually pure with the possible presence of small quantities of unknown impurities). The general conclusion conveyed by the four expert panels, therefore, was that the available scientific evidence, if not giving Fenitrothion a clean bill of health, certainly suggested that there should be little cause for alarm when considering the tiny doses to which humans were exposed. This view is caught succinctly in the following[122] (p.17):

The pesticide has been evaluated repeatedly in acute, subacute, and chronic toxicity tests including multigenerational studies. Bioassays have included evaluation of teratogenic, mutagenic, and carcinogenic potentials of fenitrothion . . . A detailed evaluation of (this) scientific literature with the major emphasis on Reye's syndrome produced little evidence to indicate that the levels of fenitrothion used in the spruce budworm control programme would elicit a toxic response in the human population in or near the application areas.

These sentiments were echoed, albeit more cautiously, in the later panels and, also, some years later by Ecobichon.[135] If one accepts the value of single-chemical studies, in which interactions are ignored, then this conclusion is plausible. However, there was evidence available to the panels that suggested otherwise.

First, the research by the Dalhousie group on pesticides and viral enhancement demonstrated clearly enough that pure Fenitrothion, by itself, did not appear to produce this phenomenon. Yet, a significant synergistic effect between DDT and Fenitrothion had been found in the Crocker group's first study.[109] One conclusion could be that bioassays of pure Fenitrothion would likely show nothing of interest. Another, even more disturbing piece of research from Dalhousie was also given short shrift. One of the technicians working with the research group had been exposed, over a period of 7 days, to a mixture of Fenitrothion and emulsifiers that she had been transporting in her car. Evidently, the container leaked and she inhaled some of the vaporized insecticide, as well as getting some of the pesticide on her skin while cleaning up the leak barehanded.[118] Within two days of exposure, she was hospitalized with a variety of symptoms including nausea, blurred vision, diarrhea, abdominal cramps, muscular weakness, mental confusion, and tremors. These symptoms responded to treatment, and she was released 16 days after exposure only to be readmitted 1 day later suffering from mental depression. Some 35 days after exposure she was finally released, although she experienced bouts of fatigue and weakness for several months thereafter. Significantly, an attempt to diet eight months after exposure led to a recurrence of symptoms, the implication being that toxin had been stored in her fatty tissues. This is contrary to conventional wisdom, which asserts that Fenitrothion is *not* stored in the body, a finding that might well have been afforded a more prominent role in the proceedings. Nonetheless, I conclude that the cautious but sanguine conclusions of the panels reflect their positivist thinking.

What I have tried to show here is that at the root of each of the panels' studies has been a series of analytical decisions that have simplified and reduced the issues at hand. While analytical scientists accept this as a necessary step to reduce an otherwise unmanageable complexity, those of a more holistic bent would likely say that the panels were not actually studying the question of pesticides and health in New Brunswick but rather a greatly reduced and inconsequential version of the problem.

Turning to the objective stance adopted by positivist scientists, it follows that the deliberations of epidemiologists in general[126] and the expert panels in particular would reflect this attitude. One common form of "objectivity" adopted by scientists is a studied disinterest in the historical and social dimensions of the problem at hand. Their ostensible concern lies only with the scientific data, the latter being the product of analytical studies conducted by appropriately trained personnel, preferably in laboratories. It follows that the rich fund of lay knowledge about spraying and illness would be seen as being both suspect and unreliable.This form of selective attention has some interesting consequences. For example, in estimating the potential exposure of the ill to spraying, the panels relied on maps of spraying operations provided by the provincial government, the distance between the person's residence and the nearest spray block being the crucial metric. However, according to local environmental groups , many of these maps were out of date.[136] There were endless complaints from citizens that the spraying operations were prone to error. Planes were said to have strayed from their allotted path, inadvertently spraying residences, bodies of water, woods workers, and even schools, or to have operated when the prevailing wind was above the maximum permitted for spraying to continue.[137-141] The seriousness with which citizens took these alleged infractions is reflected in the court actions brought against the main spraying agency.[142] Little of this contextual information is evident in the panels' deliberations, however, even though it throws doubt on the accuracy of provincial maps.

One might assume that, in analytic science, data collection is routine and uncomplicated. In practice, however, nothing is further from the truth. Take, for example, the process of estimating chromosome damage in mutagenicity studies where one looks for aberrations in the genetic material. Frequently, scientists disagree over the presence or absence of breaks, whether *gaps* should be counted as *breaks*, and whether there are more chromosome fragments than usual in the material being studied.[113,143,144] The importance of these subjective judgments is apparent in the deliberations of the Spitzer panel.[122] Referring to a study that appears to have shown the genotoxic capabilities of Fenitrothion, the panel notes:

The data are recorded as chromosome breaks, but appended documents reveal that in fact, chromosome gaps and breaks were added together. While there may be some association between gaps and chromosome aberrations or breaks, the genetic consequences of gaps are not understood and it is not really clear what they are. Many investigators in this field do not count gaps in the chromosome aberration assay. If gaps were ignored, of course, Fenitrothion would be negative in this assay" (p. 48).

At the root of data collection, therefore, is this basic subjectivity. One can find a pesticide to be genotoxic, or not, depending on the judgment made about the significance of chromosome gaps. The more general point, however, is that these methodological issues cannot be readily resolved by objective means, leaving fundamental value disputes at the core of analytical methodology. This uncertainty reduces the faith one can place in objective data in attempting to resolve issues such as the one at hand.[145]

In sum, four expert panels were convened by the provincial government to review the available scientific evidence on the issue of environmental chemicals and health and to conduct additional studies where necessary. Using standard analytical methods, none of the panels were able to find any significant associations between forest spraying and Reye's syndrome, cancer, or birth defects, although there were sufficient indications of an increased incidence of some of these problems to warrant further study. The political effect of these findings was to reassure the general public that neither was government policy exposing them to untoward danger nor was there a pressing need for any substantial change in that policy. Meanwhile, the antispray groups, unable to mount an effective challenge to the panels' findings and having seen their decade of struggle come to naught, faded from the scene. Of interest here is whether this outcome was justified in light of the evidence amassed by the

panels. I would suggest that the analytical approach to epidemiology and toxicology taken by the panels suffers from many of the problems associated with technical rationality and objective-analytic thinking. As a result, a narrow conceptualization of the problem was adopted, one that sought evidence on the etiology of relatively rare and severe illness, rather than on the more relevant but daunting problem of general ill health. Similarly, in using the traditional analytical strategy of isolation, the more significant problem of interaction between pesticides and other environmental factors remained unexamined. To add to this list of shortcomings, it is also clear that the panels' attempts at being objective resulted in some disinterest in lay knowledge and a preference for authoritative information. I would suggest, therefore, that the panels did not really study the question of pesticides and health but rather some minor, possibly inconsequential, subset of the larger problem. Furthermore, the reductive methods employed virtually guaranteed that nothing of significance would be found.

Case Study Summary

After 1985, the SBW debate simply dropped from public view. It was as if there was a collective sigh of relief from both the general public and the government. Few stories on the SBW appeared in the local media, and the Concerned Parents, to all intents and purposes, disbanded, its former members moving on to other environmental issues. Contributing to this turn of events was the rapid and inexplicable decline in SBW infestation during the 1980s and the consequent marked reduction in the spray program.[146] However, with Ontario and Quebec reporting a recent increase in infestation, it is feared that a new epidemic may reach NB in 1999.[147] These predictions come at a time when the federal government, after a careful review of available evidence, has finally decided to ban the use of Fenitrothion in aerial spraying operations, effective December 1998.[148] The province, therefore, faces this new threat without its main pest management tool. In light of this renewed threat, one wonders whether the situation will be managed differently from 1952, or will the same scenario be replayed? When the Minister of Natural Resources referred, in 1995, to Fenitrothion as "the only way we had to combat" the SBW,[149] one hears echoes of similar statements in 1952 about DDT. What, then, has been learned after four decades of problem-solving experience with one of forestry's most persistent problems?

Among the professional forestry community, one sees the expression of mixed feelings about the era of confrontation and the role of environmental interest groups. This is well expressed by Baskerville[9] when he concludes that the alleged link between Reye's syndrome and pesticide spraying was a "scary false alarm" brought on by the use of scientific information out of context:

The insecticide issue demonstrated the influence of special interest groups or nongovernmental organizations in forest policy. These groups choose an issue and tend to isolate it from the context in which the issue is embedded in reality. They then push 'the answer' based on this out-of-context view . . . The fact that a small group can create and maintain a dominant view of an issue in the media is seen by some as a danger to reasoned development in a technological society (p. 58).

Baskerville does go on to say that, at times, special interest groups can lead in the recognition of problems but, in general, have a tendency to stop things. The "things" he refers to are, I assume, the policies considered to be important by the iron triangles that controlled forestry and natural resource management in the past. Environmental interest groups most certainly oppose many such Imperial policies.

Those who espouse technical rationality as a route to salvation continue their studies much as before. With the removal of Fenitrothion from the pest management armory, and the inconsistent performance of Bt, the province is experimenting with MIMIC, a hormone that interferes with the larvae's molting process.[150] In an era of concern about the effects of estrogen mimics, any widespread use of this insecticide is likely to produce renewed fury among environmental groups and a resur-

gence of environmental conflict. Nevertheless, pesticide spraying remains the official strategy for dealing with the problem. Similarly, impact studies continue to produce the kind of ambivalent results that, as in the past, fail to reduce decision-making uncertainty. For instance, although attempts to replicate the Dalhousie toxicological studies by independent laboratories failed to reproduce the viral-enhancing effects found in the original studies,[151,152] the depression of cholinesterase by Fenitrothion is now a well-established phenomenon.[153,154] However, recent research on pesticides has begun to explore their more subtle, neurobehavioral effects, thereby opening up an even greater area for conjecture and controversy. To add to this morass, it is now clear that illnesses such as Reye's syndrome can be triggered by a variety of environmental factors, such as food additives, agricultural pesticides, emulsifiers in paint, and so on.[132,155] Some success has been achieved in identifying aspirin as a major cofactor in the genesis of Reye's syndrome,[156,157] while a decrease in administering aspirin to children appears to be correlated with a significant decline in the illness.[155] In addition, inborn errors of metabolism have been linked with the onset of the disease, suggesting that some children may be genetically predisposed to develop the condition in the presence of the other factors.[156,158] Reye's syndrome, therefore, can result from any number of permutations of these numerous factors, something that standard epidemiological methods find difficult, if not impossible, to disentangle. The effect of forest pesticides on public health, therefore, is a trans-science problem, one that can be couched in scientific terms but remains unanswerable by science.[159] The implications of this conclusion for policy making remains to be explored in the NB context.

Thus, environmental groups chose to make their stand on an issue that was insoluble and unwinnable. This experience may have diminished the credibility of the Arcadian perspective, which was never very strong in the first place. A fully fledged Arcadian position has always been and remains a marginal position in NB. Many of those involved in recent forestry debates in the province have more pragmatic goals. For instance, the small-woodlot owners have engaged the political-corporate elites in a lengthy battle to establish themselves as the primary suppliers of wood fiber to the pulp mills.[160,161] However, this protracted struggle would seem to be in pursuit of a more equitable slice of the Imperial pie, rather than for Arcadian motives. It is only in marginal groups, such as the Conservation Council of NB, that one finds innovative proposals for restructuring the forest industry base on Arcadian principles.[162] However, their impact on policy discourse is minimal in a province where the great mass of people show little interest in any fundamental reappraisal of the Imperial belief in economic growth.

As the SBW epidemic declined in the mid-1980s, the province settled back into its usual routine.[21] The provincial government is not contemplating any radical changes in industrial policy but is, instead, making every effort to ensure the survival of its industrial base.[163] At the same time, efforts are being made to develop value-added manufacturing and alternative employment for New Brunswickers independent of its traditional reliance on primary production.[164] One could conclude, therefore, that political-economic rationality has proved effective in allowing the establishment successfully to defend its policies against the demands of the Concerned Parents and other environmental groups. In a broader sense, however, the incremental changes implemented over the last four decades have really not addressed the systemic problems in NB forestry. In addition, the unwillingness of the New Brunswick establishment to draw interest groups into the policy process in a significant way during the 1970s and early 1980s may have been consistent with the behavior of other provincial governments of the time, but it also reduced the amount of mutual learning that could have taken place. Recently, there have been some modest efforts to remedy this situation through incorporation of lay groups into forestry planning (e.g., the Fundy Model Forest) and policy discussions (e.g., Round Tables on Environment and Economy). Although this involvement of the informed public in decision making is a laudable first step in problem-solving reform, it pales in comparison

to the changes that are needed to overcome the many personal and institutional barriers to adaptive problem solving. The next chapter looks at the shortcomings of interest-group politics in more detail.

References

1 Janis, I. 1989. *Crucial decisions: Leadership and policymaking in crisis management.* New York: The Free Press.

2 Briassoulis, H. 1989. Theoretical orientations in environmental planning: An inquiry into alternative approaches. *Environmental Management* 13:381–392.

3 Rochefort, D., and R. Cobb. 1993. Problem definition, agenda access, and policy choice. *Policy Studies Journal* 21:56–71.

4 Catton, W. 1989. Choosing which danger to risk. *Society* 27:6–8.

5 Bosso, C. 1987. *Pesticides & politics: The life cycle of a public issue.* Pittsburgh: University of Pittsburgh Press.

6 Dryzek, J. 1987. *Rational ecology: Environment and political economy.* New York: Blackwell.

7 Gillis, R., and T. Roach. 1986. *Lost initiatives: Canada's forest industries, forest policy and forest conservation.* Westport, Conn.: Greenwood Press.

8 Kimmins, H. 1995. Sustainable development in Canadian forestry in the face of changing paradigms. *Forestry Chronicle* 71:33–40.

9 Baskerville, G. 1995. The forestry problem: Adaptive lurches of renewal. In *Barriers and bridges to renewal of ecosystems and institutions,* ed. L. Gunderson, C. Holling, and S. Light, 37–102. New York: Columbia University Press.

10 Porter, T. 1997. Protestors: Don't cut old forest. *Daily Gleaner,* May 2. Fredericton, N.B., Canada.

11 Foulds, J., and S. Manley. 1990. "Toothpicks" and the forests of tomorrow. *New Maritimes* 8:12–13.

12 Lansky, M. 1992. *Beyond the beauty strip: Saving what's left of our forest.* Gardiner, Maine: Tilbury House.

13 MacNutt, W. 1963. *New Brunswick: A history: 1784–1867.* Toronto: Macmillan.

14 Regier, H., and G. Baskerville. 1986. Sustainable redevelopment of regional ecosystems degraded by exploitive development. In *Sustainable development of the biosphere,* ed. W. Clark and R. Munn, 75–103. Cambridge: Cambridge University Press.

15 Freedman, B. 1989. *Environmental Ecology.* New York: Academic Press.

16 Blais, J. 1985. The spruce budworm and the forest. In *Recent advances in spruce budworms research: Proceedings of the CANUSA spruce budworms research symposium*, ed. C. Sanders et al., 135–136. Ottawa: Canadian and U.S. Forest Services.

17 Versteeg, H. 1984. The spruce budworm programme and the perception of risk in New Brunswick. In *Pesticide policy: The environmental perspective*, ed. Anonymous, 77–127. Ottawa: Friends of the Earth.

18 Young, R. 1988. 'and the people will sink into despair': Reconstruction in New Brunswick, 1942–52. *Canadian Historical Review* LXIX:127–166.

19 Parenteau, W. 1992. "In good faith": The development of pulpwood marketing for independent produceers in New Brunswick, 1960–75. In *Trouble in the woods: Forest policy and social conflict in Nova Scotia and New Brunswick*, ed. A. Sandberg, 110–141. Fredericton: Acadiensis Press.

20 Fairley, B., et al. 1990. Restructuring and resistance in Atlantic Canada: An introduction. In *Restructuring and resistance: Perspectives from Atlantic Canada*, ed. B. Fairley, C. Leys, and J. Sacouman, 11–19. Toronto: Garamond Press.

21 Cashore, B. 1988. *The Role of the Provincial State in Forest Policy: A Comparative Study of British Columbia and New Brunswick*. Master's thesis, Carleton University, Ottawa.

22 Wynn, G. 1981. *Timber Colony: An historical geography of early nineteenth century New Brunswick*. Toronto: University of Toronto Press.

23 Sandberg, L. 1992. Introduction: Dependent development and client states: Forest policy and social conflict in Nova Scotia and New Brunswick. In *Trouble in the woods: Forest policy and social conflict in Nova Scotia and New Brunswick*, ed. L. Sandberg, 2–21. Fredericton: Acadiensis Press.

24 Anonymous. 1995. *New Brunswick's forestry sector.* Forestry Report, 10. Canadian Forestry Service-Maritimes Region, Fredericton, N.B., Canada.

25 Parenteau, W. 1989. Pulp, paper and poverty: Then and now: Past and present in the New Brunswick woods. *New Maritimes* 7:20–26.

26 Whelton, K. 1987. The Bathurst blockade: Non-unionized woodcutters in northern New Brunswick are through with accepting nineteenth-century working conditions and poverty wages in silence. *New Maritimes* 5:3–5.

27 Canadian Press. 1994. Arson suspected in wood-harvester fire. *The Daily Gleaner,* August 9. Fredericton, N.B., Canada.

28 Webster, P. 1991. *Pining for the trees: The history of dissent against forest destruction in Nova Scotia 1749–1991*. Master's thesis, Dalhousie University, Halifax, Nova Scotia, Canada.

29 Clark, T., A. Curlee, and R. Reading. 1996. Crafting effective solutions to the large carnivore conservation problem. *Conservation Biology* 10:940–948.

30 Rochefort, D., and R. Cobb. 1994. Problem definition: An emerging perspective. In *The politics of problem definition: Shaping the policy agenda*, ed. D. Rochefort and R. Cobb, 1–31. Lawrence, Kans.: University Press of Kansas.

31 McMartin, J. 1995. *Personality psychology: A student-centered approach*. Thousand Oaks, Calif.: Sage.

32 Mahoney, M. 1976. *Scientist as subject: the psychological imperative*. Cambridge, Mass.: Ballinger.

33 Savan, B. 1988. *Science under siege: The myth of objectivity in scientific research*. Montreal: CBC Enterprises.

34 Rothman, S., and S. Lichter. 1987. Elite ideology and risk perception in nuclear energy policy. *American Political Science Review* 81:383–404.

35 Dake, K. 1992. Myths of nature: Culture and social construction of risk. *Journal of Social Issues* 48:21–37.

36 Francis, G., and H. Regier. 1995. Barriers and bridges to the restoration of the Great Lakes Basin ecosystem. In *Barriers and bridges to the renewal of ecosystems and institutions*, ed. L. Gunderson, C. Holling, and S. Light, 239–291. New York: Columbia University Press.

37 Cable, S., and M. Benson. 1993. Acting locally: Environmental injustice and the emergence of grass-roots environmental organizations. *Social Problems* 40:464–477.

38 Balch, R., and W. Reeks. 1945. Report on forest insects in New Brunswick. Department of Lands and Mines, Government of New Brunswick, Fredericton, N.B., Canada.

39 Demeritt, D. 1947. Letter to chairman of the Forest Insect Committee of the Northeast, January 11, Bangor, Maine. In Forest Insects Control Board archives, N.B. Provincial Archives, Fredericton, N.B., Canada.

40 Forest Development Commission. 1957. Report of the Forest Development Commission. Government of New Brunswick. Fredericton, N.B., Canada.

41 Swaine, J., F. Craighead, and I. Bailey. 1924. *Studies on the spruce budworm*. Department of Agriculture, Technical Bulletin No. 37, Dominion of Canada, Ottawa, Canada.

42 Miller, A., and P. Rusnock. 1993. The rise and fall of the silvicultural hypothesis in spruce budworm management in eastern Canada. *Forest Ecology and Management* 61:171–189.

43 Reeks, W. 1946. Minutes of December 12th meeting of the Advisory Committee on forest entomology and pathology of the Canadian Pulp & Paper Association, Montreal. In Forest Insects Control Board archives, N.B. Provincial Archives, Fredericton, N.B., Canada.

44 Sewall, J. 1945. Letter to members of the Forest Insect Committee of the Northeast. November 13. Old Town, Maine. In Forest Insects Control Board archives, N.B. Provincial Archives, Fredericton, N.B., Canada.

45 Paananen, D., R. Fowler, and L. Wilson. 1987. The aerial war against Easter Region forest insects, 1921–86. *Journal of Forest History* 31:173–186.

46 Marshall, J. 1946. Letter to Dr. Prince, Deputy Minister, N. B. Dept. of Lands and Mines, March 20, in Forest Insects Control Board archives, N. B. Provincial Archives, Fredericton, N.B., Canada.

47 Marshall, J. 1949. Minutes of the October 4–6 meeting of the Executive Committee of the Forest Insects Control Board, Ottawa. In Forest Insects Control Board archives, N. B. Provincial Archives, Fredericton, N.B., Canada.

48 Palladino, P. 1989. *Entomology and Ecology: The Ecology of Entomology. The 'Insecticide Crisis' and Entomological Research in the United States in the 1960s and 1970s: Political, Institutional, and Conceptual Dimensions.* Ph.D. thesis, University of Minnesota.

49 Marshall, J. 1949. Quarterly report on the activities of the Forest Insects Control Board. January 1 to March 31. Department of Mines and Resources, Government of Canada. In Forest Insects Control Board archives, N. B. Provincial Archives, Fredericton, N.B., Canada.

50 Marshall, J. 1951. Minutes of the October 26 meeting of the Forest Insects Control Board, Ottawa. In Forest Insects Control Board archives, N. B. Provincial Archives, Fredericton, N. B., Canada.

51 deGryse, J. 1947. Noxious forest insects and their control. In *Canada Year Book*, ed. Anonymous, Ottawa: Dominion Bureau of Statistics.

52 MacLean, D. 1996. Forest management strategies to reduce spruce budworm damage in the Fundy Model Forest. *Forestry Chronicle* 72:399–405.

53 Graham, S. 1951. Developing forests resistant to insect injury. *The Scientific Monthly* LXXIII:235–244.

54 Prebble, M., and R. Morris. 1951. The spruce budworm problem. *Forestry Chronicle* 27:14–22.

55 Zon, R. 1908. Principles involved in determining forest types. *Forestry Quarterly* 6:263–271.

56 Westveld, M., H. MacAloney, and J. Hansbor-

ough. 1950. Forest crop security: The right tree on the right site. *Forestry Chronicle* 26:144–151.

57 Tothill, J. 1922. Notes on the outbreaks of spruce budworm, forest tent caterpillar and larch sawfly in New Brunswick. *Proceedings of the Acadian Entomological Society* 8:172–182.

58 Swaine, J. 1928. *Forest entomology and its development in Canada.* Department of Agriculture, Pamphlet No. 97, Dominion of Canada, Ottawa.

59 Heimberger, C. 1945. Comment on the budworm outbreak in Ontario and Quebec. *Forestry Chronicle* 21:114–126.

60 Wellington, W., et al. 1950. Physical and biological indicators of the development of outbreaks of the spruce budworm. *Canadian Journal of Research, section D* 28:308–331.

61 Baskerville, G. 1975. Spruce budworm: the answer is forest management: or is it? *Forestry Chronicle* 51:157–160.

62 Irland, L. 1980. Spruce budworm: economics and management for the long pull. *Maine Forestry Review* 13:24–26.

63 MacLean, D. 1984. Effects of spruce budworm outbreaks on the productivity and stability of balsam fir forests. *Forestry Chronicle* 60:273–279.

64 Baskerville, G. 1975. Spruce budworm: super silviculturalist. *Forestry Chronicle* 51:138–140.

65 Baskerville, G. 1976. Report of the task force for evaluation of budworm control alternatives. Department of Natural Resources, Government of New Brunswick, Fredericton, N.B., Canada.

66 Adams, N. 1952. Letter to Premier Flemming. Flemming archives, N. B. Provincial Archives, Fredericton, N.B., Canada.

67 Staff reporter. 1975. Starvation of spruce budworm proposed as way to end insecticide crisis. *Daily Gleaner*, February 12. Frederiction, N.B., Canada.

68 O'Neill, J. 1949. Hazard to human life is found in growing use of insecticides. *Saturday Night*, April 5, Toronto, Canada.

69 McCormack, W. 1952. Minutes of the Upsalquitch meeting of August 6, Flemming archives, N. B. Provincial Archives, Fredericton, N.B., Canada.

70 Webb, F., and H. Irving. 1983. My fir lady: The New Brunswick production with its facts and fancies. *Forestry Chronicle* 59:118–122.

71 Bedard, J. 1952. Letter to Premier Flemming, October 24. Flemming archives, N. B. Provincial Archives, Fredericton, N.B., Canada.

72 Blenis, H. 1952. Letter to Premier Flemming, October 15. Flemming archives, N. B. Provincial Archives, Fredericton, N.B., Canada.

73 Lund, Wilk, Scott, and Goodall. 1979. Study of

alternatives to state management of spruce budworm spraying. Consultative Report. Maine Department of Conservation, Augusta, Maine.

74 Editorial. 1952. Operation budworm. *Telegraph-Journal,* May 31. Saint John, N.B., Canada.

75 Flemming, H. 1952. Address to Canadian Institute of Forestry, November 14. Fredericton, in Flemming archives, N. B. Provincial Archives, Fredericton, N.B., Canada.

76 Woodsman. 1956. On spraying the forests, *Daily Gleaner,* Fredericton, N.B.

77 Naturalist. 1956. Balance of nature, *Daily Gleaner,* Fredericton, N.B.

78 Johnson, V. 1956. Letter to Premier Flemming, October 1. Flemming archives, N. B. Provincial Archives, Fredericton, N.B., Canada.

79 Anonymous. 1962. Atlantic Advocate article discusses chemical poisons. *Daily Gleaner,* November 20. Fredericton, N.B., Canada.

80 Concerned. 1962. Says N.B. homes sprayed. Letter to Editor, *Daily Gleaner,* June 28. Fredericton, N.B., Canada.

81 Staff Reporter. 1978. Michael Wardell's impact on this area was immense. *Daily Gleaner,* May 1. Fredericton, N.B., Canada.

82 Almeida, P., and L. Stearns. 1998. Political opportunities and local grassroots environmental movements: The case of Minamata. *Social Problems* 45:37–60.

83 Staff Reporter. 1969. DDT residues are breeding bans: New Brunswick joins Sweden, Wisconsin in pesticide controls. *Financial Post,* June 25, Toronto, Canada.

84 Kettela, E. 1995. Insect control in New Brunswick, 1974–1989. In *Forest Insect Pests in Canada,* ed. J. Armstrong and W. Ives, 655–665. Natural Resources Canada, Canadian Forest Service, Science and Sustainable Development Directorate, Ottawa, Canada.

85 Irving, H. 1985. *Coping with the spruce budworm: The technology factor.* The E. B. Eddy Distinguished Lecture Series, 33–42. Faculty of Forestry, University of Toronto, Canada.

86 DeLoitte and Touche. 1992. *Economic benefit assessment of spruce budworm control in eastern Canada.* Prepared for the Forest Pest Management Caucus by Deloitte & Touche Management Consultants, Guelph, Ontario, Canada.

87 Rayner, J., and D. Peerla. 1987. The spruce budworm spray controversy in Canada: Foresters' perceptions of power and conflict in the policy process. In *Social science in natural resource management systems,* ed. M. Miller, R. Gale, and P. Brown, 213–231. Boulder, Colo.: Westview Press.

88 Fowle, C. 1988. Using information to cope with risks in the spruce budworm control program in the maritime provinces. In *Information Needs for Risk Management,* ed. C. Fowle, A. Grima, and R. Munn, 157–175. Institute of Environmental Studies, University of Toronto, Canada.

89 Boisjoly, R., E. Curtis, and E. Mellican. 1989. Roger Boisjoly and the Challenger disaster: The ethical dimension. *Journal of Business Ethics* 8:217–230.

90 Miller, A., and W. Cuff. 1986. The Delphi approach to the mediation of environmental disputes. *Environmental Management* 10:321–330.

91 Baskerville, G. 1979. Implementation of adaptive approaches in provincial and federal forestry agencies. In *Environmental policy seminar.* Institute for Applied Systems Analysis, Laxenburg, Austria.

92 Sanders, C., et al., eds. 1985. *Recent advances in Spruce Budworms reasearch.* Canadian Forestry Service, Ottawa, Canada.

93 Cuff, W., and H. Walker. 1985. Integration of forest management and pest management in the eastern provinces of Canada. In *Recent advances in spruce budworms research,* ed. C. Sanders et al., 443–464. Canadian Forestry Service, Ottawa, Ontario.

94 Runyon, K., et al. 1983. *Canusa spruce budworm program: Organization and administrative effectiveness.* May 9, Report for the Canadian Forestry Service and United States Forest Service.

95 Blum, B., and D. MacLean. 1985. Potential silviculture, harvesting and salvage practices in eastern North America. In *Recent advances in spruce budworms research,* ed. C. Sanders et al., 264–280. Ottawa: Canadian Forestry Service.

96 Hudak, J. 1991. Integrated pest management and the eastern spruce budworm. *Forest Ecology and Management* 39:313–337.

97 Irland, L., and J. Dimond. 1991. IPM and the spruce budworm: Lessons learned in Maine 1950–1985. *Forest Ecology and Management* 39:263–273.

98 Staff Reporter. 1975. Starvation of spruce budworm proposed as a way to end insecticide problems. *Daily Gleaner,* February 12, Fredericton, N.B., Canada.

99 Howard, R. 1980. *Poisons in public: Case studies of environmental pollution in Canada.* Toronto: James Lorimer & Company.

100 Miller, A., and P. Rusnock. 1993. The ironical role of science in policymaking: The case of the spruce budworm. *International Journal of Environmental Studies* 43:239–251.

101 Associate Committe on Scientific Criteria for Environmental Quality, eds. 1977. *Proceedings of a symposium on fenitrothion: The long term effects of its use in forest ecosystems.* Ottawa: National Research Council of Canada.

102 MacTaggart-Cowan, P. 1977. Fenitrothion-The long term effects of its use in forest ecosystems: Current status. In *Proceedings of a symposium on fenitrothion: The long term effects of its use in forest ecosystems,* ed. Associate Committee on Scientific Criteria for Environmental Quality, 573–577. Ottawa: National Research Council of Canada.

103 Varty, I. 1977. *Credibility of the NRCC 'Current Status' report on fenitrothion.* Maritime Forest Research Centre, Information report M-X-79, Fredericton, N.B., Canada.

104 Anonymous. 1989. A review of the Environment Canada, Atlantic Region, document: Effects of fenitrothion use in forestry-Impacts on insect pollinaters, songbirds & aquatic organismis. Ottawa: Forestry Canada.

105 Ernst, W., P. Pearce, and T. Pollock. 1989. *Environmental effects of fenitrothion use in forestry: Impacts on insect pollinators, songbirds & aquatic organisms.* Dartmouth, Nova Scotia: Environment Canada.

106 Collingridge, D., and C. Reeve. 1986. *Science speaks to power: The role of experts in policy making.* London: Francis Pinter.

107 Staff Reporter. 1969. Public 'mistrusts' forest spraying. *Daily Gleaner,* February 19. Fredericton, N.B., Canada.

108 Miller, A. 1993. The role of citizen scientist in natural resource decision making. *The Environmentalist* 13:47–59.

109 Crocker, J., et al. 1974. Insecticide and viral interaction as a cause of fatty visceral changes in encaphalopathy in the mouse. *Lancet* 2:22–24.

110 Crocker, J., et al. 1976. Lethal interactions of ubiquitous insecticide carriers with virus. *Science* 192:1351–1353.

111 Schneider, W., et al. 1976. *Forest spray program and Reye's Syndrome.* Report of the panel covened by the Government of New Brunswick, April 26, Fredericton, N.B., Canada.

112 Brown, P., and E. Mikkelsen. 1990. *No safe place: Toxic waste, leukemia, and community action.* Berkeley, Calif.: University of California Press.

113 Levine, A. 1982. *Love Canal: Science, politics, and people.* Lexington, Mass.: Lexington Books.

114 Taylor, M. 1982. Brief to the Select Committee on Environment of the New Brunswick legislature on behalf of the Concerned Parents. January 13. Fredericton, N.B., Canada.

115 Deveaux, B. 1982. Poison Mist: A special investigation into New Brunswick's forest spray programme. Radio documentary, January 3, Toronto: Canadian Broadcasting Corporation.

116 Ecobichon, D., and J. Crocker. 1978. Depression of blood cholinesterases as a marker of spray exposure. *Chemosphere* 7:591–596.

117 Rozee, K., et al. 1978. Emulsifiers as enhancement factors in virus virulence. In *International Conference on Reye's Syndrome.* June 22–23. Halifax, Nova Scotia, Canada.

118 Ecobichon, D., et al. 1977. Acute fenitrothion poisoning. *Canadian Medical Association Journal* 19:377–379.

119 Concerned Parents. 1980. News release on the findings of Dr. Kawachi, June 3, Fredericton, N.B., Canada.

120 Canadian Press. 1980. Robertson discounts claim—Concerned Parents claim spray ingredients linked to cancer. *Daily Gleaner,* May 26, Fredericton, N.B., Canada.

121 Rozee, K., et al. 1982. Is a compromised interferon response an etiological factor in Reye's Syndrome? *Canadian Medical Association Journal* 126:798–802.

122 Spitzer, W. 1982. Report of the New Brunswick task force on the environment and Reye's Syndrome. Department of Health, Province of New Brunswick, Fredericton, N.B., Canada.

123 Hatcher, J., and F. White. 1985. Report of the task force on chemicals in the environment and human reproductive problems in New Brunswick. Department of Health, Province of New Brunswick, Fredericton, N.B., Canada.

124 Spitzer, W. 1984. Report of the New Brunswick task force on the environment and cancer. Department of Health, Province of New Brunswick, Fredericton, N.B., Canada.

125 Brown, P. 1992. Popular epidemiology and toxic waste contamination: Lay and professional ways of knowing. *Journal of Health and Social Behavior* 33:267–281.defined.

126 Inskip, H., and J. Davies. 1987. Methodological aspects of epidemiological studies on groups of workers and members of the public. In *Nuclear Energy Agency workshop on epidemiology and radiation protection,* ed. Anonymous, 13–24. October 13–15. Paris: OECD.

127 Thomas, P., and R. House. 1989. Pesticide-induced modulation of the immune system. *American Chemical Society Symposium Series* 414:94–108.

128 Bertell, R. 1985. *No immediate danger? Prognosis for a radioactive earth.* Toronto: Women's Educational Press.

129 Staff Reporter. 1995. DDT linked to abnormalities. *Daily Gleaner,* June 15. Fredericton, N.B., Canada.

130 Anonymous. 1996. Reproductive anomalies: Are fears of chemicals justified? *Globe and Mail,* August 10, Toronto, Canada.

131 Canadian Press. 1977. Angry parents not satisfied about spray. *Daily Gleaner,* February 9, Fredericton, N.B., Canada.

132 Crocker, J., et al. 1986. Biochemical and morphological characteristics of a mouse model of Reye's Syndrome induced by the interaction of influenza B virus and a chemical emulsifier. *Laboratory Investigation* 54:32–40.

133 Efron, E. 1984. *The apocalyptics: cancer and the big lie: How environmental politics controls what we know about cancer.* New York: Simon and Schuster.

134 Associate Committee on Scientific Criteria for Environmental Quality 1985. *Strengths and limitations of benefit-cost analyses applied to the assessment of industrial organic chemicals including pesticides, Monograph III. Extrapolation of toxicological data from laboratory studies to the human situation.* NRCC No. 23909. National Research Council of Canada, Ottawa, Canada.

135 Ecobichon, D. 1990. Chemical management of forest pest epidemics: A case study. *Biomedical and Environmental Sciences* 3:217–239.

136 Staff Reporter. 1981. Spray map review urged. *Daily Gleaner,* July 17, Fredericton, N.B., Canada.

137 Canadian Press. 1981. Tree planters sprayed. *Daily Gleaner,* June 5, Fredericton, N.B., Canada.

138 Concerned Parents. 1977. Bloomfield area: Depositions from sprayed citizens. Fredericton, N.B.: Concerned Parents of New Brunswick.

139 Kotzwinkle, W. 1971. You are a worm. *Mysterious East* 19:11–16.

140 May, E. 1982. *Budworm battles.* Halifax, Nova Scotia: Four East Publications.

141 Staff Reporter. 1976. Minto school children complain of another spray incident. *Daily Gleaner,* May 26. Fredericton, N.B., Canada.

142 Burrows, J. 1978. Friesen wins suit against aerial spraying. *Daily Gleaner,* May 18, Fredericton, N.B., Canada.

143 Hays, S. 1987. *Beauty, health and permanence: Environmental politics in the United States, 1955–1985.* Cambridge: Cambridge University Press.

144 Paigen. 1982. Controversy at Love Canal. *Hastings Center Report* June: 29–37.

145 Marcus, A. 1988. Risk, uncertainty, and scientific judgement. *Minerva* 26:138–152.

146 Staff Reporter. 1992. Budworm spray program cut back. *Daily Gleaner,* February 26. Fredericton, N.B., Canada.

147 Llewellyn, S. 1996. Budworm alert on. *Daily Gleaner,* September 27, Fredericton, N.B., Canada.

148 Interdepartmental Executive Committee on Pest Management. 1995. *Registration status of Fenitrothion insecticide.* Information Division, Agriculture and Agri-Food Canada, Decision Document, E95–01, Nepean, Ontario, Canada.

149 Ermen, D. 1995. Province will stop using Fenitrothion in budworm fight. *Daily Gleaner,* April 20. Fredericton, N.B., Canada.

150 Staff Reporter. 1995. Province to spray for budworm. *Daily Gleaner,* May 9. Fredericton, N.B., Canada.

151 Brookman, D. 1984. Assessment of the potential of insecticides, emulsifiers, and solvent mixtures to enhance viral infection in cultured mammalian cells. *Applied and Environmental Microbiology* 47:80–83.

152 Menna, J. 1985. Effect of emulsifiers on influenza type A virus infection, in vivo and in vitro studies. *Journal of Toxicology and Environmental Health* 16:441–448.

153 Bouma, M., and R. Nesbit. 1995. Fenitrothion intoxication during spraying operations in the malaria programme for Afghan refugees in North West Frontier Province of Pakistan. *Tropical and Geographical Medicine* 47:12–14.

154 Fakhri, Z. 1993. Cholinesterase assessment as a result of fenitrothion exposure: a survey in a group of public health workers exposed to an organophosphorus pesticide. *Occupational Medicine* 43:197–202.

155 Zamula, E. 1990. Reye Syndrome: The decline of a disease. *FDA Consumer* 24:21–23.

156 Glasgow, J., and R. Moore. 1993. Current concepts in Reye's syndrome. *British Journal of Hospital Medicine* 50:599–604.

157 Hurwitz, E., et al. 1987. Public health service study of Reye's Syndrome and medications: Report of the main study. *Journal of the American Medical Association* 257:1905–1911.

158 Rowe, P., D. Valle, and S. Brusilow. 1988. Inborn errors of metabolism in children referred with Reye's Syndrome. *Journal of the American Medical Association* 260:3167–3170.

159 Weinberg, A. 1972. Science and trans-science. *Minerva* 10:209–222.

160 Staff Reporter. 1991. Woodlot owners take concern to candidates. *Daily Gleaner*, September 13, Fredericton, N.B., Canada.

161 Staff Reporter. 1992. Judge won't rule on woodlot owners' complaint. *Daily Gleaner*, January 7, Fredericton, N.B., Canada.

162 Conservation Council of New Brunswick. 1994. Public Lands in Public Hands, Managing Crown Forests in the Public Interest, A Conservation Council Proposal. Conservation Council of New Brunswick, Fredericton, N.B., Canada.

163 Hanton, E. 1992. N.B. unveils forestry aid plan. *Daily Gleaner*, January 24. Fredericton, N.B., Canada.

164 Llewellyn, S. 1997. Mill owners told to keep pace or lose Crown wood. *Daily Gleaner*, March 13. Fredericton, N.B., Canada.

5

Pluralistic Competition

The upsurge of militant environmentalism in the 1970s introduced a variety of new psychosocial elements into the problem-solving process, making the market place of ideas much more diverse and turbulent. Particularly troublesome for the iron triangles, who had enjoyed a monopoly on decision-making power for so long, was the militant anti-industrial, antigrowth, proenvironmental stance of many of these newly organized groups. In other words, after decades of quiescence, the Arcadian worldview was beginning to reassert itself. As a result, the relatively orderly world of the decision maker was invaded by advocates of worldviews they could barely understand using styles of thinking (subjective-holistic) they were unable to tolerate. On the positive side, this period of upheaval rescued elements of Arcadian thinking from oblivion, thrusting them into public awareness where they have remained.[1] On the other hand, the period was one of bad-tempered bickering during which the problem-solving process was engulfed in posturing rather than improved in quality. Debate seldom reached beyond short-term preoccupations to the more fundamental issues of sociopolitical change and social justice but became mired in the details of parochial conflicts. As a result, the period between 1970 and 1985 was one of confrontation and litigation, which hindered, rather than facilitated, innovative policy making. What I shall argue in this chapter, therefore, is that although this period of interest-group politics may have achieved some local gains, it did not result in more adaptive forms of problem solving.

Problem Recognition

Many of the new players in environmental controversies were women who introduced a different gender perspective into deliberations that had been conducted largely by men. This was particularly evident in disputes over toxic wastes, where women activists constituted the majority of those involved. The psychosocial experiences of these women activists were quite different from that of the scientific experts and bureaucrats whom they began to confront, making communication difficult in the extreme.[2] Excellent examples of grassroots problem recognition can be found in the two classical cases at Love Canal[3] and Woburn, Massachusetts.[4] Both involved controversy over the perceived health effects of toxic wastes on residents living in close proximity to chemical waste dumps. In both instances, housewives who had had no previous involvement in environmental affairs led the citizens' charge to have their concerns recognized by authorities. The questions are what psychosocial factors caused them to do something so apparently out of character, and how did these factors influence the problem that they saw?

At Love Canal (an abandoned canal used as a dump site in upper New York State), Lois Gibbs had been aware for some time about the concerns felt by her neighbors over their proximity to the Canal but felt that her family lived far enough away not to be alarmed.[3] However, her calm turned to concern after reading a newspaper article about the neurological and respiratory effects of chemicals known to be in the Canal. Her 6-year-old son had developed asthma and convulsions after starting to attend the primary school that was built on top of the Canal itself. When her attempts to have her son transferred to another school were rebuffed by school authorities, she became sufficiently incensed to begin the long battle to have her concerns taken seriously. Initially, the problem that she recognized revolved around concerns with her son's school and the need to make it safe for the children. As she became more involved, however, and learned more about the problem, it grew into something much more comprehensive.[3] In a strikingly similar scenario, Anne Anderson's domestic life in the small Massachusetts town of Woburn was shattered when her young son was diagnosed with leukemia.[4] She and others had been having trouble with their discolored and unpalatable drinking water for some time but only began to suspect a link between contaminants in the water and her son's illness as she talked to neighbors about what appeared to be a high incidence of childhood leukemia in their neighborhood. Her concerns languished for a number of years until the problem of water contamination was brought to the attention of the media by a series of fortuitous events. The discovery of barrels believed to contain toxic waste dumped along a local watercourse led to the further discovery that two of the wells providing city water were contaminated with organic compounds known to cause cancer in laboratory animals.[4] While the State took the precaution of closing the two wells in question, additional investigation by authorities began to reveal considerable industrial pollution in Woburn. This, of course, was of interest to the media, the resulting storm of attention galvanizing citizen action to explain the leukemia clusters that afflicted the area. Anne Anderson played a significant role in all this, having, by this time, lost her son to the illness.

Why, then, did these and other housewives play such a prominent role in events? Neither had been involved in environmental matters prior to the outbreak of controversy, which suggests that, at the outset, their activism was not rooted in a well-developed environmental ideology. They were not Arcadians struggling to impose their ideological demands on the Imperial authorities. Rather, they were concerned with more parochial issues to do with the health of their own family and that of their neighbors. Brown and Ferguson[2] argue that the motivations of women activists can be understood best in terms of their roles as primary caregivers within the family and the way in which their adoption of an ethic of caring has shaped their lives. Cultural tradition affords women the role of protectors of their children, especially against threats from toxic wastes. Indeed, it is the threat of damage to the reproductive process that seems to be at the root of such toxic waste activism.[2]

Although many instances of grassroots activism do not stem from environmental ideology, other social beliefs and values appear to play a significant role in motivating women activists. For instance, both Love Canal and Woburn are working- to lower middle-class communities. The residents were dependent on the local industries. A central element in the blue-collar worldview is that they see themselves as hard-working, law-abiding citizens with strong patriotic ties to their country. Just as they are willing to work and, when necessary, fight for their country, they expect in return to be treated fairly by those in power.[2,5,6] If, through no fault of their own, their health is put at risk, then they expect to be compensated. When this doesn't happen, their sense of fair play is violated, which results in an upsurge of outrage.[5] Such a sense of outrage, even shock, comes across very strongly in Lois Gibbs' account of her first exposure to the politics of risk assessment.[3] With regard to Anne Anderson, her family pastor recalls: "Anne cut everybody else off. She did it in the name of (her son) and the fact that he was so sick and dependent on her. It became absolutely necessary for her to find some reason for why the kid was so sick. She had to find an enemy, a reason, something to focus her rage on for afflicting her son"[7] (p. 25).

Making the transition from housewife to activist is not an easy thing to do. Women activists have to overcome a number of gender and social class barriers, not the least of which is gaining sufficient confidence in their perception of the environmental problems that they believe afflict them. The persistent message from society is that they are housewives, first and foremost, and should remain so. Any attempt to step outside the bounds of family contravenes these cultural norms and results in varying degrees of displeasure from many of those around them.[2] For instance, Lois Gibbs recalls that her first efforts at collecting health information from her neighbors were extremely trying because she had had no previous experience of knocking on strangers' doors to obtain some very personal information from them.[3] In addition, some residents were beginning to fear the effect of the controversy on their property values. Others simply didn't want to think about the possibility that their pleasant neighborhood might have been damaging their health all these years.[3,4,8] Lois Gibbs, therefore, met with considerable animosity in her endeavors, not all of which emanated from some of her neighbors. In addition, women activists often face domestic problems as the internal balance of their family life is upset by their political activity. Conflict, separation, and divorce is a common consequence.[2]

While problem recognition by individual citizens is influenced by aspects of their personality, confirmation of their views on the matter, and the development of plausible stories, are developed within group contexts. Brown and Mikkelsen[4] have found that activists in contaminated communities follow a predictable path in developing a collective understanding of the problem they believe they face. In the beginning, individuals become concerned about the health effects of pollutants. As they exchange concerns with others, they begin to see that they share a common per-

ception of the problem and eventually form a group to give themselves some political credibility in persuading authorities about the validity of their concerns. Eventually, they obtain the official medical and other studies they believe will settle the matter, but usually such studies don't. In response, community groups seek their own information and use it in litigation and public relations to effect a change in official policies. Collective action also helps women activists make the transition from home to public arena. In attempting to make this shift, the support of a group of like-minded citizens helps in the development of the confidence to keep going.[2]

One does have to be wary, however, of the effects of group processes on problem recognition. While collaborative learning does hold out the prospect of a better understanding of their predicament, there is always the possibility that members of such groups merely reinforce one another's misperceptions. Presumably, community groups can follow both directions depending on circumstances. Renn et al.[9] refer to the processes involved as the "social amplification" of risk. That is, a group's perception of risk can be amplified or attenuated by the influence of a variety of psychological and social factors acting within and on the group. For instance, the feelings of injustice mentioned earlier, coupled with the distrust of science, government, and business commonly found among citizens,[10,11] can cause a group to become fixated on redressing perceived wrongs even though there is little actual evidence in support of their claims. Indeed, group behavior can take on an ugly edge, when anger and panic fuel extreme actions. On one occasion during the Love Canal controversy, a crowd of angry residents held two EPA officials hostage by detaining them for a few hours in the residents' association office.[3] Actions of this kind lead Schwartz et al.[12] to wonder whether hysterical contagion is at work in at least some community groups faced with stressful situations. However, they conclude that while the behavior of citizens might at times be mistaken for hysteria, "in general the public's response is neither irrational nor hysterical. It is for the most part a logical and reasonable response to both the information available and to the behavior of the organizations responsible for protecting community health" (p. 72). In other words, when citizens are trapped in a situation from which they cannot extricate themselves, where both media coverage and organizational prevarication seem to add plausibility to the perceived threat, then citizen responses seem in no way irrational.[8,12]

Conflicting Ways of Knowing

The conflicts that have arisen over problem recognition between grassroots activists and the establishment stem, in part, from their different "ways of knowing."[2] While women activists' perceptions of risk (from toxic dumps) is grounded in their personal immersion in family and community, scientists and officials are inclined to take a more detached, *rational* view, one that may ignore the actual lived experiences of those more intimately involved with pollutants. Consequently, the gap between citizen common sense and scientific skepticism adds to confusion in environmental controversies. A major part of the problem lies in the adoption of a reductive approach to risk assessment, by experts. They tend to ignore, or perhaps overlook, the significance of the psychosocial context within which decisions are made. Thus, by adopting a positivist approach to epidemiology, for instance, technical estimates exclude many aspects of risk considered to be crucial by the affected citizens. As one scientist explained, "We deal only with numbers; we're scientists," which Levine[3] (p. 85) took to mean that "scientists must avoid being emotionally swayed in their professional judgments by the sight, smell, and sounds of suffering." However, in addition to the scientific facts of the case, those at risk from toxic contamination are also worried about such matters as the potential loss of property values, the responsibility of looking after loved ones, the loss of control over their lives, being stigmatized as belonging to a contaminated community, and the general uncertainty underlying their situation.[8] In other words, lay estimates of risk are inclined to include a wide range of

social concerns that are missing from the more scientific estimates of risk. At the same time, when citizens look at the latter, they do so with a jaundiced eye because, as mentioned earlier, there is great distrust of the establishment (government, business, and science) in contaminated communities.[10] Often, scientific findings are at such odds with citizens' own experience of health problems that it is difficult to place any faith in official pronouncements. This is due in part to a misunderstanding, on the part of citizens, of the caution and skepticism that is an essential feature of science. It may also stem from the perception by citizens that politics may also be influencing scientific objectivity. This may not be too farfetched a possibility for, in commenting on the role of science at Love Canal, Savan[13] (p. 59) observes: "Any immediate response . . . by the New York State Health Department would have cost a great deal of money, and would have set a dangerous precedent for a state with more than nine hundred chemical dumps—many of them toxic—within its boundaries. So it would have made tactical sense for Health Department scientists to err on the side of scientific fastidiousness and to avoid raising hopes or fears that they would be unable to act on or allay."

The mutual incomprehension between grassroots activists and the scientific and managerial establishments, which often occurs during controversies, reflects their adoption of different ways of knowing (cognitive styles). Activists are more inclined to use a subjective-holistic style whereas many experts operate within the objective-analytic modes. As we have seen in chapter 2, subjective-holistics make little attempt to separate thoughts from feelings, nor do they view subjective, intuitive, experiential knowledge with the same disdain afforded it by the objective-analytic personality type. An unfortunate consequence of these differing styles is that women activists' perception of problems are frequently dismissed as the products of undisciplined minds—something that one would expect to find in housewives. Thus, rather than seeing some merit in the ideas generated by subjective-holistic people, members (male) of the establishment use cultural stereotypes about hysterical women to dismiss activists' views on the problem at hand.[2] Because science and objectivity are associated culturally with masculinity and expertise, it seems that scientifically trained experts have difficulty in accepting the possibility that other (more feminine) ways of knowing can be useful.[2] This is not to downplay the role of science in problem recognition. Scientific monitoring can play a significant role in detecting problems. For instance, it was a study by the International Joint Commission on the Great Lakes that found unacceptably high levels of pesticides in Great Lakes fish that, it turn, led to the Love Canal story. A subsequent investigation by New York State's Department of Conservation (DEC) identified Love Canal as one possible source of the contamination. As a result, the story was picked up by the local *Niagara Gazette* that began a series of stories on the history, contents, and possible health effects of the toxic waste.[3] Subsequent studies found at least 200 different chemicals in Love Canal, including "more than 13 million pounds of lindane (benzene hexachloride), more than 4 million pounds of chlorobenzenes, and 400,000 pounds of dioxin-contaminated trichlorophenol, all highly carcinogenic compounds"[14] (p. 6). Similarly at Woburn, the chance discovery of chemical barrels dumped along the Aberjona River led to the testing of two city wells that were found to contain unacceptably high levels of trichloroethylene and tetrachloroethylene. It was the credibility of scientific information that sparked official problem recognition. However, local citizens had known all along that something was wrong. At Love Canal, for instance, residents were only too aware of the inconvenience posed by proximity to the dump site. From the early 1950s and 1960s, complaints were commonplace. In addition to unpleasant odors that emanated from the waste site, there were problems with skin irritation among children and pets who played on the fields covering the canal. Chemical leachate in the form of dirty black ooze was also a problem in basements of houses backing on the canal. At that time, however, awareness of the potentially serious health effects of chemical effluents was not realized, and residents tolerated the inconvenience as part of the price

to be paid for living and working in the midst of a thriving chemical industry.[3]

On the other hand, one should not assume that risk assessment by grassroots activists is invariably correct. As explained in chapter 2, it is well known that intuitive thinking is prone to cognitive errors, especially those induced by emotional involvement. People are convinced they see, for instance, correlations and patterns that to others are illusory in nature. For instance, at a very early stage, Anne Anderson became convinced that her son's problems stemmed from their contaminated water supply even though many of those around her were skeptical.[7] Yet the relationship between toxic contamination and the health problems of Woburn and Love Canal remains uncertain. Subsequent scientific studies at Love Canal have been unable to demonstrate such links unequivocally,[15,16] and the litigation brought by the concerned parents of Woburn that would have dealt with the matter of health effects was settled out of court.[7]

Establishment Problem Recognition

Political and corporate elites are in an unenviable position. They are leading players in a socioeconomic system that has been identified by environmental activists as the source of our woes. Yet these same elites are held responsible by the silent majority for keeping the system working efficiently so that material well-being is maintained. One can see why bureaucrats and managers may feel beleaguered by what seem to be powerful environmental lobbies.[17,18] As a result, there is an understandable tendency to engage in defensive maneuvering, a form of self-protection expressed in terms of both selective attention to problems and an incremental caution.

One of the more unfortunate aspects of this self-protective behavior is the tendency to recognize only those problems that are perceived to be manageable. For instance, in their study of organizational problem recognition, Lyles and Mitroff[19] (pp. 114–115) found that:

There was a striking absence of explicitly stated problems dealing with intergroup conflict, motivation, leadership, etc., although these were components of all the problems presented. One reason might be that the managers wanted to present only problems that their organizations could deal with. Human problems are often too complex and seemingly unsolvable, and managers have been trained to avoid feelings and intuition ... It seemed that human problems were too sensitive and would expose too much of the organization if they were discussed.

In addition, it is reasonable for agencies to proceed with caution in responding to possible environmental risks and problems. They cannot respond to every anxious protestation from alarmed citizens. Government officials work within a worldview informed by political, legal, and economic considerations, not the least of which is the need for some credible evidence to justify action. One might sympathize with their plight at Love Canal: "The Health Department officials were concerned, but they were also committed to defending their actions with solid, indisputable scientific evidence ... Their insistence on what they termed a 'conservative' scientific approach was also geared to the inevitable political battles for funding any actions indicated by their results. And they were all too well aware of how much money might be required: according to a State Health Department estimate, there were about a hundred chemical dump sites in Niagara County alone, and Love Canal was by no means the largest or most dangerous"[13] (p. 58).

The need for caution did not temper the reaction of at least some citizens to efforts of officials to wrestle with the Love Canal dilemma. It seems that no matter what officials did, they were perceived as bungling and prevaricating[3] (p. 24):

The more that officials met with residents, the more negative feelings and relationships developed. When professionals presented raw data, it confused people. When they tried to interpret the data in down-to-earth terms, describing risks as some number of deaths in excess of the usual number expected, people interpreted that to imply *their* deaths and their children's deaths. When they tried to calm people by saying that, despite all serious possibilities, there was no evidence of serious health effects, the officials were seen as covering up, since no health studies had been done. Authorities trying to

coordinate multiple government and private agencies were seen as wasting time in meetings . . . What officials saw as preliminary studies conducted to assess the situation were viewed by residents as wasting resources on repetitious research projects rather than doing something helpful. When they took action quickly or tried to do everything at once, for everyone, they overloaded the facilities, made errors, and were faulted for bungling."

Problem recognition, therefore, is not some objective process in which problems simply emerge from the jumble of everyday life. Rather, they are constructed by individuals working within the confines of their own biases and self-interests and are further refined in the crucible of social conflict where numerous sociopolitical pressures determine whether the problem, or some part of it, is to be recognized. At Love Canal, for instance, a potential health problem was recognized by New York State's commissioner of health soon after the scientific evidence of leakage from the site had been established. Much of the ensuing controversy stemmed from later stages of problem solving—establishing the seriousness of the threat and deciding what to do about it. On the other hand, at Woburn, while chemical contamination of the water supply was also officially recognized at an early stage, its relationship to the leukemia clusters was not. In both cases, 20 years of turmoil and controversy have not resulted in an unequivocal recognition of any serious health problem due to contamination. In essence, the cases remain unproven.

One wonders what was gained in these examples of interest-group politics. Both Love Canal and Woburn do seem to have become symbols in focusing national attention on a ubiquitous problem—chemical waste sites. As Brown and Mikkelsen[4] (p. 42) put it: "In Woburn, the litigant families . . . set in motion an extensive process that is a model for many communities contaminated by toxic wastes. They pioneered the detection of the leukemia cluster; they spurred government to investigate and clean up (the contaminated sites); they brought the contamination to national attention." On the other hand, while all of this is laudable, the problem recognized in both cases was one of how to "clean up" contaminated communities and determine appropriate compensation for victims. The problem that emerged was not a more fundamental questioning of the Imperial system. Thus, citizen activists seemed not to be asking questions about how they could adopt Arcadian lifestyles of voluntary simplicity but, rather, sought social justice within the prevailing Imperial mode.

Problem Definition

To reiterate a point made earlier, all problems are socially constructed. That is, once a problem is recognized, its further elaboration involves subjective choices about how the problem is to be mentally constructed (or framed) and how it is to be distinguished from other problems (bounded). Information is usually sought in developing a better understanding of the problem being delineated, and as this accumulates, the framing and bounding of the problem may be modified. At the same time as these enterprises are underway, deciding what might be done about the problem (goal setting) becomes a priority. Thus, *problem definition* is a complex mix of procedures conducted by different parties who, of course, have conflicting interests or ideologies. The ensuing turmoil may be unfortunate because the confrontation between interest groups is likely to hinder the level of cooperation needed to frame problems in adaptive ways. As a result, the victors in these power struggles may direct attention toward the wrong problem.

Problem Framing

Asking the right question is crucial in constructing a problem definition. Unfortunately, what is considered to be the *right* question is a matter of ideology, and problem framing provides an opportunity to pursue ideological agendas. The emergence of interest-group politics, therefore, has intensified the battles between coalitions espousing versions of the Imperial and Arcadian worldviews. Bardwell, for instance, describes how differences in problem framing led to conflicting approaches to a water problem in Colorado.[20] One conception of the problem (Imperial) was that a water

shortage existed, and the solution was one of procuring more water through building a dam. Those who opposed dam building preferred to focus on the problem of the excessive use of water and to develop solutions based on conservation (Arcadian). These contrasting problem definitions led to radically different solutions to what are really distinct problems. Bardwell believes that the concerns of all parties were addressed, however, by reframing the problem as: "How much water do we really need?"[20] However, this question would also be interpreted differently by proponents of the two ideologies.

Similar ideological differences underlie disputes over transport policies in the United States, especially over what Coughlin[21] calls the *traditional* (Imperial) and *green* (Arcadian) perceptions of traffic congestion. The traditional perception of the problem is one of inefficiency in traffic flows and insufficient capacity of the highway system, one that results in costly delays that are detrimental to the economy. Solutions are couched in terms of technical fixes involving better electronic control of traffic flows and increased highway construction. In contrast, greens see traffic congestion as a symptom of a country that has embarked on an unsustainable trajectory of economic growth and is suffering the consequences. Highway construction and other technical fixes do not address the real problem, which is the promotion of rampant individualism. Traditional transport policies actually subsidize private automobile use with its attendant overconsumption of energy and subsequent pollution. The sensible solution, greens believe, is to curtail highway construction and suburban development with a concomitant promotion of public transit.

Because the arguments on both sides of the ideological divide are socially constructed, the debate often takes the form of a conflict between two, often equally plausible, stories. Thus, "the 1980–82 California Medfly Controversy provides an especially well-documented case study of a policy issue sufficiently complex and uncertain that its proposed 'solutions' were generally recognized to be stories about what might possibly happen rather than predictions about what would probably happen"[22] (p. 253). Two disparate stories were used by opposed groups in promoting their causes. On the one hand, "a powerful coalition of politicians, government officials, and agribusiness interests" offered their Aerial story in favor of the aerial spraying of pesticides (Imperial). Proponents of the Ground story on the other hand, a diverse group of scientists and policymakers, fought to limit pesticide use while encouraging biological control methods (Arcadian). Roe concludes that it was the intelligibility and "readability" of the Aerial story, not its scientific merits *per se*, that proved decisive in influencing policy making.[22]

Story construction is an intriguing political dance in which the various parties "attempt to wrap themselves and their policy objectives in the symbols of legitimacy . . . Thus a group's real power should be measured by its skill in defining and redefining an evolving issue to its advantage as circumstances warrant"[23] (p. 29). One of the key elements in this political struggle is the extent to which a group can enlist the political support of others by expanding the number of issues being addressed. For instance, in the battle over the future of the Siskiyou National Forest[23] (p. 31):

it was the Siskiyou environmentalists' strategy to try to expand the issue at every turn, since they were the group seeking to change the status quo and to do so required intervention by a broader public. The local timber interests, on the other hand, sought to maintain the status quo . . . Accordingly, timber interests sought to assure the public that, despite the environmentalists' hysterics, everything was fine down in the woods . . . While this message aimed at defusing environmentalist charges never wavered, the timber interests, nonetheless, found themselves increasingly unable to keep the issue from expanding. They had no choice, therefore, but to play the expansion game themselves, turning to wider audiences to plead their case in an attempt to gain the edge in defining the issues . . . The central goal of either side's efforts toward issue expansion was to gain as many allies and sympathizers as possible.

Between 1983 and 1992, therefore, the environmentalist camp expanded the issues under contention from an initial concern with the prevention of herbicide spraying to include protection of roadless areas, a series of preservation issues, and, finally, to much broader matters dealing with global warming, citizen's rights, genetic diversity, and forest service reform. The basic thrust of this story centered on "linking the health of the forest to the well-being of all people" (p. 33). In contrast, the timber interests started with concerns about protecting local access to the timber commodity before moving on to preservation of family values, free enterprise, and the economic health of the broader community. The timber story, therefore, emphasized the threat posed by environmentalists to economic growth and human progress.[23]

In addition to enlisting support for their disparate causes by enlarging the problem definition, the two sides also used additional tactics to persuade their audiences of the merit of their stories. The environmentalists, for instance, made widespread use of photography, taking "great advantage of the gruesome visual ugliness inherent to clearcuts. By widely distributing pictures, typically of the jagged, smoking, stump-filled wreckage of a clearcut, often juxtaposed with a photo of a beautifully intact forest . . . the environmentalists exploited what had to be their most potent symbols for all they were worth"[23] (p. 35). On the other hand, according to Davis, their opponents engaged in their own symbol manipulation. "Perhaps even more important than how a group presents an issue is whether that group can create an aura of legitimacy around itself while simultaneously discrediting its opponents . . . While passing themselves off as a 'cross-section of Americana,' . . . timber interests in the Siskiyou, with rhetoric that strongly resembled the Vietnam debate two decades earlier, often described their adversaries with fairly negative 'hippie' stereotypes . . . As the conflict in the Siskiyou intensified, so did the timber interests' assault on the environmentalists' character. The standard timber term 'radical preservationists' soon degenerated in pro-timber rhetoric from hippie throwbacks into people-hating fanatics and even terrorists"[23] (pp. 36–37).

Thus, in the context of interest-group politics, problem definition becomes a jousting match in which protagonists engage in political maneuvering to ensure that their particular version of reality captures the public's attention. In the era of iron triangles, this process was relatively simple. Only one message filtered down to the unsuspecting public. Now that we are in the era of interest-group politics, the public is presented with a *tower of babel*, a barrage of conflicting messages about what we are supposed to believe. This is unlikely to lead to adaptive problem solving.

Knowledge Production

While problem *recognition* requires only enough understanding to conclude that a problem might exist, problem *definition* is an attempt to take this understanding to the point where something can be done about it. The general difficulty faced by all concerned is how to determine when enough is known about the problem to justify action. Those who subscribe to the rational-comprehensive model argue for postponement of remedial action until comprehensive knowledge is acquired. In contrast, incrementalists are willing to act in the absence of complete understanding.[24] One sees this difference of opinion in arguments over the *precautionary principle*, the notion that we must act now on environmental problems, such as global warming, despite an absence of scientific consensus on their nature.[16] This dispute is fueled by the feeling, among members of the public, that calls for yet more research are simply political maneuvers to avoid hard decisions or self-serving tactics by scientists to obtain more funding.

Although information is sought by all those involved in dealing with an environmental problem, the type and quantity of information collected depends on the cognitive style employed. This would not create any difficulty were it not for the fact that most of us are unaware of the weaknesses of our preferred styles and tend to use them inappropriately. For instance, the exclusive use of analytic-reductive (positivist) thinking by natural and social scientists leads to some serious flaws in problem solving.[25] Thus, a major assumption underlying objective-analytical thinking is

that a more comprehensive understanding of any problem can be achieved by examining its parts in ever-increasing detail. The resulting fragmentation of problems into a bewildering, and ultimately uninterpretable, array of snippets of information calls into question the usefulness of endless analysis. This problem with the analytic route to understanding is conveyed cleverly in the film *Blow Up* in which a photographer accidentally captures a scene in which there appears to be a body lying behind a bush. However, the more closely he looks at the photographic image by enlarging it, the more obscure it becomes and the less able he is to decide what he is looking at. The parallels with scientific analysis are obvious. As the intensity of analysis increases, understanding of the problem in question recedes, lost in the accumulating detail. An interesting case in point is provided by Smil.[26] One would think that deciding how much nitrogen fertilizer to apply to a farmer's field is a relatively simple problem, one that can be easily resolved by appropriate analysis. However, Smil argues that the complexity of the nitrogen cycle makes prediction difficult. The amount of nitrogen in a field is determined by a very large number of factors in dynamic relation to one another, including interactions between the farmer's actions, soil structure, soil chemistry, soil microflora and fauna, atmospheric gases, and water regimes. Thus[26] (pp. 9–10):

A farmer wishing to know the nitrogen story of his field would not only have to set down a long equation describing the dynamics of this intricate system—but he would have then to quantify all the fluxes . . . If the farmer would have numbers ready to be attached to this alphabet-full array of key variables he could fertilize with precision, with minimized losses and with the lowest costs. In practice, he knows exactly only one variable—the amount of synthetic fertilizer he plans to apply to the field.

Smil's point is that "even on this small scale reliable understanding still eludes us . . . it is beyond our capability to chart and to forecast accurately nitrogen fluxes of a single crop field" (p. 11). Even if we were to turn loose a horde of empirical scientists on to that benighted field, they might spend a working lifetime collecting data and still be unable to improve appreciably the accuracy of forecasts. If this is so, with such a tiny ecosystem, what prospect is there that analytical thinking will provide the understanding necessary for managing ecosystems on a larger scale? Smil suggests that the major strength of analytic schemes, their generality and abstractness, is also their major weakness. They are difficult to apply, in a meaningful way, in concrete situations where more specific local knowledge is at a premium. Thus, analytic thinking, at least in its scientific guise, produces abstract knowledge that may be of little use in specific situations.

It is possible that the holistic cognitive style might be more appropriate in these circumstances. For instance, Smil[26] suggests that traditional farmers may not have had a detailed knowledge of the nitrogen cycle but through trial and error, and long experience in a particular location, had developed a considerable degree of control over their environment. However, the cognitive errors to which intuitive knowledge is prone make lay assessment of problems particularly vulnerable to distortion. For instance, as we have seen, many environmental controversies revolve around whether or not there is some connection between a cause (pollution) and an effect (illness). Proclaiming any such link requires some caution, however, because we live in a multicausal world. Yet, as human beings, we need to identify patterns in the world around us to foster our (often mistaken) belief that we can understand and control what happens to us. Unfortunately, this may result in seeing patterns of cause and effect where no pattern really exists. Winter[27] (p. 205) suggests that: "In the arena of environmental concerns, the principle of regression toward the mean leads us to interpret essentially random events as meaningfully related to some human action." She goes on to quote an example from Lee[28] (pp. 71–72) in which he points out:

Regression to the mean predicts . . . that after a species has been declared endangered it will tend to become more abundant. This is not an effect at all, but a reflection of the fact that the human decision to declare a population in bad trouble is based upon its being *in extremis*. To the extent that that condition is caused by a variety of factors . . . some of them will fluctuate in the next year, and the fluctuations

will tend to bring the population up. In the early years of the Columbia Basin program, before any of the rehabilitation measures could be carried out, there was a resurgence of salmon populations from the historic lows of the late 1970s . . . It took a special effort of political will *not* to take credit for this change, even though there was as yet no cause to which such an effect could be attributed.

While all of us, regardless of cognitive style, are prone to these and other cognitive errors,[27] those who prefer holistic styles (and make judgments on the basis of hunches, personal experience, and intuition) are particularly vulnerable to such distortions. To reiterate the obvious conclusion, some judicious mix of analytic and holistic styles would compensate for the weaknesses of each. Unfortunately, a variety of personal and political factors militate against this, as explained earlier. When ideology and politics start to intrude, cooperation in overcoming cognitive weaknesses is curtailed.

This is regrettable when it precludes cooperation between lay people and experts. Brown and Milliken[4] make the argument, for instance, that ordinary people with extensive local experience of the effects of contamination have made important contributions to the recognition of environmental problems that otherwise might have gone unnoticed. An interesting case in point is that of the *swale hypothesis* at Love Canal. Lois Gibbs, along with other members of her Homeowners group, had been collecting information on the incidence of illness among their neighbors. "Late one night, she hit on the idea of marking the information . . . on a street map, with pins to indicate homes with medical problems. As she worked on this project she thought she discerned a pattern of disease outbreaks concentrated in certain areas and along narrow paths. Might this physical arrangement coincide with the old drainage ditches (swales) that used to cut through the neighborhood?"[13] (p. 55). The prospect thrown up by the swale hypothesis was that ill health might be associated with proximity to the swales or wet areas around the Canal, while those living in dry areas might be less effected even though they might live closer to the toxic dump. Accounts of subsequent events diverge at this point. One version asserts that Gibbs presented the swale hypothesis to the state official in charge of the Health Department study who showed some interest in it. However, while Health Department scientists tested a variety of hypotheses in later research,[16] initially they "planned to test the very reasonable hypothesis that chemicals were seeping away from Love Canal at a more or less even rate. Logically, they compared the miscarriage rate on different streets, at increasing distances from the canal"[13] (p. 57). In contrast, the Homeowner's study, conducted by Dr. Beverley Paigen, sought to test the swale hypothesis. Thus, two studies were eventually offered to the public, each of which studied the possibility of health problems due to exposure to toxic effluent but were based on very different assumptions. If the swale hypothesis was correct, the design of the official study would not pick up relevant effects, because wet and dry areas would cancel one another out, as it were. Initial findings from the official study indicated no causative links between contamination and illness, whereas the Paigen study appeared to confirm an increased incidence of illness among families living in the wet areas.

Clearly, the swale hypothesis was the result of an intuitive insight in which a graphic display of clusters of illness was linked with another piece of apparently unrelated information about old drainage channels. These intuitive leaps are characteristic of holistic cognitive styles that, in this case, resulted in an intriguing hypothesis about the patterns of illnesses. In contrast, the official study design appears reasonable but conventional, something one associates with the more objective-analytic modes of thought. The reception of the swale hypothesis by officials seemed polite but restrained until the Homeowners released the results of their unofficial study to the media. At this point, official opinions became rather terse, the data in the Paigen study being dismissed as "totally, absolutely and emphatically incorrect" by one official and by others as "information collected by housewives that is useless"[3] (p. 93). Brown and Mikkelsen[4] would see this as an example of the resistance by professionals to popular intrusion into their domains, which "threatens not only the division

of knowledge and power between lay persons and professionals, but also the corporate system that produces the environmental hazard" (p. 151).

Data Interpretation and Integration

Empirical data and other forms of information only become meaningful when they are integrated into a conceptual framework of some kind. In the absence of this integration, they remain a jumble of uninterpretable data points. Finding effective ways of integrating data has proven to be a continuing headache, however, because of the volume and disparate nature of the information generated by the problem-solving process. While the integration of ideas into meaningful wholes is a central human preoccupation in which we all engage, the dilemma faced by environmental professionals is particularly difficult. All professionals work within the limits imposed upon them by their cognitive styles, personal ideologies, disciplinary specialization, and sociopolitical context. Their personal mental model of the problem at hand is, therefore, likely to be highly circumscribed. Difficulties arise when those with widely disparate mental models are required to cooperate in order to solve a complex environmental problem. This multidisciplinary process is a form of intellectual mayhem in which participants attempt, as best they can, to reconcile their conflicting mental models. Often, this is achieved only after the more powerful impose their will and their problem conception on the rest (chapter 3). The unsatisfactory nature of expert groups has led to the development of more formal ways of interacting through which it was hoped that interpersonal politics might be circumvented.[29] In this section, I shall focus on only one of these methods, the use of quantitative modeling, dealing with the more qualitative approaches to information integration in chapters 7 and 8.

Quantitative modeling of environmental systems is the domain of objective-holistic (OH) personality types. It follows that modeling would reflect many of the strengths and weaknesses exhibited by this kind of person.[30] Objective-holistic individuals seem to enjoy creating models, conceptual systems, and imaginative fantasies that incorporate their perception of the objective world. In all cases, the primary ingredient is a holistic vision of the system in question into which data, collected by objective-analytic personality types, might occasionally be fitted. Although the strengths of these models and schemes is that they provide a way in which disparate information can be integrated into a meaningful picture, their weakness is that the picture may be seriously flawed.

One of the cognitive problems with the OH personality types is that their intellectual reach may exceed their grasp. The development of global and other large-scale models is a case in point. They have tended to suffer from a variety of problems associated with the question of scale, which Smil explains at length.[26] For instance, the global model created by Meadows et al.,[31] in their controversial study on the *Limits to Growth*, used "excessively large, and fundamentally meaningless aggregates, such as 'pollution,' linked by no less abstract generation rates and multipliers" (p. 17). Aggregation of this kind is the result of too little information and computing power. For instance, the spatial resolution of large-scale climate models (number of data points) is too coarse to reflect reality in an adequate way. "Grids of one degree of latitude by one degree of longitude and forty vertical levels would bring us much closer to the reality" (p. 22). However, to achieve this would impose a prohibitively expensive and impossible level of data collection on researchers. In addition, so little is known about the dynamics of ocean currents, cloud formation, and atmospheric chemistry that it is not clear which data should be collected or what one might do with them once available. Even if such data were to become available, they still would not be useful without an increase in the speed of computers by several orders of magnitude.[26] It is for these reasons that climate modeling is sometimes viewed as dealing with a *trans-science* problem, one that gives the appearance of being soluble by orthodox science but which, in reality, is not.[32]

In my earlier discussion of objective-holistic personality types (chapter 2), I suggested that the desire for mastery that underlies their behavior is expressed in the development of fanciful schemes and an attempt to force reality to fit these schemes. While this is useful in that it provides us with an abundance of theories and conceptual schemes, model makers and users may forget that they are working with fantasies that may have drifted far from reality.[26,33] *Delusions of grandeur*, the feeling that one is capable of controlling nature, provide the ultimate high for some professionals.[26] However, recent understanding of complex systems belies these assumptions of mastery. It is now evident that the nonlinearity of ecosocial systems means that their behavior can be both uncertain and chaotic.[34] Small changes in the initial conditions of a system can lead to divergent paths of development. Under such conditions, predictions of systems behavior are difficult to say the least.[26] It is unfortunate, therefore, that many ecological models are based on older notions of ecosystem stability, which appears to be more of an ideological than a factual assumption.[35]

Objective-holistic types also have difficulty, as do we all, in comprehending human unpredictability. For instance, some years ago, Shell curtailed its use of big, quantitative models because staff were spending so much time generating data that they failed to anticipate the upheaval in oil prices caused by OPEC in 1973.[36] It was concluded that a more qualitative understanding of geopolitics might be preferable to time spent on number crunching. Human passions, therefore, give the lie to accurate prediction. As Smil[26] asks, Who could have predicted Saddam Hussein's firing of the Kuwaiti oil fields, China's increasing use of refrigerant chemicals, or the collapse of the Soviet Union? Even at a more prosaic level, the incorporation of psychosocial information into biophysical models is problematic. Often, the kind of psychosocial information that is amenable to this process is of little interest to social scientists and may add little to one's understanding of the problem. For instance, in a recent modeling exercise in which I was involved, the economist in the group explained how it would be useful to include a metric that accounted for the price of land in terms of its distance from the provincial capital. The farther away the piece of property was from Fredericton, New Brunswick, the cheaper it could be expected to be. Although, in a gross sense, this does apply to many properties, it tells you little about the family who will not sell their farm under any circumstances because their ancestors arrived at that very spot with other Empire Loyalists after the Revolutionary War in 1783. It is difficult to incorporate these more subtle, idiosyncratic psychosocial elements into a model, just as it is difficult to persuade OH types that such information is important.

This difficulty extends beyond modeling to a more ubiquitous resistance to psychosocial information among natural resource professionals. One cannot expect to "integrate" information when the very people who need this kind of information avoid it. Clark et al.[37] (p. 943), for instance, suggest that many professionals prefer to define environmental problems as primarily technical in nature so that they are amenable to their preferred disciplinary solutions:

Most recent definitions of the carnivore conservation problem in North America focus on single species and on ecological and management concerns. They rarely address valuational, social, institutional, or other factors that have contributed to the problematic ecological and management situations . . . Aside from brief references to the sociopolitical context of grizzly bear conservation, the U.S. Fish and Wildllife Service . . . defined the problem in terms of ecological variables such as deteriorating habitat and human-induced bear mortality . . . (in offering a solution, the Service) emphasized such biological factors as secure and sufficient habitat (which) cannot ensure grizzly conservation over the long term both because it does not . . . address an array of factors identified as essential for understanding and solving the problem.

Data collection and integration, therefore, are handicapped by a variety of psychological and social constraints. The traditional positivist methods used by professionals,which emphasize rampant empiricism and quantitative

modeling, continue to dominate problem-solving efforts even though their limitations are increasingly evident. Among environmental groups and a minority of professionals, therefore, there is growing interest in qualitative (soft systems) modeling and other more informal approaches to data integration (which is discussed in chapter 7). In part, this shift in opinion, though modest, is due to the pressure from previously marginalized groups to have their views included in the decision process.

Alternative Generation

Generating promising solutions to a problem is a potentially creative phase of problem solving, one that holds the promise of innovation and adaptive change. However, creativity is prized in only some sectors of society, such as the commercial world where new ideas may become a source of enormous wealth. Matters are different in environmental management. To begin with, actions are constrained by the practical limits imposed by ecological cycles and processes. In addition, the culture of natural resource management is more conservative, less prone to flights of fancy than one sees elsewhere in the commercial world.[38,39]

Even though grassroots organizations have sought to be part of the decision process, the construction of alternative solutions to problems remains a responsibility of scientific and managerial professionals.[40] It follows that their preferred modes of thought are likely to have a serious impact on the kinds and quality of solutions offered to decision makers. Because the Imperial majority prefer to see planning as a technical exercise,[41] technical rationality continues to occupy a significant role in planning and management,[42] despite planning theory's recognition of the importance of the political aspects of planning.[43] Technical rationality combines the use of instrumental-analytic thinking with a belief in mastery and control (chapter 2). Professionals who adopt this approach to problems are field independent in perception in that they make a sharp distinction between relevant and irrelevant aspects of a problem. While technical facts are relevant, questions of value and sociopolitical context are viewed as being peripheral to the main agenda—although this would be more true of scientists than managers, as the latter are drawn into the political sphere by the demands of their work. The semantic networks (mental models) built with these facts are logical templates, stripped of irrelevant psychosocial context. Any reasoning based on these models is convergent, serial, and tightly focused on finding so-called correct answers.[44] It follows that, in generating solutions, natural resource professionals who use this way of thinking are likely to produce technical fixes aimed at narrowly conceptualized problems while avoiding, as far as possible, the psychosocial aspects of the problem situation. In addition, convergent thinkers are not renowned for their intellectual playfulness but, instead, prefer a more sober, intensely practical approach to thinking.[45] What they consider relevant to a problem then is what their disciplinary training and work experience tells them is important. As a consequence, problem solutions are unlikely to deviate from current professional practice to any significant extent. Nor are they likely to contravene prevailing agency policy or organizational culture.

Where innovative solutions are offered by professionals, it is likely that they take the form of a technical breakthrough. Pacey,[46] for instance, describes the way in which the pursuit of cutting-edge technology is a product of what he calls the "technocratic master value" that permeates the culture of engineering. The same master value encourages a distaste for the technically mundane tasks involved in routine servicing activities or in the less interesting prospects of simple, low-tech solutions to problems. If Pacey is correct about this preoccupation with technical prowess, then there are unfortunate implications for the solution of environmental problems. Smil,[26] for instance, argues that successful adaptation to environmental dilemmas requires not some spectacular technical invention but, rather, the political will to implement relatively simple technologies that are already known. Pacey[46] suggests that past attempts to control malaria

with DDT is an excellent example of the sacrifice of modest, psychosocial strategies in favor of the more impressive technical fix:

One spectacular example . . . is provided by the effort to control malaria in India during the 1950s and 1960s. In a massively remarkable campaign, the walls of every dwelling in the subcontinent were sprayed with DDT to kill mosquitoes that entered people's homes. This had an immediate impact on transmission of the disease, whose incidence fell to a very low level. But with this achieved, it was necessary to sustain surveillance work, to nip any new outbreaks in the bud . . . But with malaria apparently defeated, it proved difficult to carry out these tedious maintenance tasks with conviction . . . too much reliance was placed on the insecticide and too little on parallel mosquito control measure and the organization necessary to sustain the programme (pp. 38–39).

As a result, malaria has re-emerged on a massive scale and in even more virulent forms. Nevertheless, the professional community presses ahead with esoteric technical research while researchers and officials have to be reminded that such research should not take resources away from more socially based control strategies,[47] the need for which is evident.[48] Thus, the solutions to environmental problems generated by those most responsible for coping with them tend toward technical fixes, while more mundane responses to the problem that require innovative psychosocial behavior is not something toward which technical professionals are inclined. In the present era of competitive pluralism, however, it is precisely these latter strategies that are promoted, often vociferously, by environmental interest groups. The stage has been set, therefore, for a confrontation between two quite different conceptions of adequate problem solutions.

Many environmentalists of the Arcadian persuasion are inclined to use "ecological rationality" in thinking about environmental problems and their solutions.[49] This combines a primary concern for environmental integrity with a subjective-holistic style of thinking. As seen in chapter 2, the latter cognitive style makes use of emotional reactions to events in making intuitive judgments about problems, something that many objective professionals find difficult to tolerate. The result can be innovative problem solutions that, to the professional mind-set, are decidedly fanciful and impractical. A case in point is the recurring dispute over the use of clear-cutting as a harvesting technique in forestry. A clear-cut area is an affront to many environmentalists and members of the general public, who see it as an ugly wound on the forest landscape. Their arguments against clear-cutting are both intellectual and emotional (aesthetic), with the latter playing an important role. Environmentalists want clear-cutting stopped, or reduced in size to areas smaller than 100 ha. This, they believe, would solve some of the problems created by past excesses. However, professional foresters are bemused by this. First, they cannot understand why an agrarian landscape, which has been carved out of a forest, is not subjected to the same criticism, since a farmer's field is, in essence, a former clear-cut. Second, clear-cutting is essential in some circumstances. For instance, in a case with which I'm personally familiar, a small-woodlot owner in New Brunswick decided to use selection harvesting as a more ecologically sensitive way of treating a stand of balsam fir. Having conducted the thinning exercise, a subsequent windstorm flattened the remaining trees in the stand that had been exposed to windthrow by the harvesting process. The resulting jumble of trees was difficult, even dangerous, to salvage. Ironically, therefore, an initial clear-cut would have been the more technically sensible, but not politically correct, approach.

The confrontation between the professionals who are responsible for generating solutions to problems and organized environmental interest groups often takes place within a variety of forums such as workshops and public hearings. In the mid-1980s, for instance, toward the end of the period of most active confrontation, a workshop was held in New Brunswick to find a solution to the spruce budmoth (no relation to the budworm) infestations of white spruce plantations. This insect had emerged from obscurity as a minor inhabitant of the natural forest to become a voracious pest of the local conifer plantations. The threat was serious enough to warrant the con-

vening of a workshop involving representatives of industry, government, academia, and regional nongovernmental organizations. As one might expect, the composition of this workshop lent itself to a polarization of views between those with Imperial (industry and government) and Arcadian (some nongovernmental organizations) views. One purpose of the workshop was to overcome these antagonisms and work constructively toward innovative solutions to the problem.

Recall that (chapter 4) the province of New Brunswick had been engaged in a vitriolic dispute over forest management since 1975. In 1984, a decade of bitter dispute over the aerial spraying campaign against the spruce budworm was just beginning to abate. The prospect of having to protect the spruce plantations in a similar fashion risked igniting this furor once again. The delicacy of the situation was very apparent to participants, as was the urgency in finding some judicious solution. However, the scars from this decade of confrontation were brought to the workshops. Many of the players knew one another and had heard each others' pitch many times before, with the result that the opening maneuvers, as in so many similar situations, took on the quality of a "ritualistic verbal dance" in which each group of players "reaffirms its particular faith"[50] (p. 166). The apparent lack of trust between various factions was a major impediment to effective problem solving.[51]

In framing the problem to be addressed by the workshop, the steering committee grappled with the question of scale. Should the problem be defined narrowly, focusing discussion on pest management in plantations, or be more inclusive, dealing with fundamental questions about the role of plantations in forest policy? In the event, the steering committee opted for a narrow problem definition, one that limited the exercise to a consideration of the specific problems being experienced in certain New Brunswick plantations. Opinion was divided on the validity of this decision. Most of the government and corporate participants in the workshop viewed the choice as being eminently sensible, whereas representatives of environmental groups were disturbed by what they saw as the elimination from consideration of alternatives to plantation forestry. This fundamental difference of opinion hindered communication throughout the whole of the exercise. From the perspective of the bureaucratic and corporate participants, a practical problem had to be solved, whereas the broader policy issues raised by environmentalists were simply political matters. When the environmentalists repeatedly raised the matter of point of entry during the first workshop, the reaction from a number of professionals was a certain impatience. For instance[50] (p. 10):

These people (environmentalists) argue that any discussion of pest management in plantations should proceed first with evaluation of the role of the forest industry in the province, concentrating on the need for plantations in that industry, and only then to the issue of pest management. Thus, the need for plantations must be established before any discussion of pest problems can take place.

At the root of these views, the same commentators argue, is "a naive perception of forest management that has become embedded in New Brunswick society . . . The public and the environmental groups do not have, and will not acquire, a realistic understanding of the realities of forest management on 500,000 hectare forests over four decades unless someone (industry) gives them *the basis* to acquire such understanding"[50] (p. 45). It follows that, given this difference of opinion over the scale of the problem to be addressed, conceptions of adequate problem solutions tended to diverge dramatically.

Because the workshop was composed of primarily technically trained men, and in light of the avoidance of broad policy issues in the problem definition, it follows that deliberations quickly focused on technical issues while sociopolitical concerns played little part in the proceedings. One can see this most clearly in the structure of the conceptual model that was developed as part of the workshop process, a simplified version of which is shown in Figure 5.1. The solution to pest management problems was seen to lie in the conventional domain of pesticide spraying, whereas social aspects of the case have been reduced to two

issues: the risk of public exposure and the possible effects of spraying on public attitudes. Both social aspects construe the public in a passive role, at the receiving end of decisions made by others. Though the risk to the public, being quantifiable, received some attention, there is little recognition, in the conceptual model, that the public could be brought in as constructive partners in decision making. Nor is there any attention paid to the basic political assumptions at the root of professional-lay conflict.

The conceptual model of the pest-plantation interaction reflected the predominant, orthodox assumption that the exercise was to find some aerial spraying regime that would satisfy regulatory authorities and protect the plantations. Given these parameters, it follows that the model would reflect conventional wisdom on the subject. The prospects for more original thinking had been dealt an early blow by the steering committee's narrow framing of the problem. Unusual options could not be considered once the basic framework had been established. For instance, the environmentalists repeatedly asked what the role of plantations in forestry was supposed to be and why it was thought possible that monoculture plantations could be established in the first place and yet avoid pest problems. In the absence of any coherent answer, one environmentalist proposed that a possible solution to the pest problem was to dig up the plantations and plant potatoes, a facetious suggestion that, nevertheless, contained a germ of creative possibility. Such suggestions did not appear to warrant serious attention and were dismissed without deliberation. The peremptory treatment of such impractical suggestions, however, fails to recognize that creativity flourishes in the context of childish playfulness.[52] Silly ideas have to be tolerated for they may contain novel possibilities if they are allowed time to germinate. Scientific and technical training and years of experience in routine departmental and interdepartmental politics, however, do not encourage flights of fancy.[53] For instance, during the final day of the first workshop, the suggestion was made that the model simply reinforced conventional thinking and that it was

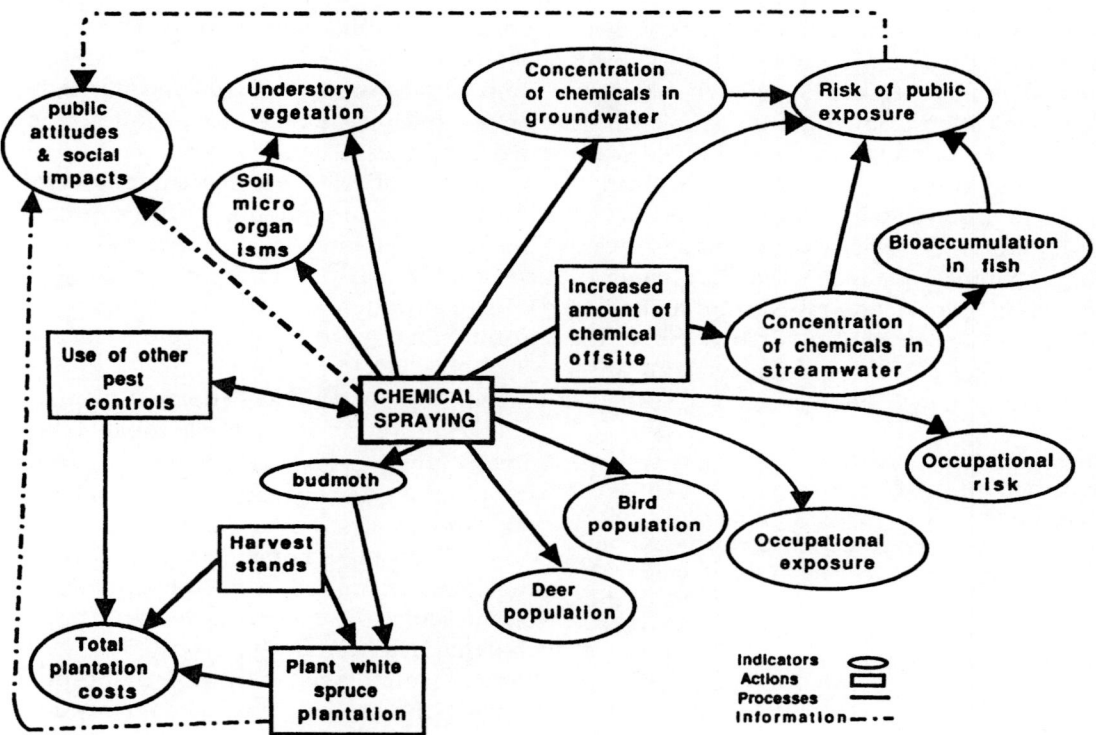

FIGURE 5.1. Pest management strategies. (Simplified version of that in Sonntag et al.[51] Used by permission of ESSA Ltd. and Environment Canada.)

unconventional wisdom that was needed in this situation. This suggestion was dismissed quickly with the comment that there were no unconventional strategies that were at the same time practical.

A major problem in generating innovative problem solutions, therefore, is that the logistical and other practical constraints on potential solutions are used too soon to dismiss seemingly impractical ideas. This sober but premature evaluation of alternatives derails creative thinking. Practical concerns are important, of course, but need to be suspended in the early stages of problem solving so that novelty might flourish.[52] Unfortunately, by virtue of their personality and training, environmental professionals are inclined to impose technical constraints at too early a stage in their search for solutions. This infuriates those environmentalists who prefer divergent ways of thinking while, in turn, the impractical

products of such divergent thought annoy their professional antagonists. None of this helps adaptive problem solving.

Decision Making (Choice)

Decision making, or choice between alternative problem solutions, takes place within a sociopolitical context in which there is unequal distribution of decision-making power (chapter 3). As we have seen, in earlier decades, the latter was concentrated in the hands of iron triangles that dominated the formulation and implementation of policy decisions. Recently, this decision-making power has been challenged by a variety of environmental interest groups. The victors in these conflicts do not have a free hand to pursue whatever policies they wish, however, for their actions are constrained by the massive inertia among the silent majority. As Bosso[54] (pp. 198–199) explains:

Policy elites, interest groups, media organizations, and other elites certainly play key roles in defining problems and setting agendas. There also is more than a little fighting among elites and groups over whose alternative construction of reality will stick. But none of this takes place in a vacuum. . . . there are a great many times when prevailing values . . . simply screen out most . . . alternative definitions of a problem. Interest groups or policy elites in the short term seemed to have little chance to alter dominant social constructions of (problems) . . . the 'received culture' seems to have had a greater role in defining the range of legitimate alternatives than any policy elite or interest group . . . it is (hard) to change strongly held public values.

Decision making is an intensely political process, therefore, one that is constrained by the conflicting demands of interest groups and, more importantly, by the looming presence of the silent (voting) majority. Within these limits, there is a lively debate over how decisions are and should be made. The era of interest-group politics that has ushered in a confrontation between differing rationalities pits several decision-making models against one another, namely, rational-comprehensive, incremental, and what might be called *principled* decision making.

From the rational-comprehensive perspective, complete information about a range of available options and their possible consequences must be achieved before a rational choice can be made by the decision maker.[55] This rational view is promoted by, and attractive to, technical and scientific professionals of the objective-analytic and objective-holistic personality types, whose espousal of technical rationality is based on a variety of assumptions, or myths, about the objectivity and impartiality of science and scientists.[55] The perpetual conflict within science caused by the irrationality and self-interest of many scientists[56] seems not to detract from this overall claim of objectivity. Rather, conflict is seen as playing a crucial role in uncovering bias and error in the scientific process.

The typical response of technical rationalists to complex problems is, as we have seen, to advocate more research on a massive scale. I've mentioned this earlier in connection with the Ecological Society of America's research agenda in pursuit of knowledge with which to combat the undesirable aspects of global change.[57] The belief is that even though such information may be difficult to achieve, it is a crucial element in making rational decisions. There have been countless examples of this same rational orientation. For instance, the response of the U.S. National Research Council to the hazard posed by environmental lead contamination to public health was to propose a comprehensive list of technical research projects that, it was believed, was necessary in understanding the problem.[55] Following this extensive data collection, it would then be possible to identify alternative control strategies and apply risk-benefit or cost-benefit analyses to "compare alternatives for control, and decide what is an acceptable level of lead in the environment for each specific population" (p. 52). Who would make these decisions about acceptibility is unspecified, a political consideration outside the mandate of the technical experts providing the basic scientific information.

There are a number of reasons why this rationalist model of decision making is ineffective in practice.[55] The idea that problems can be resolved by further scientific research ignores the simple fact that all data have to be interpreted, thereby creating the opportunity for endless technical dispute. Thus, the myth of scientific objectivity falls apart as "rival armies of technical experts," working for opposing vested interests, offer their conflicting interpretations to bemused decision makers.[55,58] Nevertheless, proponents of rational-comprehensive decision making continue to promote its veracity, as seen in a later section.

Incremental decision making differs from the rational model in that decisions are made by compromise between interested parties in a process referred to as "partisan mutual adjustment."[55] In this way of decision making, various players protect their own interests by focusing on policies that differ only marginally from what is in place. Such a process makes few demands on new information and, as a result, is not beholden to the scientific community. Typical examples of incremental decision making are to be found in the way in which the U.K. government regulated the hazard posed by lead in the environment.[55] Rather than some radical program to eliminate lead from gasoline or to strip the lead piping from domestic residences, successive governments engaged in serial removal of the hazard over a period of decades. As the volume of traffic increased, so the government reduced the lead content of petrol. At the same time, the amount of lead piping was reduced as new construction replaced older dwellings. Collingridge and Reeve[55] believe that this incremental approach is preferable to a more radical change that, in the case of lead, would have thrown society into turmoil.

Incrementalism suffers from many of the problems inherent in any form of pluralisitic social choice. Policy making becomes fragmented, as the various parties pursue self-interest. Often, decisions are in favor of those with the most power, although the very fragmentation of the decision process itself makes comprehensive power difficult to achieve. Agencies may work at cross-purposes, therefore, as they concentrate on their respective parts of the problem. In addition, incremental change does not encourage innovative measures in addressing problems.[55] As Caldwell[59] (p. 336) observes:

The risk of incremental approaches is the tyranny of small decisions. An emergency response to a critical environmental situation or event may make the larger systemic problem worse. Faced with demands for action, governments characteristically seek practical, bounded "solutions"—preferably technological with the least disruption of institutions, behaviors, or expectations. It is sufficient that a temporizing solution "work" politically. Repetition of unsuccessful but familiar incremental policies has generally been more acceptable and less politically hazardous than efforts to find permanent solutions that require major systemic changes.

Many environmentalists, especially the more radical element (Arcadian), reject both rational-comprehensive and incremental modes of decision making in favor of what, for want of a better label, might be called *principled*

choice. That is, radical environmentalists adopt what Dryzek[49] refers to as "ecological rationality," a way of thinking that espouses the prime value (principle) of ecological integrity. Thus, decisions that appear to promote ecological integrity are supported even though they may require radical action based on flimsy evidence with uncertain outcomes. Several examples come to mind. For instance, the demand by the Concerned Parents Group in New Brunswick for the immediate cessation of aerial spraying against the spruce budworm would have been a radical step with untold consequences for the forest industries.

Environmental decision making, therefore, throws these various rationalities into a crucible. Some kind of decision usually emerges from the ensuing conflict but often one that offers little satisfaction to anyone. The process by which decisions are arrived at in real-world situations is sometimes described in terms of the "garbage can" metaphor[49] (p. 127):

The garbage can itself represents an occasion on which the system must produce a decision—perhaps a sudden crisis, perhaps a situation determined by a legislative calendar. The ensuing 'choice opportunity' enables various actors to throw problem interpretations and candidate solutions into the garbage can . . . Actors, problems, and solutions move in largely independent trajectories; they mix in an arbitrary way in each garbage can, and so choice opportunities are not easily described in problem-solving terms. Solutions look for problems, rather than vice versa. The collective choices that emerge do so so in a largely unpredictable manner.

Of interest to us is the extent to which interest-group politics, especially the conflict between environmental groups and the establishment, contributes to this mess.

There are at least two views on whether decision makers in government actually listen to the opinions of interest groups. In a study of decision making in the U.S. Forest Service, for instance, Mohai[39] evaluated the two contrasting views in the context of the agency's policies regarding road access to wilderness areas in national forests. One common perspective is that "Forest Service decision-making is strongly molded by its professional ideology. The Forest Service is relatively uninfluenced by public opinions differing with that ideology because of its relatively closed organizational structure"[39] (p. 125). The ideology referred to is Imperial, focusing on sustained yield and utilitarian values. In contrast, the alternative view is that government agencies depend on their client groups for their political and financial support. Especially after 1970, the diversity of such interest groups "precludes the capture (of policy) by any one group." Thus, agency decision making is anticipatory, seeking to avoid conflict and a loss of bureaucratic influence by alienating its constituency. After reviewing a number of cases, Mohai concluded that both interpretations are valid in the sense that the Forest Service does appear to have responded to public concerns by incorporating some of their demands into the criteria for designating roadless areas. However, at the same time, the Forest Service allowed "the majority of the roadless areas to remain subject to development and utilitarian values . . . Thus environmentalists may have been allowed the 'pick of the litter' so that most utilitarian values could be preserved" (p. 154). In other words, environmentalist groups were "allowed to have a few 'high quality' areas while preserving the majority of development options." One might argue that this was a political compromise between the demands of environmentalists and industry. Thus, environmentalists are listened to on occasion, but never get all they want because of competing demands by other interest groups.

In a similar vein, Landre and Knuth,[60] in evaluating the impact of citizen advisory groups on Great Lakes planning, found that some degree of consensus over plans was possible when deliberations took place in areas relatively free of pollution and industrialization, that is, where problems were less severe. The corollary would seem to be that consensus and mutual communication is less probable when the problems being addressed are severe and the emotional climate one of distrust and hostility. Under such conditions, the attempts by governments to provide vehicles for public involvement in decision making often exacerbate these feelings of mutual distrust and frustration. For instance, at the instigation of

antispray groups, the government of New Brunswick initiated annual public hearings as part of the pesticide licensing process to determine if the aerial spray program against the spruce budworm should continue. Two presentations from irate environmentalists catch the flavor of their frustration with the hearing process:

This is the last brief I shall present before the Pesticides Advisory Board. My reasoning is this: If you do your duty, if you act rationally, if you are not at the sole service of owners and executives of pulp and paper companies, you will publicly recommend and recommend to government in private that no aerial spraying with herbicides and insecticides take place in future. If you do not so recommend, it is proof to me that appearing before this committee is totally futile.[61] (p. 67)

This whole process which we are involved in is wrong . . . there is not one woman among you . . . there is no one among you except government representatives . . . In the make-up of the Board you are without public representation . . . You make up your minds and you decide to tell the Minister . . . that that is the proper decision. Yet it is without public input except for this ridiculously short four hour meeting that you have once a year.[62] (p. 37)

Yet, contrary to the impression conveyed by these presentations, the government had previously responded to the demands of environmentalists by curtailing the aerial spraying program adjacent to inhabited areas in establishing clear set-back (no-spray) zones. Many environmentalists, however, may have felt that this was a token gesture aimed at placating public feelings while allowing industrial practices to continue virtually unabated.

The credibility of public hearings, and similar forums, is not helped when spokespeople for industry argue in hearings that, though public input is interesting, decisions should be made by experts.[63] Nor is confidence raised when, as a result of extensive hearings, an environmental impact panel recommends against an industrial development only to have its recommendations overturned by government edict.[63] Nor when studies conclude that environmental mediation "demobilizes and de-politicizes grass-roots opposition to development projects by channeling activism away from confrontation and publicly visible tactics and narrows their demands and concerns . . . Mediation in Hawaii has helped pre-empt environmental and land use conflicts from raising public awareness and generating debate over broader policy issues that may constrain development interests"[64] (p. 313).

One response of some environmental activists to what they see as a cynical disregard of their concerns is to engage in unorthodox tactics that seek to blunt the overwhelming power of corporate elites and their expert spokespeople. For instance, in a recent manual entitled *The Polluters' Secret Plan and What You Can Do to Mess It Up*, several ways of deflating experts in public hearings are outlined.[65] The following is one way of dealing with experts' requests that citizens trust them:

There's also the tried-and-true tactic of trying to get the risk assessment expert to trust YOU, by asking him to drink a glass of something you hand him. Usually, the expert won't do it, claiming he has no way of knowing what's in the glass or whether it's safe. This, of course, is the point of the exercise— since he won't take your word that a glass of water is safe, why should you take his (p. 16)?

While these political tactics are an understandable response to the continued feeling of being marginalized, none of this is conducive to collaborative problem solving. Such feelings and tactics are likely to persist, however, in the context of conventional interest-group politics. The kind of emotional turmoil generated by adversarial politics is well illustrated by Solecki[66] in his analysis of an environmental controversy in a small mill town in the southern United States. The town's major employer, a pulp mill, practiced what Solecki interprets as a form of corporate paternalism, behaving like an authoritarian but benevolent father, adopting "the role of judge and protector, acting ostensibly in the workers' best interests" (p. 7). He points out, however, that modern forms of paternalism are no longer the unilateral expressions of power by a single company but are embedded in networks of community rela-

tions. In other words, paternalistic relationships are to the mutual benefit of a large section of the community who are bound together in a system of dependency and obligation.

The managers of the mill had attempted to play a constructive role in the community by providing funds for numerous community activities, as well as making concerted efforts to reduce pollution emissions from their operation. As the problems with household water wells continued, the company "was able to diffuse local concern during critical moments of the crisis by stepping up its long-standing practice of supplying alternative sources of water , thereby 'taking care of its own' '' (p. 13). Solecki argues that this benevolence achieved two corporate goals. First, it projected "a benevolent image, distancing itself from the source of the problem while clearly asserting a role in its solution . . . A second result of corporate paternalism was that it diverted public discourse away from the problem of pollution and cast the debate as a referendum on (the firm's) presence as an employer in the county." As a result, environmental agitation about the need to stop the pollution at source was seen by many in the community as an attack on a benevolent firm that was, as a result, faced with the prospect of having to close down, thereby robbing the community of its major source of employment. To complicate matters further, many of the environmental activists were not only women but also "outsiders," leading to accusations of female overemotionality and interference from out-of-town interests. Solecki concludes that:

The paternalistic relationship between the community and the firm guided the development of the conflict and dramatically narrowed the terms of public debate. The framing of local discourse away from issues of environmental health and safety toward a focus on economic livelihood wrested legitimacy away from local environmentalists, who were cast as potential destroyers of the community . . . It can be argued that power moved through at least three pathways: (1) through the firm itself, by its maintenance of the local culture of paternalism; (2) through local elites and powerful local institutions, which pledged their allegiance to the facility through various actions and proclamations; and (3) through other local individuals who also came forward in support of (the firm) against the environmentalists . . . (As a result) this discourse of community identity became a repressive force that both reflected and reinforced the paternalistic social contract between the town and the mill. (p. 17)

In this instance, as in so many others, adversarial politics seems only to have promoted social divisions while consolidating vested interest behind the Imperial status quo. Arcadian dreams have little hope of survival under such circumstances.

In sum, decision making during this period of interest-group politics tended to be (and remains) confrontational. Although control of the decision-making process has been strongly contested, it remains firmly in the hands of political-corporate elites with the government having the unenviable task of balancing competing interests. Government's dual and contradictory roles of minimizing public concerns about the environment while promoting economic growth[64] is increasingly difficult under the intense pressures generated by environmental conflict. As a result, implementation of environmental policy is fraught with yet more conflict, as we shall now see.

Implementation

Until recently, most research on complex problem solving in the policy and planning context has focused on the policy formulation and decision-making stages. What happened afterward, whether the policy was successfully implemented, tended to receive less attention. The disparity was exacerbated, perhaps, by the separation of function between those who formulated policy and those who administered it.[67] Broadly speaking, however, conceptions of the implementation process have fallen into two camps: top-down and bottom-up versions.

Until the mid-1970s, a top-down rational model was the prevalent notion of how implementation did and should occur. In this Imperial way of thinking, political-corporate elites form policy and pass it on to an army of managers and administrators who impose the policy on a sometimes reluctant public. "The rational model is imbued with the ideas that implementation is about getting people to do what they are told, and keeping control over a sequence of stages in a system; and about the development of a programme of control which minimizes conflict and deviation from the goals set by the initial 'policy' . . ."[67] (p. 466). Even in the heyday of iron triangles, however, this top-down depiction failed to capture the byzantine nature of policy implementation. As shown in chapter 3, selective modification of and resistance to centrally generated policies occur at all levels below that of the policymakers. Increasingly, this resistance came from organized groups as we entered the era of interest-group politics in the 1970s, when those on the lower rungs of the policy ladder started to make themselves felt. This is not to say that top-down, centrally controlled implementation is impractical in all situations. Matland,[68] for instance, suggests that the global campaign to eradicate smallpox is a prime example of a centrally planned exercise that was successful because there was little ambiguity about, or conflict over, policy goals or the means to implement them. Under such conditions, administrative implementation is likely to be the most effective way to proceed. However, as the ambiguity and conflict over both means and ends increases, effective implementation must necessarily involve competing coalitions engaged in political discourse over how to proceed.[68]

The bottom-up view is that policy implementation is more than a mechanical imposition of orders from above; it is a much more complex interaction at all levels in the social system between the original policy directives and administrators, managers, field staff, and public. At all these levels, the original ideas are interpreted and modified to suit local conditions and the interests of those involved. Thus, environmental managers and professionals have some discretion in how they apply policy directives, just as the lay public select those aspects of policy with which they feel comfortable and reject the rest, at least to the extent that they can do so without being punished.

This *dilution* of policy as it descends from above is not an unmitigated disaster. It can be argued, for example, that local discretion has the effect of garnering support for a policy that might be rejected if imposed unilaterally and without concern for local sensitivities.[67] On the other hand, local control can at times subvert official policy in pursuit of personal goals and local vested interest, so that policy control is no longer exercised by those who are accountable through the electoral process.[68]

This creates something of a paradox. While there is the need to maintain some kind of central control of policy, centrifugal forces at the local level threaten to pull the policy apart.[69] Presumably, some judicious balance between central control and local autonomy is needed in implementing any policy. However, with the rise of interest-group politics, the resistance to unilateral imposition of policies increased dramatically, at least among some sectors of the population. It is doubtful whether the resulting adversarial politics has improved the problem-solving process. Karlberg,[70] for instance, suggests that the adversarial framework within which environmental issues are reported by the media has not helped matters. First, there is a tendency in the media to couch disputes in terms of simple dichotomies between opposing "highly stereotyped perspectives" (such as my use of Imperial and Arcadian ideologies). The oversimplification involved may obscure the common ground shared by the competing groups and so limit the prospect for dialogue. Second, the dramatization of conflicts by coverage of the most extreme behaviors of those involved exacerbate confrontation and contributes to the "systemic pressure to adopt extreme and confrontational modes of expression in order to engage in public discussion" (p. 24). Third, "adversarial news frames typically amplify the positional statements and demands of the dueling camps—as opposed to probing and clarifying their underlying interests and motives" (p. 24). Again, this obscures possible commonalities on which compromise could be built. Finally, adversarial politics tend to be constructed primarily in terms of economic considerations, playing on fears of immediate

economic security and leaving to one side the more long-term problems of sustainability.[70] Whether all this has resulted in more effective environmental management is a matter to which we now turn. Two examples might help clarify matters.

Resistance to official policies by professionals within the bureaucracies is a growing phenomenon, one that O'Leary[69] has studied in the context of wetlands legislation in Nevada. At issue was a long-standing dispute over how best to deal with water-quality problems induced by irrigation runoff. A bill was passed by the U.S. House of Representatives in 1990 giving the Department of Interior (DOI) the power to take various steps in mitigating the effects of this agricultural pollution on natural wetlands. Ironically, the bill "was not supported by the U.S. Department of Interior . . . (but) was supported, however, by a vast 'web' comprised of environmental groups, Indian tribes, chambers of commerce, hunters, trappers, and conservation groups"[69] (p. 444). Central to the success of this heterogeneous group was the assiduous organizing activity of a small group of "organization deviants" within the U.S. DOI and the State of Nevada Department of Wildlife. These critics of official policy were willing to embark on a potentially career-damaging course of action, contrary to the expressed policies of their organizations, because the goal of saving the ailing wetlands "became more important than pleasing their superiors, upholding departmental policy, or appeasing potentially hostile interest groups"[69] (p. 447). The strategy they adopted was to attempt to change policy by working circumspectly outside of their respective bureaucracies that they felt had had difficulty in looking objectively at the problems with their policies. In an interview, one of the dissenters explained that he had come to this decision after spending 20 years banging his head against a wall within the agency. As a consequence, they launched into a program of activity aimed at forcing a public review of water-use policy through public exposure of salient issues. This was achieved by an exhaustive attempt to forge coalitions between grassroots organizations, working the media and lobbying political representatives at all levels of government. Several of the dissidents openly acknowledged that they manipulated the press by among other things providing the media with "gruesome photographs of fishkills and deformed birds," as well as feeding information to "aggressive young reporters on the way up." Another public relations tactic used successfully was to look for crises to publicize their position. At one point, a massive fish-kill hit one of the wildlife reserves seemingly due to selenium toxicosis, a product of irrigation runoff. Although it was discovered later not to be the culprit, the event was used by the dissidents as a symbol to bring their agenda to the attention of policymakers.[69] On another occasion, a U.S. senator who was proposing the construction of a set of dykes, which in the opinion of the dissenters was not needed, was contacted and asked if he would like a tour of the wildlife refuge. The trip was justified to superiors in Washington by telling them that he had requested the tour. An exceptionally innovative approach to accumulating water rights was also taken. The dissidents hit upon the idea of persuading owners of unused water rights to donate them to the wetlands while claiming, in the process, a tax deduction.[69]

These and the many other tactics used by dissidents in pursuit of their agendas were only possible because of their willingness to take professional and career risks. Apparently, they were ready to do so even in the face of subsequent stress and uncertainty. One dissident was quite aware that "the government punishes risk takers" but felt that his whole life couldn't be centered around a paycheck[69] (p. 459). Nevertheless, "they were often ostracized for going against the scientific and political conclusions of their peers . . . some colleagues refused to speak to them. When they attended meetings with groups that were formerly thought of as the 'enemy,' some other government workers called (them) traitors"[69] (p. 460). The ensuing loneliness resulted in one of the dissidents having to take a medical leave due to mental exhaustion. Thus, although they were politically successful, they did not emerge from their ordeal personally unscathed.

As an example of adversarial politics, however, what was actually achieved here? On the face of it, we see a bottom-up resistance to an official policy that was successfully overturned thanks to the organization of resistance by dissident bureaucrats. As O'Leary[69] points out, environmentalists and conservation groups (Arcadians) would most likely see this as a courageous action by caring public servants committed to the protection of the environment. On the other hand, those of a more Imperial mind-set might well view their actions as the kind of outrageous insubordination that leads to policy chaos, "since they took secret actions against the wishes of their superiors" (p. 462). O'Leary[69] makes the additional point that one's reaction to this story might have been different if the dissidents had been "anti-Black, anti-Jew skinheads who used the tools to undermine federal civil rights actions. Or what if they were anti-environmental bureaucrats bent on destroying wetlands . . . against the wishes of their superiors?" (pp. 464–465).

Adversarial politics, therefore, and the tactics that have been used in pursuing environmental goals, raise some pressing procedural and ethical problems. No policy can be successfully implemented if groups at all levels decide to oppose it. On the other hand, top-down and implementation merely sows the seeds for subsequent insubordination. The obvious remedy is to include minority opinion more effectively into policy making at the outset—a matter to which I shall return in later chapters. For the time being, however, I'm more interested in examining the shortcomings of adversarial politics. Consider the next example.

The controversy surrounding hydrodevelopment on the Franklin River in Tasmania is a classical example of a development versus conservation dispute.[71,72] In one camp were proponents of industrial development who saw hydrodevelopment as a means of supplying electricity to the hinterland. Opposing this was a heterogeneous collection of conservation and wilderness groups determined to stop the further destruction of the pristine wilderness of southwest Tasmania. Early skirmishes in the 1960s, at a time when there was relatively little environmental awareness or concern, had been resolved in favor of development, with a dam on the lower Gordon River and the flooding of Lake Pedder, a jewel in the wilderness crown. While the conservation groups that had been formed to protect the Franklin system were unsuccessful in protecting Lake Pedder, the event spurred grassroots organizations that were successful in getting a national park created in the southwest region.[72]

At the center of development proposals was the state Hydroelectric Commission (HEC), a semiautonomous public corporation over which the state government had limited authority.[72] Supporters of the HEC "included industrial consumers (who benefited from any oversupply of electricity by being able to buy the surplus cheaply) and unions (who members relied on dam building for their jobs)"[72] (p. 208). When the HEC proposed further dam building on the Franklin system, a clash between industrial development and the grassroots conservation organizations intensified, eventually consuming the whole of Tasmania in a controversy with which the political system could barely manage to cope. In the face of a stark choice between development and conservation, a compromise was suggested by the state cabinet, in the form of an integrated strategy of energy generation and conservation. Energy was to be generated in a number of ways, including a dam built in an alternative location, one that spared the Franklin. This failed to satisfy the conservation groups who took the position that no dams were acceptable. Meanwhile, exciting new archeological finds in the Franklin River valley put additional pressure on the state government to curtail the HEC plan to build more dams. Finally, the government decided to put the matter to a referendum vote, one that did not include a "no-dam" option, to the chagrin of conservationists who advocated a spoiled-ballot strategy. As it transpired, the majority of legitimate votes were in favor of the Gordon-below-Franklin dam, which was contrary to the government's proposal for an alternative site. As a result, parliament reconvened and the government fell to a new party that had promised to build at Gordon-below-Franklin

and that subsequently authorized damming the Franklin itself. At this stage, conservation groups had no other recourse but to take their cause to the national stage where they found a majority of voters in mainland Australia in favor of protecting the Franklin. They were lucky in that not only was the new national government open to green ideas but it had recently acquired additional powers to control environmental affairs. Eventually, in the face of looming elections, the national government was able to defuse the situation by offering the state of Tasmania substantial compensation if it declined to build the Franklin dam, which it did.[72]

Again, on the face of it, we see an important victory for the conservation movement and an illustration of the way that grassroots organizations use adversarial politics to block the implementation of policies with which they disagreed. Yet they were able to do so only by fortuitous political circumstances outside of the state and in the face of widespread local concerns about the state's future economic prospects. Many citizens in this economically vulnerable state saw hydrodevelopment as vital to economic well-being and wilderness preservation as an unaffordable luxury.[72] The adversarial framework seemed not to persuade the majority about the virtues of the conservation position nor provide them with economic alternatives other than the dubious pleasures of tourism.[72] Karlberg[70] may be correct, therefore, in suggesting that adversarial politics is so preoccupied with position taking that the more fundamental issues are ignored.

In conclusion, while there have been some positive effects of adversarial politics, the system that generated the need for such politics remains much as before. Adversarial politics are the product of a system of decision making in which grassroots involvement in the crucial early stages of policy making is minimal. While it is understandable that some professionals might conclude that environmental groups seem only to stop things,[17] if policies are presented as a *fait accompli*, how else can dissenting groups respond other than with vigorous opposition? The emotional heat generated is hardly conducive to scholarly debate over the finer points involved in environmental issues nor to adaptive problem solving. This is unlikely to change as long as decision making is dominated by Imperial values and corporatist politics. The prospects for significant change, under such circumstances, are dealt with in chapter 8.

Evaluation

I believe I'm becoming repetitive in making the point that although interest-group politics have opened environmental problem solving to wider public involvement and scrutiny, the resulting changes in conventional modes have been cosmetic rather than substantive. Instead of belaboring the point, therefore, I shall keep this final section short.

Because of the complexity and unpredictability of ecosystems referred to in chapter 1, evaluation of the impacts of human interference on natural systems is both difficult and controversial. Not only is it difficult to separate signal from noise in such complex situations, with the result that immediate impacts are obscured by extraneous information, but the distant effects of any human action mean that relevant impacts are overlooked. To complicate matters further, programs are often implemented in ways that diverge from the original policy goals. In addition, new political masters may enter the policy arena eager to make their mark on policy before the impacts of previous efforts have been ascertained. In forestry, for instance, several management strategies may be implemented during the life cycle of a commercial forest, making evaluation of any single policy impossible.[17]

In previous decades, when iron triangles controlled the evaluation process, it was possible to maintain a narrow focus on Imperial goals. Management strategies were deemed successful to the extent that they enhanced production values or met a restricted range of program objectives. For instance, the aerial spraying of pesticides to control the ravages of the spruce budworm in New Brunswick has been judged a success in terms of its effectiveness in protecting the commercial wood sup-

ply. When evaluative criteria are extended to include the broader impacts of spraying pesticides on environmental and public health, however, the efficacy of the program is thrown into doubt. Thus, when environmental interest groups started to make their voices heard on environmental issues in the 1970s, the narrow positivist criteria commonly used by scientists and managers to evaluate management programs became less tenable.

Needless to say, the adversarial atmosphere of pluralistic politics is hardly conducive to a calm, measured evaluation of the impacts of any management program. Nor does it encourage a willingness to admit error by any of the parties involved. Instead, defensive maneuvering and posturing may take the place of a simple recognition that past efforts were misguided. As a result, efforts to evaluate programs are fraught with animosity and tension for reasons that are far from mysterious. Few people enjoy having their best efforts scrutinized by others who may not be sympathetic to their goals or are unaware of the constraints within which they are working. It is unfortunate, therefore, as Marcus[73] (p. 150) observes, that "those scientists and engineers who work in special monitoring units are likely to face considerable hostility and resentment" from those whose work they are evaluating. After all, program evaluation can uncover varying degrees of incompetence and ineffectiveness that reflects badly on those involved, notably the managers and civil servants who "may be reluctant to admit error because of fear of blame"[73] (p. 151).

We learn to protect ourselves from the admission of malfeasance early in life when fear of punishment takes precedence over any sense of moral obligation. Later in childhood, *guilt* (transgressing one's moral codes) and *shame* (being humiliated) come to the fore as psychological mechanisms that control our behavior. Avoiding both guilt and shame are central ambitions in later life, as we strive to maintain some semblance of pride in ourselves and our individual achievements. To do so, we engage in various coping strategies that, to put it bluntly, involve persistent self-deception and other forms of self-lies to protect our rather fragile egos.[74] In this difficult world, some degree of self-deception is necessary to maintain psychological "health," for without it we may face the torment of psychological honesty,[75] a nightmarish existence for all but the hardy. To add to these personal problems is the fact that we are inclined to believe, with some justification, that admission of error can be a form of professional and political suicide. Even the admission of the simplest error can be an excruciating, humiliating experience for many professionals and politicians who are seldom willing to shatter their facade of invincibility with such confessions. Yet, learning from error is a central tenet of adaptive management. Anything that hinders this learning process is likely to be counterproductive. Thus, the open recognition of error is crucial in the evaluation of management programs just as failure to do so limits the ability to learn from past mistakes. How this unfolds in the real world is well illustrated by the recent history of attempts to manage the cod stocks off the coast Atlantic Canada. I draw heavily from Blades[76] in what follows, using his interpretations liberally in outlining the story.

Atlantic cod has been the mainstay of the economy of Newfoundland for 400 years, as well as an important element in the well-being of fishing communities in Canada's Maritime provinces. For hundreds of years prior to World War 2, fisheries technology was relatively simple with demands on the groundfish stocks being manageable. However, all this was to change in the postwar period with "the tremendous growth and resulting overcapacity of the harvesting sector—domestic and foreign—fishing the waters of the northwest Atlantic. The unbridled expansion of fishing power involves an increase in the number and size of vessels that pursue cod and other groundfish, as well as improvements in the industry's ability to search, chase and catch its valuable prey"[76] (p. 47). At least one contributing factor to this harvesting overcapacity, and the subsequent collapse of the cod stocks, was the impact of successive governmental policies. In formulating fisheries policies over the years, the government of Canada has been faced with the persistent economic weakness

of the Atlantic provinces. One way to overcome this, it was hoped, was to promote the development of the fisheries sector through improving the technical proficiency of the domestic fleet. Loans were provided for the building of more modern vessels in both the inshore and offshore fisheries, with the intent that Canada would become a dominant force in the increasingly competitive North Atlantic fisheries. At the same time, grants and subsidies were offered to attract private capital investment in the fish processing capacity of the region. As a result, there developed an enormous fishing fleet of more than 30,000 vessels "ranging in size from a dory powered by a small outboard motor to a large factory freezer ship"[76] (p. 47). This policy of expansion was predicated on the continuing existence of large stocks of cod and other groundfish on which the cod, and humans, fed. Information on the health of these stocks was provided to policymakers in the Department of Fisheries and Oceans (DFO) by the agency's scientists. As we shall see, in the face of declining stocks, managers and politicians overestimated the health of the cod stocks.[76]

During the late 1960s and early 1970s, signs of overfishing outside of Canada's 12-mile limit had encouraged the federal government to extend its jurisdiction over the whole continental shelf. Accordingly, a 200-mile limit was imposed in 1977. At the same time, the DFO adopted a quota management system in which annual harvest was reduced from 35 to 20 percent of groundfish biomass. This more conservative approach was based on calculation of fish mortality and estimates of biomass derived from harvesting records and other scientific data obtained by cruise ships operated by DFO researchers. Estimating the mortality rate of groundfish stocks is a difficult business, one that entails determining the amount of fish landed, as well as discarded at sea, together with degree of predation, adverse environmental changes, and the number of young fish "recruited" into the harvestable stock.[76] For a while in the late 1970s and early 1980s, these new management arrangements seemed to be working well with good catches and the prospect for higher quotas in the offing. However, "by 1984 inshore northern cod catches off Newfoundland had dropped significantly, and fishermen were becoming more and more vocal about declining stocks. Inshore fishermen were among the first to challenge the government perception that the northern cod stocks were healthy"[76] (p. 38).

According to Blades,[76] what had happened was that the quota management system had been derailed by "conflict between industry sectors and interference from politicians working to protect the interests of the fishermen and sectors they represented . . . Despite evidence that the resource was no longer responding to management measures and that high catch rates were suppressing stock growth, a series of fisheries ministers, under duress from fishermen's groups and lobbying from powerful offshore companies often set quotas at levels above those recommended by scientists" (p. 111). This is not to say that the federal government had failed to act at all. The quota system itself was an attempt to retrieve the situation, just as a moratorium on fishing licenses sought to curtail the number of fishers reliant on the dwindling resource. However, it seems that federal politicians were unwilling to take more draconian measures for fear of creating an economic disaster in the Atlantic region. Needless to say, the inevitable collapse of one fish stock after another began to occur in the 1990s, resulting in the virtual closure of the groundfish industry and the economic calamity all involved had feared. A moratorium on northern cod was imposed in 1992, with other closures following a year later. Approximately 50,000 fishermen and processing plant workers were affected directly, together with an additional 47,000 indirect jobs.[77] Of interest to us is the subsequent autopsy that has taken place over the collapse of this previously abundant fishery and the extent to which responsibility has been accepted by the various players. Blades[76] argues that all the major players contributed something to what happened. Thus, politicians, DFO bureaucrats and scientists, the fisheries corporations, and independent fishers all played a significant role.

Even as late as 1989, senior government officials were denying that the cod stocks were

in crisis, a position that was inconceivable to the other players in the drama. Instead, "rather than accept sole responsibility for mismanagement of the resource, government focused on research into environmental circumstances that had contributed to the decline"[76] (p. 38). What Blades is saying is that, rather than paying attention to past policies that had encouraged overcapacity in the industry and that were based on overoptimistic perceptions of fish stocks, government explanations for the shortages emphasized such factors as inclement weather and predation by seals. Even after a sobering appraisal of the situation in 1989 by the Canadian Atlantic Fisheries Advisory Committee, an independent scientific panel, and their recommendation that the annual harvest be cut in half, only a modest reduction was announced by the federal fisheries minister. Thus, "the official line from the DFO remained optimistic even during the worst years, when the offshore companies faced financial ruin; when the domestic fishery had to contend with failing groundfish stocks; and when the closure of the groundfish fishery appeared inevitable. The DFO chose to emphasize the overall strength of the Atlantic fishery . . ."[76] (p. 119).

When he turns to deficiencies in the policy process, Blades[76] recognizes a number of problems, two of which will be mentioned here. First, he suggests that, until quite recently, the federal fisheries policy-making process was controlled by what we would call an iron triangle composed of Cabinet ministers, senior bureaucrats in the DFO, provincial bureaucrats, and special interests. This policy community, he argues, has been "insulated from input by individual fishermen and the general public, except in crisis situations, when past policy is exposed to attack" (pp. 121–122). On the other hand, the large fisheries corporations play a major role in policy deliberations because government policies for development of the Atlantic provinces hinged on continuing corporate vitality. The net effect is that, while DFO gives "appearance of inviting input on policy formation from all sectors of the domestic industry, independent fishermen and independent fish processors realize they have little impact on government decision making ... The bureaucratic inertia inherent in the system effectively stifles independent input" (p. 122). This turned out to be most unfortunate because fishers possess a fund of valuable information on the ecology and health of the fish stocks with which they are so intimately involved.[78] In other words, the policy process is relatively closed, opening up only in times of crisis when it is impossible to engage in measured debate so that errors can be identified and rectified.[76]

Second, Blades[76] believes that fisheries scientists had made a number of errors in their estimates of cod stocks. According to the Harris Report, an independent review of the state of the northern cod stocks, their overestimates were the result of several technical errors. Earlier overoptimistic projections by fisheries scientists had seemingly been based on inadequate mathematical modeling of the stocks, particularly the way in which fish mortality had been estimated. The formula used "did not recognize the need to maintain a sufficient number of older aged spawning members in the cod population. The ... formula simply measured the biomass comprised of all year classes of exploitable age and paid no particular attention to the health of the sexually mature age group" (p. 116). In addition, the research data obtained by scientists in their research surveys were subject to error because of the highly migratory nature of the stock. Further, the reliability of the data used by scientists on the annual catch were skewed by, among other things, inaccurate reporting of catches, particularly by some offshore fishers. The consequence of these errors, Blades argues, was an overoptimistic assessment of the health of the cod stocks,[77] such that "managers and politicians, eager to hear an optimistic forecast, naturally subscribed to the promising reports and set northern cod total allowable catches at the upper end of the range"[76] (p. 129).

Blades[76] takes the position that as time went by and it became more obvious that something was terribly wrong with the stocks, the majority of fisheries scientists did not speak out because "the scientific community had been coopted by its political leaders. Senior bureaucrats and politicians did not want to hear bad news, so those scientists who would sound the alarm bells were given no audience. Even scientists with tenure were reluctant to speak out because their careers were dependent upon government funded research grants" (p. 129). Those few DFO scientists who did speak out, by making their concerns known in both scholarly and public forums, experienced some predictable problems. Two former DFO scientists, in particular, are currently at the center of a political and legal storm arising in part from their recent statements about the use of science in the fisheries policy process. For instance, Hutchings et al.[79] argue that the scientific uncertainty and differences of opinion that accompany all scientific work were not dealt with adequately in DFO policy making. Indeed, they go so far as to say that public discussion of this uncertainty has been curtailed by DFO officials:

One can conclude that constraints imposed by the DFO stifled efforts to undertake, or to discuss publicly, such analyses of scientific uncertainty. These constraints took various forms. Prominent among them was the withholding of research survey data from DFO scientists who did not normally participate in the annual stock assessments ... Scientists were also explicitly ordered then, as they are today, not to discuss 'politically sensitive' matters (e.g. the status of fish stocks currently under moratoria) with the public ..." (p. 1202).

An increasing number of concerned scientists, primarily from outside the agency, have stepped forward with their views on the fisheries fiasco, charging suppression of dissenting opinion and manipulation of scientific data by DFO bureaucrats.[80,81] Several DFO officials have reacted strongly to such charges, with one reported as saying that the paper by Hutchings et al.[79] represents a "platform from which to launch an unprofessional and unsubstantiated attack. They have maligned the reputations of hundreds of dedicated, hard-working scientists and managers ... "[82] As the controversy has escalated, the atmosphere has become even more vitriolic, culminating in legal action being instituted by two senior DFO managers against one of the more outspoken scientists and the newspaper that reported his, and other scientists', comments (*DFO News Release*, June 30, 1997).

To bring this sorry story to a close, given the current turmoil in fisheries management in Canada, beset with controversies on both the east and west coasts, it is difficult to imagine how the parties can engage in adaptive problem solving. The atmosphere is too fraught with charges and countercharges, recriminations, and distrust. Even when the storm abates, it is likely that the policy-making structures that led to problems in the first instance will still be in place. One might conclude, from these various examples, that the adversarial politics that typifies environmental management is a reaction by marginalized groups to exclusion from a meaningful role in decision making, an attempt to force open the policy-making door to let in a wider variety of players. While this has had the useful effect of opening up discourse on a variety of environmental issues, it is doubtful whether it has, or will, result in the political changes that must precede adaptive problem solving. As powerful elites fight a rearguard action to protect the status quo by offering token concessions here and there, there is no real change in the policy-making process or in who holds the power to control events. Meanwhile, too much time is wasted on emotional altercation and confrontation. Is there any practical alternative?

References

1 Eder, K. 1996. The institutionalisation of environmentalism: Ecological discourse and the second transformation of the public sphere. In *Risk, environment and modernity: Towards a new ecology,* ed. S. Lash, B. Szerszynski, and B. Wynne, 203–223. London: Sage Publications.

2 Brown, P., and F. Ferguson. 1995. "Making a big stink": Women's work, women's relationships, and toxic waste activism. *Gender & Society* 9:145–172.

3 Levine, A. 1982. *Love Canal: Science, politics, and people.* Lexington, Mass.: Lexington Books.

4 Brown, P., and E. Mikkelsen. 1990. *No safe place: Toxic waste, leukemia, and community action.* Berkeley, Calif.: University of California Press.

5 Cable, S., and M. Benson. 1993. Acting locally: Environmental injustice and the emergence of grass-roots environmental organizations. *Social Problems* 40:464–477.

6 Couch, S., and J. Kroll-Smith. 1985. The chronic technical disaster: Toward a social scientific perspective. *Social Science Quarterly* 66:564–575.

7 Harr, J. 1995. *A civil action.* New York: Vintage.

8 Hallman, W., and A. Wandersman. 1992. Attribution of responsibility and individual and collective coping with environmental threats. *Journal of Social Issues* 48:101–118.

9 Renn, O., *et al.* 1992. The social amplification of risk: Theoretical foundations and empirical applications. *Journal of Social Issues* 48:137–160.

10 Cvetkovich, G., and T. Earle. 1992. Environmental hazards and the public. *Journal of Social Issues* 48:1–20.

11 Dake, K. 1992. Myths of nature: Culture and social construction of risk. *Journal of Social Issues* 48:21–37.

12 Schwartz, S., P. White, and R. Hughes. 1985. Environmental threats, communities, and hysteria. *Journal of Public Health Policy* 6:58–77.

13 Savan, B. 1988. *Science under siege: The myth of objectivity in scientific research.* Montreal: CBC Enterprises.

14 Hoffman, A. 1995. An uneasy rebirth at Love Canal. *Environment* 37:5–9, 25–31.

15 Whelan, E. 1993. *Toxic terror: The truth behind the cancer scares.* Buffalo: Prometheus.

16 Wildavsky, A. 1995. *But is it true? A citizen's guide to environmental health and safety issues.* Cambridge, Mass.: Harvard University Press.

17 Baskerville, G. 1995. The forestry problem: Adaptive lurches of renewal. In *Barriers and bridges to renewal of ecosystems and institutions,* ed. L. Gunderson, C. Holling, and S. Light, 37–102. New York: Columbia University Press.

18 Webb, F. and H. Irving. 1983. My fir lady: the New Brunswick production with its facts and fancies. *Forestry Chronicle* 59:118–122.

19 Lyles, M., and I. Mitroff. 1980. Organizational problem formulation: An empirical study. *Administrative Sciences Quarterly* 25:102–119.

20 Bardwell, L. 1991. Problem-framing: A perspective on environmental problem-solving. *Environmental Management* 15:603–612.

21 Coughlin, J. 1994. The tragedy of the concrete commons: Defining traffic congestion as a public problem. In *The politics of problem definition: Shaping the public agenda,* ed. D. Rochefort and R. Cobb, 138–158. Lawrence, Kans.: University Press of Kansas.

22 Roe, E. 1989. Narrative analysis for the policy

analyst: A case study of the 1980–1982 Medfly controversy in California. *Journal of Policy Analysis and Management* 8:251–273.

23 Davis, S. 1995. The role of communication and symbolism in interest group competition: The case of the Siskiyou National Forest, 1983–1992. *Political Communication* 12:27–42.

24 Robison, W. 1994. *Decisions in doubt: The environment and public policy.* Hanover: University Press of New England.

25 Miller, A. 1985. Technological thinking: Its impact on environmental management. *Environmental Management* 9:179–190.

26 Smil, V. 1993. *Global Ecology: Environmental change and social flexibility.* London: Routledge.

27 Winter, D. 1996. *Ecological Psychology: Healing the split between planet and self.* New York: Harper Collins.

28 Lee, K. 1993. *Compass and gyroscope: Integrating science and politics for the environment.* Washington, D.C.: Island Press.

29 Olsen, S., ed. 1982. *Group planning and problem solving methods in engineering management.* New York: Wiley.

30 Miller, A. 1987. Psychopathology amongst environmental professionals. *The Environmental Professional* 9:111–120.

31 Meadows, D. H., *et al.* 1972. *The limits to growth.* New York: Universe Books.

32 Rastetter, E. 1996. Validating models of ecosystem response to global change: How can we best assess models of long-term global change? *BioScience* 46:190–198.

33 Arney, W. 1991. *Experts in the age of systems.* Albuquerque: University of New Mexico Press.

34 Waldrop, M. 1992. *Complexity: The emerging science at the edge of order and chaos.* Harmondsworth: Penguin.

35 Botkin, D. 1990. *Discordant harmonies: A new ecology for the twenty-first century.* New York: Oxford University Press.

36 Beck, P. 1982. Corporate planning for an uncertain future. *Long Range Planning* 15:12–21.

37 Clark, T., A. Curlee, and R. Reading. 1996. Crafting effective solutions to the large carnivore conservation problem. *Conservation Biology* 10:940–948.

38 Magill, A. 1988. Natural resource professionals: The reluctant public servants. *The Environmental Professional* 10:295–303.

39 Mohai, P. 1987. Public participation and natural resource decision-making: The case of the RARE II decisions. *Natural Resources Journal* 27:123–155.

40 Schnaiberg, A., and K. Gould. 1994. *Environment and society: The enduring conflict.* New York: St. Martin's Press.

41 Forester, J. 1989. *Planning in the face of power.* Berkeley: University of California Press.

42 Dalton, L. 1986. Why the rational paradigm persists—the resistance of professional education and practice to alternative forms of planning. *Journal of Planning Education and Research* 5:147–153.

43 Brooks, M. 1993. A plethora of paradigms? *Journal of the American Planning Association* 59:142–145.

44 Udwadia, F. 1986. Management situations and the engineering mindset. *Technological Forecasting and Social Change* 29:387–397.

45 Miller, A. 1991. *Personality types: A modern synthesis.* Calgary: University of Calgary Press.

46 Pacey, A. 1983. *The culture of technology.* Cambridge, Mass.: MIT Press.

47 Tangley, L. 1987. Malaria: fighting the African scourge. *BioScience* 37:94–98.

48 Kondrachine, A., and P. Trigg. 1995. Malaria: hope for the future. *World Health* 48:26–27.

49 Dryzek, J. 1987. *Rational ecology: Environment and political economy.* New York: Blackwell.

50 Baskerville, G., and P. Duinker. 1986. Pest management in plantations: An institutional analysis. In *Pest management in plantations: A consultative approach, ed.* N. Sonntag et al. Vancouver, B.C., ESSA Ltd.

51 Sonntag, N., *et al.* 1986. *Pest management in plantations: A consultative approach.* Vancouver, B.C.: ESSA Ltd.

52 Isaksen, S., and D. Treffinger. 1985. *Creative problem solving: The basic course.* Buffalo, N.Y.: Bearly Ltd.

53 Wenk, E. 1987. *Trade Offs: Imperatives of Choice in a High-tech World.* Baltimore: Johns Hopkins University Press.

54 Bosso, C. 1994. The contextual bases of problem definition. In *The politics of problem definition: Shaping the policy agenda, ed.* D. Rochefort and R. Cobb, 182–203. Lawrence: University of Kansas Press.

55 Collingridge, D., and C. Reeve. 1986. *Science speaks to power: The role of experts in policy making.* London: Francis Pinter.

56 Mahoney, M. 1976. *Scientist as subject: the psychological imperative.* Cambridge, Mass.: Ballinger.

57 Lubchenko, J., et al. 1991. The sustainable biosphere intiative: An ecological research agenda. *Ecology* 72:371–412.

58 Ozawa, C. 1991. *Recasting science: Concensual pro-*

cedures in public policy making. Boulder, Colo.: Westview Press.

59 Caldwell, L. 1995. Environment as a focus for public policy. College Station: Texas A &M University Press.

60 Landre, B., and B. Knuth. 1993. Success of citizen advisory committees in consensus-based water resources planning in the Great Lakes Basin. Society and Natural Resources 6:229–257.

61 Labossiere, J. 1982. Presentation to the Pesticide Advisory Board: Public hearings on December 14th, Newcastle, N.B., Canada.

62 Coombs, D. 1985. Presentation to the Pesticide Advisory Board: Public hearings on November 18th, Woodstock, N.B., Canada.

63 Richardson, M., J. Sherman, and M. Gismondi. 1993. Winning back the words: Confronting experts in an environmental public hearing. Toronto: Garamond Press.

64 Modavi, N. 1996. Mediation of environmental conflicts in Hawaii: Win-win or co-optation? Sociological Perspectives 39:310–316.

65 Collette, W. 1989. The polluters' "secret plan". Arlington, Va.: Citizens' Clearing House for Hazardous Wastes, Inc.

66 Solecki, W. 1995. Paternalism, pollution and protest in a company town. Political Geography 15:5–20.

67 Parsons, W. 1995. Public policy: An introduction to the theory and practice of policy analysis. Aldershot, U.K.: Edward Elgar.

68 Matland, R. 1995. Synthesizing the implementation literature: The ambiguity-conflict model of policy implementation. Journal of Public Adminitration Research and Theory 5:145–174.

69 O'Leary, R. 1994. The bureaucratic politics paradox: The case of wetlands legislation in Nevada. Journal of Public Administration: Research and Theory 4:443–467.

70 Karlberg, M. 1997. News and conflict: How adversarial news frames limit public understanding of environmental issues. Alternatives 23:22–27.

71 Birkeland. 1990. Tasmania: The story. Presented at Clayquot Sound Wilderness Conference, September 8, Vargas Island, B.C., Canada.

72 Doyle, T., and A. Kellow. 1995. Environmental politics and policymaking in Australia. South Melbourne: Macmillan.

73 Marcus, A. 1988. Risk, uncertainty, and scientific judgement. Minerva 26:138–152.

74 Goleman, D. 1985. Vital lies, simple truths: The psychology of self-deception. New York: Simon and Schuster.

75 Smail, D. 1984. Illusion and reality: The meaning of anxiety. London: Dent & Sons.

76 Blades, K. 1995. Net destruction: The death of Atlantic Canada's fishery. Halifax, N.S.: Nimbus Publishing.

77 Neis, B., and S. Williams. 1997. The new global right, gender and the fisheries crisis: Local and global dimensions. Atlantis 21:47–62.

78 Neis, B. 1992. Fishers' ecological knowledge and stock assessment in Newfoundland. Newfoundland Studies 8:155–178.

79 Hutchings, J., C. Walters, and R. Haedrich. 1997. Is scientific inquiry incompatible with government information control? Canadian Journal of Fisheries and Aquatic Science 54:1198–1210.

80 Enman, C. 1997. Science silenced to hide 'disasters': Top biologist says DFO suppressed evidence of calamitous errors. Ottawa Citizen, June 27, Ottawa, Canada.

81 Enman, C. 1997. 36 scientists: End the suppression: Manifesto calls for the restoration of integrity within DFO. Ottawa Citizen, July 4, Ottawa, Canada.

82 Canadian Press. 1997. Fish experts draw fiery response from Ottawa: Scientists who accuse Fisheries officials of deception prompt blistering denials. Globe and Mail, June 25, Toronto, Canada.

6

Single Visions

The persistent ineffectiveness that typifies our dealings with complex environmental problems has resulted in a bewildering array of conflicting recommendations about how matters might be improved. Fortunately, there is some underlying pattern to this chaos, with proposals dividing along ideological and disciplinary lines (Table 6.1). Contrasting Imperial and Arcadian pathways are clearly evident, as both sides in the argument propound different routes to disparate goals. Within each ideological pathway, advocates of each perspective emphasize one substantive domain (environmental, social, or psychological) while underplaying the other two domains. This is not to say that proponents of each viewpoint ignore the remaining domains completely, rather, their consideration of the latter is often cursory. The result is at least six broad perspectives on necessary reforms to environmental problem solving. My argument is that, when used as the sole basis for problem solving, each perspective is a form of "single vision," a singularly inadequate foundation for adaptive behavior. In support of this contention, we should look at the pros and cons of each single vision in turn.

Environmental Visions

This group of reforms is offered by those who believe that environmental problems can be resolved, or at least contained, by refining en-

6.1. Single Visions

Domain	Imperial	Arcadian
Environmental	Scientific ecosystem management	Ecological coexistence
Social Psychological	Eco-Hobbesian Behavioral engineering	Ecosocialism Self-development

vironmental management methods. Attention is focused primarily on the environmental symptoms of problems, rather than their psychosocial causes. Although the latter are acknowledged, they take second place behind the more important technical issues involved in understanding and manipulating the biophysical environment. Beyond these commonalities, conceptions of environmental management split into Imperial and Arcadian camps.[1-4] More specifically, this ideological division reflects undercurrents within ecological theory that offer contrasting perspectives on or metaphors about nature. The Imperial approach to reform of environmental management is founded on the mechanistic and cybernetic viewpoints characteristic of ecosystems ecology. A variety of large-scale, systems-based methods stem from these assumptions,[5] the most prominent being referred to as *ecosystem management*. In contrast, Arcadian reforms are based on older, organic metaphors in ecology that depict nature as a complex organism akin to the human body.[6] The aim of Arcadian management is to promote the development of these natural communities rather than managing them in an Imperial sense. Accordingly, Arcadian management might best be referred to as *eco-community development* or, perhaps, *eco-dwelling*. Whatever the term used, the distinction being referred to is that between anthropocentric and ecocentric forms of environmental management.[1]

Ecosystem Management

Like all ideas at the center of intellectual ferment, ecosystem management means different things to different people. However, all versions share a common interest in trying to overcome the deficiencies of traditional reductive management practices.[7] Ecosystem management, therefore, is based on the assumption that environmental problems do not respond well to methods that are fragmented by discipline and jurisdiction.[1,8] The remedy, it is argued, is to attempt large-scale management based on a scientific understanding of ecosystem functioning. On the face of it, proponents of ecosystem management appear to be offering yet another version of the rational-comprehensive approach to problem solving that has been so thoroughly discredited by those who study policy-making processes. However, the situation is more complex than it appears. Some advocates appear to brush aside qualms about difficulty in understanding complex systems; others are more humble in the face of uncertainty about ecosystem functioning. Ecosystem management, therefore, is an uneasy mix of rational-comprehensive pretensions with a more pragmatic realization of the need for incremental learning. Thus, the common features underlying most conceptions of ecosystem management are as follows:[3,9,10]

Ecosystems. The basic management units are ecosystems, seen as dynamic systems interacting with one another across broad spatial and temporal scales as they cut across political and administrative boundaries.

Adaptive Management. Practices emphasize adaptive, flexible management strategies based on progressive learning through monitoring, data collection, and experimentation.

Collaborative Decision Making. It is assumed that the agencies and organizations involved in ecosystem management will change their practices to facilitate cooperative management across jurisdictional boundaries and to allow the inclusion of the public in a significant way.

Ecosystems

The current interest in large-scale management is not new but has precedents dating back to the 1950s when systems ecologists promoted "Big Ecology" as a means of resolving environmental problems.[11,12] Their enthusiasm stemmed from the development of systems ecology as an alternative to the previously dominant vision of ecology based on the study of discrete ecological communities wending their way toward stable climaxes.[13] Difficulties in defining and setting boundaries around these latter entities, as well as the anthropomorphic assumptions underlying the notion of communities as "superorganisms,"[14] led to the development of a more mechanistic view of natural systems as ad hoc collections, or loose assemblages, of species spread along continuous gradients.[15] These *ecosystems* were described in terms of energy and food exchanges at varying trophic levels within the system. At midcentury, ecosystem ecology was essentially an "energy-based economics of nature" concerned with the productivity and efficiency of energy transfers—a form of bioeconomics.[6] As such, it fitted easily into the pervading political culture that, at the time, displayed rampant Imperial ambitions. Perceptions of nature as an "automated, robotized, pacific factory" that could be controlled and managed with the help of the new system ecology and its big computer models were very attractive to political elites who were becoming aware of the environmental consequences of postwar industrial expansion.[6,14] When this promise of managerial control failed to materialize, interest in ecosystems management appears to have lapsed, along with funding for this kind of ecology. At least part of this decline in interest was due to yet another change in political climate away from big government toward a more free-market economy in the 1970s and 1980s. Recently, however, growing concerns with global change and environmental deterioration have led to a revival of interest in management on a large (ecosystem) scale. In the intervening period, ecosystem ecology has moved on from the mechanistic era of the 1960s toward a more cybernetic view of natural systems, one that eschews older notions of stability and permanence, assuming, instead, that uncertainty, change, and complexity are more realistic characteristics of ecosystems.[16] One sees again, therefore, political-corporate elites espousing large-scale management, just as they did in the 1950s and 1960s, under the rubric of sustainable development and ecosystem management. Most natural resource agencies now officially endorse ecosystem approaches to management, even if they have difficulty in deciding what that might be.[3] At the same time, ecologists and managers promote ecosystem management made possible, they claim, by the advent of new, more powerful tools.[17,18] The result is an animated debate over the prospects for large-scale management, with scientists differing over the ability of science to develop an adequate understanding of any ecosystem. The debate is an example of the much older epistemological dispute over the relative merits of reductive versus holistic forms of thought.[19-21] Although the most sensible resolution of the dispute is to accept that both forms of thought are essential, conflict between reductive and holistic thinkers continues. In the context of ecosystem management, one can see this in disagreements over at least two issues: (1) problems in defining ecosystems and (2) problems in acquiring knowledge.

Definitional Problems

The term *ecosystem* can be applied to systems of virtually any size within the natural world, from a dunghill to a watershed.[3] This means that what is considered to be an ecosystem is very much in the eye of the beholder, a subjective judgment based on what one is interested in studying. Scientists react differently to

this definitional uncertainty, in ways that reflect their underlying epistemological assumptions. For instance[22] (p. 35):

> It is the observer-dependent nature of the study of self-organizing systems which is the most difficult point for traditional reductionist science to understand . . . The response of traditional science to this is that ecosystems don't exist, since we cannot come up with an observer independent way of defining them. One consequence of this logic is that ecosystem research is not considered proper 'scientific' research by most North American granting agencies and is not a fit topic in American ecological journals.

Scientists differ, then, over how important they believe definitional precision to be. Underlying this parting of the ways are differing opinions on the nature of conceptual categories.[23] That is, any conceptual category, such as an ecosystem, can be viewed in the classical sense as a discrete entity that should be distinguished from others on the basis of a precise list of descriptive criteria. In this way of thinking, unless one can define the entity under consideration with clarity and precision, then one cannot proceed with scientific study. However, there is an alternative view in which conceptual categories are seen as fuzzy concepts, defined by a prototypical example rather than a list of defining characteristics. Prototypical categories are fuzzy concepts containing many instances that bear only passing resemblance to the defining example. Holistic thinkers are more liable to see merit in and be willing to work with this kind of fuzzy category.[24]

Fitzsimmons,[15] for instance, while accepting that boundary setting may be of limited import in research, argues that it is crucial in matters of policy. In his view, the ecosystem concept is useful in research but disastrous as a basis for policy decisions precisely because ecosystems are arbitrary subdivisions in a continuously shifting pattern of species. Attempts to devise ecosystem maps, for instance, assume an arbitrary appearance as individuals include their personal concerns into the area under consideration. Thus, "the area labeled the Greater Yellowstone Ecosystem expanded from some 5 million acres in 1979 to 19 million acres in 1990" as a result of researchers adding their areas of interest to the debate over the management of this region.[15] In contrast, some scientists are inclined to brush aside these definitional niceties arguing that it is simply a rehashing of the old continuum versus category debate in ecology.[25] Others believe that[26] (p. 229):

We think it is fruitless to argue about the conceptual and physical existence of ecosystems . . . Like it or not, the concept of ecosystem management isn't going to go away anytime soon, because something like it is necessary to address the vast natural and cultural wreckage of exploitation-focused, single-resource approaches to resource management that has accompanied European colonization of the globe.

With this mind-set, one has to do the best one can while working within the context of definitional uncertainty. "Management objectives define the ecological system of interest and the appropriate scales of ecological organization that need to be addressed"[25] (p. 210). In part, these alternative views reflect a willingness to work with fuzzy concepts within science. It is unlikely that there will be a resolution to this difference of opinion because the two viewpoints are in a dialectical relationship to one another. Each viewpoint complements the other as those working with fuzzy concepts proceed in a pragmatic manner by satisficing and approximating, while their more reductive colleagues attempt to restrain the worst excesses of holistic thought by subjecting the products thereof to the rigor of reductive analysis.

Reliable Knowledge

Ecosystem management implies the use of large-scale analyses, often at the landscape level. Achieving an understanding of the structure and dynamics of natural systems at this level requires "intellectually massive integrations."[27] The question is whether science can cope with the enormous complexity involved and produce reliable knowledge of use to managers. As intimated earlier, views are divided into pessimistic and optimistic camps.

The more pessimistic (or realistic?) view is that ecological science is strictly limited in the extent to which it can achieve the level of understanding needed to predict ecosystem responses to changing circumstances. In particular, the intellectual tools available to ecologists are thought to be inadequate. Laboratory-based research is of limited use in providing the kind of information needed to understand ecosystems, the alternative being large-scale field experiments. However the "cost, logistical constraints, and lack of replication make ecosystem-level experiments prohibitive," to the extent that they have "played a minor role in the acquisition of reliable knowledge about ecosystem function."[4] As Hilborn and Ludwig[28] point out, difficulty with replication and control of ecosystem-level experiments are serious problems. Field experiments may take a decade or more to complete, even longer when one considers the problem of replication "for issues like old-growth forest, marine mammals, or global climate change." In addition, "ecologists often rely on natural experiments to understand the dynamics of ecosystems," which raises the issue of inadequate controls and the subsequent weakness of inferences from findings. To make matters worse, ecosystems change continuously, so even the most meticulous experiments may be irrelevant because of changing circumstances.[28] It follows that "there is general agreement that ecology is not a predictive science and that our ability to predict is poor" due in large part to "our inability to verify and validate predictive models"[4] (p. 259). It seems that the presence of ecological surprises, critical thresholds, chaos, and increasing stochasticity as scale increases severely limit the predictive capabilities of ecological models.[29]

A more optimistic view of the potential contribution of science in general, and ecology in particular, to large-scale management is held by other scientists.[30-34] For instance, a significant role for traditional reductive science is seen by Mooney and Sala,[35] who believe that problems with replication and control of large-scale experiments are being overcome. In their view, numerous excellent studies have been conducted in research on global change and in such projects as the Sustainable Biosphere Initiative. Although Foran and Wardle[36] take the matter further by arguing that "all problems are potentially solvable by reductive science," others suggest that if ecological research is to reflect the complexity and chaos of ecological systems, there needs to be an epistemological reformulation of scientific methods beyond the mere recognition of a need for holistic and systems thinking. While the direction of this reformulation has been sketched out, the actual details remain to be worked out. Thus, Costanza[37] talks of a "post-normal" science that is "edge-focused," meaning that science should seek to delimit the boundaries within which ecosystems can operate.

In sum, there is considerable uncertainty within the scientific community over the prospects for the generation of reliable knowledge about ecosystems. The more pessimistic point out that, in the unlikely event that fundamental epistemological and methodological problems could be overcome, the historical record suggests that natural resource policymakers take little notice of scientists anyway.

Adaptive Management

Although optimists and pessimists may differ over their views on the prospects for attaining comprehensive understanding of ecosystems, all would agree that management decisions are usually made under conditions of great uncertainty. One of the main problems faced by advocates of ecosystem management, therefore, is how to proceed in the face of the complexity and unpredictability of ecosystems. The method of choice for many is "adaptive management," an approach that attempts to apply scientific rationality to the problem of uncertainty, by adopting an explicitly experimental approach to the design and implementation of natural-resource and environmental policies.[38] That is, management is seen as an ongoing experiment designed to facilitate continuous learning about the ecosystem in question. Halbert[38] contrasts this with traditional management in which policy implementation is less sensitive to changing environmental circumstances. Management by experiment is simply not contemplated by those for whom learning from failure (admitting error) is anathema (Table 6.2). While traditional management may accumulate knowledge in bits and pieces,

using this to make marginal adjustments in an incremental fashion, this is not what is meant by adaptive management. Rather, a much more active testing of management actions is undertaken through deliberate experimentation.[38] Where experiments on the ground become prohibitively expensive or logistically impossible, use of simulation modeling is substituted as a means for exploring the consequences of management practices.[38]

Adaptive management has many precursors, but its recent development is frequently attributed[38] to the work of Holling and his co-workers on "Adaptive Environmental Assessment and Management" (AEAM).[39] The design and planning phase of AEAM brings together "managers, researchers, policy people, and the concerned public from a variety of disciplines and responsibilities in the process of developing integrated management-research plans."[40] Responsibility for organizing the exercise rests with a small steering committee composed of representatives of the major players. The process itself involves a combination of workshops, literature reviews, and an institutional analysis (an examination of the social and organizational context within which policy proposals coming from the workshops would be implemented). The aim of the workshops is to develop a detailed understanding of the nature of the problem at hand, suitably expressed in a conceptual model or computer simulation of its basic technical and social elements. This exercise may take up to a week of intensive analysis by a selected group of 30 or so participants and involve a preliminary attempt to develop a conceptual overview followed by more specialized analyses in small work groups. Model building is one of the desired outcomes of this process that, in turn, allows exploration of alternative policy options by running simulations. Proponents of AEAM make it clear that the aim of the workshops is not model building per se but rather the development of communication and understanding among the participants. If a useful model is produced, then that is fine, but the major intent is to create an atmosphere in which constructive thinking can take place about the consequences of alternative policy options. Holling[39] suggests that the

TABLE 6.2. Differences between traditional and adaptive management.

	Traditional management	Adaptive management
Uncertainty	Uncertainty is rarely acknowledged explicitly; it is assumed that the policy is correct and it is not tested.	Uncertainty is explicitly recognized and the policy itself is "tested" by treating management as an experiment.
Link between science and management	Link absent: vague process for how science will be used to change policy	Direct link: science is used to directly inform policy and management
Management implemented as experiment	No	Yes
Implementation	Does not use controls and replications in the implementation of its management program	Uses controls and/or treatments in the implementation of its management program
Type of learning	Incremental: manages to maintain status quo, learning from failure unacceptable	Sudden shifts occur, status quo subject to change, learning from failure acceptable

From Halbert.[38] Reprinted with permission from "How adaptive is adaptive management? Implementing adaptive management in Washington State and British Columbia." *Reviews in Fisheries Science,* 1:261–283. Copyright CRC Press, Boca Raton, Fla, © 1993.

modeling exercise plays the ironical role of taking care of what he calls the "technical trivia," thereby releasing participants to explore their underlying assumptions in the process of examining policy options. An interesting feature of AEAM is the overt perception of policy development as a continuing learning exercise, one in which any policy changes that are implemented can serve as a source of experience through provision for monitoring and evaluation of the consequences of change.[41-43]

In principle, therefore, AEAM is a workshop-based exercise that attempts to enhance communication among conflicting interests through involvement in collaborative analyses and simulations. In other words, it attempts to establish an atmosphere conducive to the discussion of complex policy issues. Subsequent experience with AEAM, in a wide variety of settings, has left a mood of cautious optimism about its prospects as an integrating and conflict resolution process.[41,43-46] There are, however, a number of significant problems with this and other approaches to adaptive management.

Traditional environmental management is conducted by professional managers. Innovations like adaptive management may seek to widen involvement in the management process, but the bulk of participants are likely to remain technical professionals of one kind or another. It is not surprising, therefore, to find that the policies and plans that are generated may be biased toward technical concerns. For example, numerous recent examples of adaptive management are exercises in which interdisciplinary teams of managers and scientists develop technical management plans in the absence of significant public involvement.[47-49] Even adaptive management exercises that attempt to integrate wider social and political concerns are frequently subverted by the aversion of technical professionals to social and political matters.[50]

Adaptive management also relies upon the use of systems models to enhance the rate at which participants develop an understanding of the ecosystem under consideration.[43] However, as mentioned earlier, systems models have serious problems that limit their effectiveness in this regard. Scientists frequently disagree about the facts of the case and over the assumptions on which systems should be structured.[43] There is also a heavy emphasis on incorporating variables that are amenable to quantification, thereby downplaying nonquantifiable aspects of the problem situation. Under such circumstances, the more intangible psychosocial aspects of the situation, as well as indigenous/anecdotal knowledge, receive short shrift. As a result, the openness and free-flow of information that adaptive management is supposed to encourage is hindered. Thus, in their review of the use of AEAM in New Brunswick, McLain and Lee[43] (p. 444) concluded that there was a tendency on the part of modeling teams to adopt a narrow conception of who were the important decision makers and implementers of policy:

In New Brunswick, for example, the modeling team assumed that federal and provincial foresters and politicians were the key actors and marginalized representatives of the environmental movement from the model development process. ..The New Brunswick case also illustrates the danger of relying on one powerful stakeholder to handle monitoring and evaluation: the temptation to hide information unfavorable to the data keeper's interests is very strong.

Under these circumstances, it is no wonder that the question of trust becomes paramount. Adaptive management is conducted in local situations where players often know one another and share a history of conflict over attempts to manage the resource in question, with suspicion and distrust being the common result. In the case of pest management in New Brunswick (mentioned in the previous chapter), the scars from a decade of confrontation were brought to the adaptive management workshops. Sonntag et al.[40] (p. 32) observed that "lack of trust between interest groups is the major barrier to improving the ability of industry, government and public to manage the forests of New Brunswick," a state of affairs exacerbated by ideological differences over the direction of forest policy.[51] In New Brunswick, as elsewhere, one side is happy with the political-economic status quo, while the other wants radical change toward more participatory democracy. Managers believe that they already live in a participatory democracy and are perturbed when environmentalists claim otherwise, whereas the latter are puzzled by the animosity that arises when they pursue what they consider to be legitimate political and environmental reforms. Provincial managers and bureaucrats, for instance, consider that it is their mandate to carry out the will of the people as determined by their elected politicians

and expressed in the policies sanctioned by the provincial cabinet.[52] It follows that environmentalists would be seen as an unrepresentative minority who wish to usurp the established democratic process. As a result of these various difficulties, the open atmosphere that is so crucial to the adaptive management process is difficult to develop.

Turning to the implementation of adaptive management strategies, one finds considerable resistance to the notion that all plans should be treated as experiments. Hilborn et al.[53] argue that to understand the dynamics of systems and to better understand optimal levels of sustainable harvest, it may be necessary to exploit the resources beyond what is thought to be sustainable harvest levels. This, however, creates difficulties for managers[38] (p. 273):

One of the reasons why adaptive management may have some fundamental difficulties in becoming effectively implemented into public management programs is because the experimental nature of adaptive management implies the need to take "risk-prone" actions. Public agencies typically manage for some ill-defined equilibrium to maintain the status quo and can generally be characterized as risk-averse. This presents a major dilemma for managers who need to take risk-prone action in order to learn more about the system, when they are enmeshed in a risk-averse setting.

Agencies may also have difficulty in sustaining their commitment to experimental approach to management because of political pressures to resume traditional harvesting practices. A case in point is the attempt to adopt an experimental approach to management of ocean perch in British Columbia during the 1980s. "The program involved a 5-year period of specified overfishing in order to test assumptions about stock dynamics, biomass and productivity estimates, aging methods, and the values of population parameters. The original experimental design required that after the initial period of overharvesting, a period of no harvesting would follow in order to continue testing the assumptions about population dynamics and to allow the stocks to recover"[38] (p. 270). However, it was difficult to adhere to the second portion of the experiment because the fishing industry was reluctant to curtail harvesting.[38] Thus, it is doubtful whether proponents of adaptive management have either the technical means to achieve their Imperial ambitions or the political power to see their management plans carried to fruition.[4,38,54]

Collaborative Decision Making

The adaptive orientation at the core of ecosystems management is said to be a process that "combines democratic principles, scientific analysis, education, and institutional learning to manage resources sustainably in an environment of uncertainty"[55] (p. 676). In other words, ecosystem management has the avowed aim of combining the adaptive search for reliable scientific information with a broader, more collaborative way of using that information both in decision making and implementation. Unfortunately, the scientific pursuit of knowledge about ecosystems through experimentation is often accompanied by a preference for rational-comprehensive decision making. In contrast, nonscientific players in the drama may prefer other forms of rationality that lead to more sociopolitical or value-based decision making. Combining the various rationalities is, as we have seen, extremely difficult. While planners may operate at the ecosystem level using their newly acquired scientific understanding, implementation of programs at the local level is a much more fragmented process requiring extensive social interaction, negotiation, and consensus building.[46] There is an uneasy relationship, therefore, between need for technical information at the ecosystem level and the more local sociopolitical processes involved in implementation.

Advocates of ecosystem management recognize that it cannot be effective unless there is extensive collaboration among the many players and jurisdictions involved.[7,56] At the same time, many also recognize that traditional management structures do not encourage the degree of collaboration required by the ecosystems approach and argue for reform in current decision-making processes. Unfortunately, while ecosystem management has been formally championed by all the major natural resource agencies,[57] there has been little change

in the underlying institutional structures needed to implement it.[43] For instance, agencies continue to show "information flow pathologies," in the sense that information sharing both within and between agencies is frequently restricted.[58] In addition, the public are excluded from meaningful participation in the decision process through, for example, the preoccupation of agencies with technical rationality.[9] Indeed, rather than moving toward more open, participatory decision making, many natural resource agencies are shifting in the opposite direction toward a more centralized command and control system.[59]

There are few signs that advocates of ecosystem management are giving these problems with collaboration the attention they deserve. In this regard, Grumbine[9] concludes that the majority of biologists who write about ecosystem management pay little attention to the need for organizational change. Nor do the major natural resource agencies reviewed by Grumbine mention any substantive organizational restructuring. In addition, agency managers seem to underestimate the need for and consequences of a change in power relationships among the various players. Apparently, they are "unaware of the radical implications of creating (more equal) partnerships" with other players.[9] Thus, the problem of integrating public values into decision making remains problematic and the ideals of adaptive management unfulfilled.[60]

In conclusion, there are few examples of the successful implementation of adaptive ecosystem management.[38,43,61] In fact, a gaping void exists between current accomplishment and hopes for the large-scale control of the natural world.[18,62] This would come as no surprise to those who reject the technical rationality and rational-comprehensive planning at the root of such ambitions. They would dismiss ecosystem management as yet another example of anthropocentric, Imperial illusions of control. In mapping out ambitious visions of ecosystem control, advocates of ecosystem management seem to ignore the many examples of resource overexploitation that recur despite scientific advice to the contrary.[63] In the absence of a more critical sociopolitical consciousness, therefore, the notion of ecosystem management may simply be a rhetorical device used to promote preferred research agendas but which misses the fundamental point—the need to attend to the question of restraint in both resource consumption and human population growth.[63,64]

Arcadian Management

Reforms generated within traditional Imperial management try to move natural resource management away from the classical utilitarian paradigm toward a broader, more diverse set of goals and practices that promotes a holistic perspective on such matters as biodiversity, conservation, and sustainability. A typical example of the proposed Imperial reforms is that of "New Forestry" in which[65] (p. 193):

Classic regulation is abandoned; in its place are generally-stated concerns for biodiversity, ecological system complexity, aesthetics, protection of all indigenous species of flora and fauna . . . clean air and water, respect for those who find spiritual values in the forest, a re-emerging 'ethic of nature', holistic or systems-oriented approaches to management, coupled with older but not abandoned desires for commodity production, economic prudence and humanistic concern for rural communities disrupted by the vicissitudes of a rapidly-changing forestry praxis.

While these proposed reforms seek to move environmental management away from its traditional utilitarian concerns, in practice the shift is modest. New Forestry, and ecosystem management in general, remains anthropocentric and Imperial in nature. There is little mention, as we have seen, of sociopolitical change or the urgent need for restraint in consumption and human population growth. At the same time, ecosystem management retains the Imperial goal of the control and subjugation of nature. For instance, architecture has been proposed as a role model for New Forestry with silvicultural practices and the manipulation of forest stands being likened to an artist painting pictures.[65] Such images retain the older notion of instrumental control that, from an Arcadian perspective, has led to our

current problems. It follows that Arcadian reforms take a different, more radical route.

The Arcadian tradition adopts a more humble posture toward nature, one that emphasizes restraint, noninterference and living with, rather than dominating, nature. Because less control of natural processes is sought, the intent of management is to find ways of living with nature that enhance ecosystem health and integrity. Those who advocate such an orientation are aware that these goals of health and integrity have also been adopted by Imperial ecosystem managers but take some pains to distance themselves from the latter. For instance, a typical example of the Arcadian approach to environmental management is that of *ecoforestry*[66] (p. 38):

Ecoforestry decentralizes authority and preserves all forms of diversity. It must not be confused with sustainable development or ecosystem management. Both of these are based on translating the imperial, anthropocentric models into other terminology; they then use only palliative measures. They are based on the assumption that we must continue to control the forests and ecosystems and manage them for our benefit.

Drengson's[66] vision of ecoforestry is one of "dwellers in the land," individuals living in close contact with their home domain, carefully observing and learning the "inherent wisdom" of nature. An attitude of humility is deemed to be essential in trying to understand what the forest can supply us, much as indigenous peoples have done for millenia. The question asked by an ecoforester is not "how much can we take from this land and forest to maximize production and profits?" but rather "what must we leave . . . if we are not to interfere with its many functions and processes?"[66] (p. 37). Efforts are made to reduce disruption by "minimizing road construction, using low impact appropriate technologies, and by removing only trees which natural indicators reveal have been selected for removal by Nature"[66] (p. 38). This latter principle runs through many descriptions of ecoforestry that seeks to lessen human impacts by diversification of the species harvested, the idea being that the more diverse and value added are the products derived from an area, the less it has to be managed.[67,68]

At the stand level, the practice of ecoforestry is vernacular, that is, site-specific, intuitive, experiential.[69] It is assumed that close contact with the land allows the development of an intuitive understanding of how harvesting and silviculture should take place.The knowledge involved is tacit, making it difficult to transmit to others, except through observation of those adept in the art. Thus, beyond some general principles, there are no established prescriptions for ecoforestry because practice will differ from one place to another. It is not surprising, therefore, that Imperial forest managers, used to more explicit descriptions of management practices, commonly view ecoforestry as more like a religious movement, one that is emotionally rather than factually based.[70] However, this is a reaction that is typical of those who prize technical rationality and who tend to dismiss intuitive, lay knowledge as the product of subjective, undisciplined minds. This is unfortunate because local knowledge, formed after decades of experience in a specific location, has much to offer. Yet there are, as one might expect, several problems with ecoforestry and Arcadian management practices.

First, all professional practice involves some intuitive judgment, even among the most fervent advocates of technical rationality. However, there is no reason to believe that intuitive judgment is necessarily an adequate basis for action, given the many cognitive errors and emotional biases that can distort judgment. One of the strong points of positivist thinking is that there is an explicit attempt to subject intuitive thought to the rigors of further analysis. One of the major problems facing ecoforestry and other forms of Arcadian management, then, is whether intuitive understanding is a sufficient basis for sensible action.[70] Consider the case of the grassland ecologist who was convinced that the vigorous, fibrous grasses that flourish on the sandy soils of the southern Canadian prairies were able to do so because these soils retained moisture to a far greater degree than had hitherto been acknowledged. Noth-

ing would shake him from this belief even though it was quite eccentric in light of received wisdom that clay soils are most likely to retain moisture under drought conditions. Here was a case where intuitive insight needed exposure to rational analysis. As far as I can ascertain, however, a belief in ecoforestry and experiential management does not preclude an interest in technical forestry, nor does it imply an inability to think in rational terms. It might be more appropriate to say that the humble ambitions of Arcadian management are well served by the use of local knowledge that, nevertheless, needs to be treated with caution, even skepticism.

Second, ecoforestry (and other forms of Arcadian management) is physically and intellectually demanding. Selection harvesting using small machinery or horses is both inconvenient and hard. It requires the development of new skills, constant attention to the state of the forest, and difficult judgments about what should be taken and what should remain. It is so much easier to have a contractor clear-cut the area and invest the proceeds in a mutual fund. In addition, spending a lifetime in the woods, plagued by mosquitoes and blackflies, is not everyone's idea of a sensible use of one's time. To do so implies a love of the land and a desire to do what one can to sustain ecosystem integrity and health. The problem for advocates of ecoforestry is whether there are enough individuals with these beliefs to make a difference. Although advocates of ecoforestry clearly are committed to Arcadian management,[69] it is doubtful whether their enthusiasm will be shared by more than a small minority.

Finally, advocates of ecoforestry have yet to work out in detail how local control of the forest could be incorporated into the broader environmental decision-making process.[70] Thus, the ecoforestry literature contains little discussion of the necessary social and political changes, especially changes in land tenure, that would allow ecoforestry to be implemented.[70] As a result, the prospect for a wider adoption of ecoforestry, like so many Arcadian projects, is likely to remain an idealistic dream, an allegory used to counter the harsh reality of the Imperial world in which we live.

Conclusions about Environmental Visions

Any attempt to develop adaptive environmental problem solving must necessarily include a sophisticated understanding of the natural world. Despite their many problems, both the positivist knowledge of Imperial managers and the more local, experiential knowledge of Arcadian managers have a significant role to play in improving current methods. Some judicious integration of the two approaches to knowledge is needed. Yet even if the many obstacles to achieving this were overcome, managers would still be missing the point. Lack of knowledge about the natural world is not the limiting factor in coming to terms with the problems that confront us. More important are the psychosocial barriers that result in relevant knowledge being ignored or inappropriate ambitions being pursued. In other words, environmental knowledge is useful but needs to be supplemented by a more sophisticated understanding of the psychosocial world if environmental management is to be anything other than a way of "avoiding the negative externalities of the development process."[71]

Social Visions

Emphasis on the social roots of environmental problems is common among those who believe that environmental degradation and other problems stem from the way in which society is organized. Unlike ecosystem managers and ecoforesters, who are concerned primarily with the technical aspects of environmental management, the socially oriented reformer addresses the broader sociopolitical problems ignored by the technically minded. Social reformers, therefore, are concerned with changing social institutions and policies so as to better manage the interaction of human populations with their natural resources. At issue are such things as how best to control overconsumption and excessive reproduction, allo-

cate scarce resources, distribute wealth, and direct agriculture and industry along sustainable paths.[72] More specifically, sociopolitical approaches deal with the thorny problems of power and control: Who is to have the power to change the way we interact with nature, and what is the best way to organize how we satisfy our basic needs? It follows that the solution to environmental problems is thought to lie in social change rather than in environmental management or psychological awakening. While the latter domains are important, they are considered to be secondary to the enormous task of reshaping the structure of society.

As one might expect, opinions about the direction of social change differ according to ideology. For some, the major problem rests with the hierarchical structure of society in which elites dominate industrial production, gorging themselves at the expense of the great mass of people and to the detriment of the natural world. Others see the unrestrained habits of the common folk at the root of our dilemmas, especially the proclivity of the masses to reproduce themselves in huge numbers. Both viewpoints recognize the need for restraint but differ over who must show it and how it might be achieved. As we saw in chapter 3, preferences for different social choice mechanisms can be arranged along two important social dimensions: (1) the degree of central control and (2) an orientation toward economic growth versus restraint (Figure 3.1). Socially oriented reforms condemn the emphasis in western societies on growth and argue that no more than lip service is being paid to the alternative visions of either administered restraint or autarchy. It is these alternative viewpoints that socially oriented environmentalists promote as the only sensible solutions to environmental problems. However, ideological differences channel the pursuit of environmental and social restraint along radically different routes. For the more Imperial mind-set, the imposition of a powerful central authority with the capacity to coerce the recalcitrant is the preferred approach (administered restraint). In contrast, Arcadians are inclined to believe that radically decentralized, small-scale communities will not only heal our numerous social problems but also encourage a more benign attitude to the natural environment (autarchy). The contrast, then, is between authoritarian and democratic routes to social change.[72]

The Authoritarian Solution

Proponents of authoritarian solutions to environmental problems advocate varying levels of central control by powerful elites.[72,73] This is an old story dating back to Plato but more commonly associated with Thomas Hobbes, the seventeenth century English philosopher. Underlying his authoritarian proposals was a jaundiced view of human nature, one in which competitive, self-centered aggressiveness was seen as the natural response of most people to the struggle for material well-being. Hobbes also believed that even in the absence of material scarcity, men would still struggle to achieve power and eminence, thereby condemning society to perpetual turmoil.[74] Accordingly, the only way to avoid the political destructiveness of such human motivations was through the auspices of a despotic ruler capable of imposing order on chaos.

In the 1970s, circumstances conspired to revive these Hobbesian remedies in the face of growing alarm over environmental trends.[72] With the publication of the *Limits to Growth* study early in the decade,[75] the informed public became aware of the prospect of environmental decline on a global, catastrophic scale. The study's main thrust was that the confluence of overpopulation, pollution, depletion of renewable resources, and food shortages was pushing the earth beyond its carrying capacity for the human species. Widespread restraint, engineered by social and political change, was the only way to avoid catastrophe.[76] Although the study was subjected subsequently to intense criticism over its assumptions and methodology, it served to bring images of catastrophe into public awareness. With the onset of the OPEC oil crisis the following year, the stage was set for what Eckersley refers to as the survivalist school of environmental thought.[72]

In face of what was perceived to be immi-

nent environmental and social disaster, a number of eminent writers resurrected the remedies of Hobbes and other catastrophists. Variously labeled as eco-Hobbesians,[19] neo-Malthusians,[76] and even ecofascists,[77] their essential conclusion was that, no matter how regrettable, some form of authoritarian rule was most likely the only way in which the appetites of the masses could be restrained in the limited time available. None of these writers appears to prefer authoritarian solutions in principle, nor do they propose them as remedies for all aspects of the environmental crisis but, rather, offer them only when no other option seems to be available. For instance, one of the most prominent eco-Hobbesians is Garrett Hardin whose parable of the commons underscores the need for resolute measures to restrain resource overuse. Left to themselves, he suggests, the great mass of people would pursue self-interest in the exploitation of a common resource, while the few altruists (conservers) would suffer materially because of their self-sacrifice that would make little overall difference to conservation of the commons. Only by restricting access to the resource, or making it more costly, can some semblance of rational management be achieved.[78] Hardin suggests that this might be possible through mutual coercion, mutually agreed upon by the majority of those involved. In effect, an eco-social contract could be entered into by users of a particular resource whereby they would agree to restrain their individual use in pursuit of sustainability.[72] That is, Hardin recognizes the sociobiological principle that "actions that benefit a group will be selected for only if there is a pay-off for the individual who carries out the action" (p. 271). Thus, some degree of central coordination is implied when he suggests that[79] (p. 274):

In light of what sociobiology has taught us, we now see that Helvetius put us on the right track when he said that the 'art of the legislator consists of forcing men, by the sentiment of self-love' to do the right thing. The wise legislator writes laws that will, in fact, achieve the desired end by rewarding individuals for actions that benefit the group (of which individuals are members). This does not mean that each individual will find pleasure or profit in obeying every such law, but he should recognize that he is, in the long run, better off with such laws since they apply to all individuals.

Mutual coercion, therefore, implies a willingness to accept restraint imposed by some authority because it is seen as mutually beneficial. Where there is resistance to the legislation, then rewards are to be favored over punishments for "punishment often proves counterproductive because it motivates the subject to find a way of evading the punishment while doing the wrong thing"[79] (p. 271).

When one turns to the question of population control, Hardin's ideas have achieved widespread attention in the form of his "Lifeboat Ethic," a metaphor that depicts the affluent world in the lifeboat surrounded by a host of developing-world swimmers. Survival of those in the lifeboat (developed world) is only possible by limiting the number of swimmers allowed into the lifeboat. Thus, nothing is to be gained by letting the lifeboat be swamped by the wretched of the earth who, in turn, have to sort out their own problems. Such a triage strategy would require a strong central authority that can impose immigration controls on the developed world. Hardin recognizes that "the image disturbs many kind-hearted people, but its basic meaning—a limited universe within which practical decisions must be made—is correct, because a limited world is the only world we will ever experience"[79] (p. 280).

Fears about human survival disappeared from the public imagination in the late 1970s and early 1980s, calmed by the optimistic rhetoric of the Reagan-Bush years, only to reappear with greater intensity in the 1990s. In addition to earlier concerns with overconsumption and overpopulation, we are now faced with the uncertainties of global change. Not only are we required to struggle with environmental problems at home but are now informed that the inability of many countries to cope with the doubling of their populations will result in their environmental and sociopolitical disintegration that, in turn, will spread beyond their borders as mass migration threatens the security of other nations.[80-82]

Zimmerman[83] fears that one of the consequences of this global turmoil will be the rise of ecofascism. One of the basic premises of this latter ideology is the need for individuals' rights to give way to the collective good. In other words, individuals are to be submerged into an organic whole. It is only when such ideas become fused with nationalistic, xenophobic politics, however, that they take on fascist overtones. Zimmerman concludes that although the National Socialism of Nazi Germany was one such political movement, he sees no evidence of such ecofascism in current environmental politics in the United States, which is a blessing because "for peoples armed with modern weapons, a move in this direction would probably lead to unimaginable ecological and social disasters" (p. 232).

However, the promise of incisive action to resolve ecosocial problems remains an attractive image for those disillusioned by the inevitable confusion of democratic politics. No doubt this explains the continuing support for a more technocratic form of government within the scientific and managerial community.[84] Thus, the ecosystem managers discussed earlier, who are concerned with bringing large natural systems under managerial control, have their counterpart in the social systems managers and urban planners who believe that ecosocial problems are amenable to rational-comprehensive logic.[85] One can detect elements of this technocratic way of thinking among those environmental and social professionals who argue that solutions to our various problems are available, but the political will to implement them is missing. Similarly, the same professionals often argue that part of the solution to our many problems is to educate the public into more appropriate ways of thinking that, by coincidence, are not unlike the views held by the professionals themselves.

Examples of centrally planned economies and their consequences are to be found in the former Soviet Union and eastern Europe where both production and consumption of goods were centrally controlled. In all these countries, however, industrial production was geared toward economic growth rather than restraint, while consumption was constrained more by shortages than by any ecological rationality. As a result, environmental degradation was extensive, as we have subsequently discovered.[19] In the western democracies, as mentioned earlier, there are numerous elements of central control in the form of federal and state regulation of industrial production, but, as Redclift[71] notes, such regulations serve mainly to ameliorate the worst excesses of industrial production rather than changing the fundamental nature of that productive process. As a result, "examples of ecologically successful administered systems are less easily identified. The managers of public lands and regulators of pollution in the USA (and, for that matter, in other Western industrialized countries) have their accomplishments, but also some very severe shortcomings. Ecological values are not generally the prime concern of such agencies"[19] (p. 95).

Thus, there are some environmental thinkers and professionals who believe that the most effective way to cope with our ecosocial problems is to restructure our problem-solving efforts along more authoritarian, technocratic lines. This would require a more central role for bureaucratic problem solving and, most likely, the imposition of solutions by coercive measures. The question is, of course, to what extent this is likely to work. Perhaps the best way to answer this question is to focus on what Dryzek considers to be two key elements in administered systems: the promise of rational coordination and the effectiveness of coercion.[19]

The Promise of Rational Coordination

The technocratic, authoritarian option is attractive because it promises a coordination of effort in the face of the fragmented, decentralized complexity of modern society.[78] Successful bureaucratic problem solving, however, assumes that it is possible to achieve both consensus on goals and compliance in the implementation of policies at all levels in the bureaucratic system. As we saw in chapter 3, however, the real bureaucratic world does not conform to these functional ideals. For instance, policies are subverted at all stages of their im-

plementation as professionals interpret policy guidelines in terms of their personal viewpoints and professional interests. Innovative policies tend to disturb established structures of vested interest within and between agencies, and they are resisted accordingly. Similarly, any policy that falls outside standard operating procedures is likely to be resisted.[19] Innovative policies are unlikely to be generated in the first instance, however, because in the competitive bureaucratic world, primary attention is paid to protecting the agency's standing rather than taking risks with new, politically unpopular policies. At the same time, attempts to overcome the fragmentation caused by allocation of different aspects of problems to different agencies creates an unmanageable cognitive burden.[19] As a result, the dream of effective rational-comprehensive planning remains a figment of the technocratic imagination.

The Effectiveness of Coercion

There is no reason to believe that a technocratic regime would be benign. Ophuls[86] argues that it might be possible to have a "well-ordered and well-designed state" in which rule by a "natural aristocracy" was limited by constitutional means. At the same time, however, he recognizes that there remains the "exceedingly difficult problem of legislating the temperance and virtue needed for the ecological survival of a steady-state society without at the same time exalting the few over the many and subjecting individuals to the unwarranted exercise of power or to excessive conformity to some dogma" (p. 227). Unfortunately, the prospect for technocratic rule taking an authoritarian direction is strong given what Pacey[87] refers to as the "technocratic value system" underlying the technocratic worldview. This mind-set is "singlemindedly insistent on an unambiguous view of progress, of problem-solving, and of values . . ." (p. 127). What Pacey[87] is arguing is that the single-mindedness of technical experts, often in pursuit of technical virtuosity, coupled with a certain disregard for the social consequences of their innovations, results in a politically dangerous combination in which technical experts may "sometimes use their specialized knowledge in ways that make them hijackers of power, not its servants" (p. 128). As a consequence, technocracies could well become totalitarian states based on a system of organization foreshadowed in such endeavors as the Los Alamos project to produce the first atomic bomb.[88]

Even if some kind of totalitarian rule were to be established, however, the prospects for imposing solutions on environmental problems by coercive methods would not be strong. The great mass of people are willing to tolerate some degree of privation for short periods given the promise that things will get better, but the message that living standards are to remain stable or even get worse is unacceptable.[89] Attempts to enforce frugality in the consumption of resources and reproductive restraint through coercive means will work for a while but will meet increasing resistance as cultural norms reassert themselves.[90] Further, the record of state control of the corporate world is not encouraging for proponents of central planning. The history of forest exploitation in Canada is an illustration of how difficult it has been for provincial states to regulate the behavior of powerful industrialists and corporations.[91,92]

In conclusion, bureaucratic, or centrally planned forms of problem solving are useful in stable situations where there is little need for adaptation to rapid ecosocial change. Traditional natural resource agencies experience difficulty, however, in the face of wicked problems. There is no reason to believe that an even more imperious organization would fare any better. It is unfortunate, therefore, to find that one response to turbulent times has been an apparent increase in the incidence of command and control management strategies.[59]

The Democratic Solution

In stark contrast to authoritarian centralization and bureaucratic problem solving are reforms emanating from the socialist wing of the environmental movement. In what follows, I draw extensively on Eckersley's[72] characterization of these ecosocialist reformers.

Like all socialists, ecosocialists lay the blame for our environmental problems squarely on the shoulders of rapacious capitalism, which, in its obsession with endless expansion, ignores such externalities as environmental degradation. Thus, the capitalist class is seen as being indifferent both to the well-being of nature as well as to the working classes who provide the sinew of industrial production. There is, for instance, an extensive literature depicting the past 200 years of resource use as one of increasing exploitation by monopoly capitalism leading to resource degradation and continued rural poverty.[92-96] The traditional socialist response to this inequity is to seek social justice by resting control of the means of production from monopoly capital through augmenting state control and channeling a greater share of national wealth to the proletariat. In this scheme of things, increased industrial production is considered to be a means of creating greater wealth and alleviating the poverty of the working classes. Unfortunately for the socialist cause, the problems and excesses of state socialism as practiced in the Soviet bloc has not endeared traditional socialism to a wider public[72] (p. 136):

These problems include bureaucratic corruption and bribery; underemployment of labor; gross economic inefficiencies; one-party dictatorships; intolerance toward political dissent; widespread political intimidation and oppression; economic stagnation, and ecological devastation. In this new, post-Cold War era, ecosocialists must convince an increasingly skeptical public that it is possible to deliver economic planning that is at once democratic, ecologically responsible, coherent, and responsive to consumer demand.

In response to this unfortunate legacy, ecosocialists have taken a rather different tack. Instead of pursuing social justice through state-controlled, economic growth, they promote democratic (community-based) self-government while, at the same time, pursuing ecological restraint. It is this insistence on restraint that distinguishes them from social democrats with whom they share a number of characteristics. Thus, ecological sanity is thought to rest on the good sense of working people cooperating with one another in small-scale, self-reliant, democratic communities. Local self-reliance would entail disengagement from international capitalist systems, such as free-trade agreements, as well as avoiding any system that exploits third-world resources for first-world benefit.

The balance between social justice (Red) and ecological (Green) concerns among ecosocialists varies. Though both goals are thought to be important, too much preoccupation with environmental preservation issues is seen as distracting attention from the more fundamental sociopolitical origins of environmental degradation. The neo-Malthusian preoccupation with population control, for example, is thought to miss the point, which is that social inequalities and injustice are at the root of population problems and their resolution must take precedence over population-control programs. Thus, ecosocialism offers a mixture of resource conservation with human welfare concerns that assumes that the domination of nature is wrong because it leads invariably to the domination of people.[72] This social justice emphasis is viewed with some trepidation by other radical environmentalists, however, who believe that traditional ecocentric goals are becoming submerged in a growing anthropocentric concern with social justice.[97] In other words, Red ecosocialism is starting to dominate Green ecosocialism. As with all radical groups who are trying to develop alternatives to the status quo, there is little unanimity on how the ecosocialist agenda might be implemented.[72] Instead, one sees a great variety of efforts to foster local autonomy within the context of State regulation through community management projects. Because these endeavors are promoted by ecosocialists as problem-solving strategies that will promote social justice and ecological restraint, it will be useful to examine some examples in more detail.

In its ideal form, *community forestry* can be defined as "forest management conducted by local people, to benefit local people. It is characterized by local control in decision-making and fosters the economic independence of the community"[98] (p. 16). Social justice concerns are sought through community empower-ment and participation in decision making in the context of community economic development, while ecological restraint is pursued through sustainable, ecologically sensitive forest practices. Ideally, the community in question would own the forest to be managed and have the power to establish "local control in order to produce and retain local benefits."[99] A variety of organizational structures can be used to manage community forests (and fisheries), but some form of local council, composed of trusted local citizens, seems to be the most popular.[100] The implementation of community-based management, however, faces a number of formidable practical barriers.

It is, for instance, just as difficult to delimit a human community as it is an ecological community or ecosystem. If a community is to manage a forest or fisheries resource, it is important to decide who is to be included in it and how decision-making power is to be distributed. One way of resolving this problem in Canada has been to recognize municipalities as the responsible group.[70] Unfortunately, this doesn't solve too many problems because of the instability of many resource-based communities. As Marchak[70] points out, community management may be difficult where there is limited social cohesion:

To create a community project, there must be a socially cohesive community . . . Since many of the forest-dependent towns now are artificially created locations for a resident labour force to large companies, they are not, as a matter of course, socially cohesive. Transience rates are high if the economy is sufficiently buoyant to sustain job changes. When it is in a downturn, these towns experience various other signs of social tensions, such as domestic violence, high suicide rates, alcoholism, and family break-ups.

Community forestry is possible on a variety of land tenures, both private and public. However, the ideal is a situation where the community in question owns clear title to the land, rather than some limited access to harvesting rights or some other limited bundle of rights.[70] Unfortunately, few municipalities in Canada, or elsewhere, own any significant area of land that they could put under community management. Prospects for community forestry, therefore, require some considerable changes in land-tenure arrangements so that municipalities have access to land.[70] At present, forest lands in Canada are owned either by the Crown (the state) or by private individuals and corporations. Crown land is usually leased to major corporations for lengthy periods, a tenure arrangement on which they base their long-term planning and which they protect with considerable determination. One would not expect, therefore, to see substantial proportions of Crown land allocated to municipalities any time in the near future. When one turns to private land, it is difficult to see why private owners would put there land under community management, although there are some efforts to do so in the form of land banks.[101] One innovative solution to some of these tendentious land-tenure battles is to adopt forest zoning regulations that divide Crown land into industrial, wilderness, and community forests.[70,102,103] However, this change requires a substantial reduction in the land allocated to industrial forestry and, as a result, would meet formidable political resistance.

The management of a community forest by a forest council or board representing all the varied interests in the community may seem appropriate in theory, but in practice it runs into a number of problems. Most notable is the lack of real decision-making power by those who attempt to manage community resources. Numerous instances are available of community endeavors that founder because of the limited power of local managers to implement their plans in the face of government or corporate control of the resource.[104,105] As a result, many community-based management projects are not really self-management but comanagement endeavors.[106,107] The former refers to genuine local control of a resource in the absence of external pressures from government or the corporate world. As such, pure forms of community self-management of resources are most often seen in geographically isolated areas where it has been possible to avoid or resist the intrusion of the modern industrial world. Given the pervasive reach of the industrial world, most instances of community management are comanagement arrangements in which community efforts coexist with state control, while retaining varying degrees of autonomy.[106] The main problem for any group interested in developing community-based management in the developed world is that they must try to find some niche within a system of resource exploitation organized to facilitate large-scale production under the control of large corporations. Attempts to alleviate this situation requires a sophisticated understanding of power and social relations in rural communities. The expression of power is not a simple matter of a powerful group imposing their will on a powerless mass but, rather, a much more complex web of rights and obligations in which the powerful operate with the (at times begrudging) consent of the masses.[104] To break out of these established ways of thinking and behaving, therefore, is extremely difficult, especially if there are economic and social penalties for doing so. In the province of New Brunswick, fear of the latter is thought by some commentators to act as a barrier to the development of community forestry[98] (p. 23):

There also appears to be a fear on the part of provincial government decision makers to change from the present 'efficient' system (and) move toward a concept that remains relatively untested in the province. Other barriers exist which have more to do with the political culture of New Brunswick than the government itself. After many years of corporate domination in the forestry sector, some communities appear to be reluctant to challenge the present system and push for community power. As the forest industry is often the only industry in many towns, people are frightened that vocal opposition could mean being laid off, or reduced job opportunities for family members. Critics of the forest industry in local communities have become social outcasts as well.

Even if there were to be changes in political arrangements that allowed local councils to emerge, the social justice aims of ecosocialists would only be achieved if there is genuine participation by all interests in subsequent decision making. Unfortunately, this is not always the case,[108] for their are times when local leaders begin controlling the forest in their own interest by allowing access only to their political supporters[104] (p. 89). In addition, private or community ownership of a resource does not necessarily ensure wise use of that resource, as we have seen in the overharvesting of private woodlots in New Brunswick in response to an increase in the price of pulpwood.[109] However, one of the potential advantages of local management is that it allows closer monitoring and policing of the resource than possible by professionals based at some distance from the community. It is possible, therefore, to restrain abuses by individuals. Pinkerton[106] mentions the way in which "harbor gangs" control access to and police any abuse of the lobster fishery in parts of Maine. Anyone who defies local custom and intrudes on another's rights may find his or her gear destroyed, or worse. The point is not that one would want to encourage such summary justice but that close monitoring is possible. Offshore monitoring is less feasible for local fishers, however, which suggests an important role for the state.

In conclusion, there are few examples of community management in Canada, although interest seems to be growing in the forest, fisheries, and agricultural communities.[70,98,100,110] Yet despite some successes in community control,[105] the ecosocialist vision remains unfulfilled.

Conclusions about Social Visions

The central theme propounded by social reformers, that an adaptive response to ecological problems requires a fundamental change in the nature of society, is basically sound. Both authoritarian and democratic reformers underscore the point that it would be unwise to muddle along with conventional problem-solving systems. Yet, both sets of reforms make dubious assumptions about human nature. Authoritarian reforms are predicated on the assumption that an elite will be able to provide solutions that would be successfully implemented by an efficient bureaucracy in the face of popular resistance. There is little provision for enlisting the support of those most affected by policies. In contrast, democratic reformers assume widespread common sense and de-

cency that would come to the fore under different political arrangements. Yet, it is hard to see how the teeming urban masses of modern cities would willingly adopt such frugal, ecologically sophisticated lifestyles. Perhaps the most glaring insufficiency among social reformers, however, is the absence of any detailed program for managing the environment. Effective management practices are assumed to follow social change, but what these practices might be remains to be determined. In fairness to some social reformers, however, those ecosocialists who are engaged in community management are certainly trying to combine social and ecological innovations.

Psychological Reform

Psychologically inclined reformers of current problem-solving systems argue that adaptive problem solving can only be achieved if we direct our attention to individual human beings and their psychological processes. While it is acknowledged that we must attend to the social and ecological context of problem solving, psychological reformers maintain that even the most inventive of social and ecological plans will flounder unless those involved are willing and able to carry them out. In other words, the human face of environmental management and planning requires more than the token attention that it receives today.

The two environmental ideologies, because of their differing stances on human nature and its vicissitudes, offer contrasting perspectives on a psychological approach to solving environmental problems. As we have seen, the Imperial approach to problem solving is one of control by political-corporate elites aided by scientific and managerial experts. Because the implementation of problem solutions requires compliance by the masses, the latter being perceived as both irrational and inclined to resist elite control, psychological experts working within the Imperial system propose various forms of behavioral engineering to shape the behavior of the public. From the vantage point of behavioral engineers, the use of psychological techniques to manipulate the behavior of others is considered to be a justifiable response to the needs of legitimate authorities. Thus, behavioral engineers do not question the political status quo but seek to improve its efficiency.

The Arcadian approach is based on a different, essentially romantic, view of human nature. As we have seen, human beings are considered to be basically decent and cooperative. Beneath the frenetic, unwholesome search for material wealth, it is thought, lies a more humane individual waiting to emerge. This can only be achieved by creating the right psychological context in which people can shed the behaviors brought on by a seriously dysfunctional Imperial culture. Psychologists of the Arcadian persuasion, therefore, advocate self-development rather than behavioral change engineered by some self-appointed elite. Their task is the creation of conditions under which people may find their true self, which is assumed to be prosocial and proenvironmental. This self-discovery will reveal to the individual their natural connection with other living things in the form of an ecological unconscious, a kind of species memory waiting to be reawakened. At the very least, it involves a form of consciousness raising in which we begin to understand ourselves in relation to the world around us. The assumption is that by looking inward at ourselves we will find something we like and that this awareness will cause us to want to behave in a more compassionate way toward other people and nature.

The two views on how best to encourage proenvironmental change in people's behavior are the product of two conflicting schools of thought in psychology: behaviorism and humanism. From the perspective of the orthodox behaviorist, overt, observable behavior should provide the basic data for scientific psychology. Because, it is claimed, the inner workings of the human mind cannot be observed or measured with any certainty, then attention should be focused on understanding how overt behavior is influenced by external stimuli. Most important is determining how such stimuli might trigger specific behaviors and which contingencies, or consequences, sustain the behavior over time. Behavioral science,

therefore, strips human behavior of what some see as our most human qualities, matters of thought, feeling, and motive. The role of the latter in understanding human behavior is replaced by a mechanistic view of life in which people are perceived as reacting mechanically, rather than thoughtfully, to external forces. It follows, from this mind-set, that behavior change would be seen as a matter of manipulating the contingencies, the external forces that control behavior. Whatever is happening inside the person is essentially of little significance to this process. Any thoughts, feelings, or resentments about being manipulated are of little consequence if, through appropriate rewards and punishments, the desired behavior can be promoted. Indeed, a common assumption of behaviorists is that by persuading someone to engage in a desired behavior (such as recycling), there may be a concomitant change in the way the person thinks about the matter, and their attitudes may undergo modification. Thus, behavior change comes before attitude change.

Humanists take the opposite position, believing that only through a fundamental internal change can there be any significant, and lasting, shift in a person's behavior. In other words, only through transformation in the way a person thinks, feels, and strives can there be any meaningful and lasting improvement in environmental behavior. These internal processes are at the root of what is referred to as attitudes, beliefs, and values. It follows that humanists, unlike orthodox behaviorists, believe that it is possible to study these internal processes and that, rather than being some futile exercise in introspection, self-reflection may actually result in a change in behavior. Therefore, attitude change must necessarily come before behavior change. Behaviorists and humanists also differ over the question of ethics. A strategy that involves manipulation by an elite to further ends determined by that elite is considered to be fundamentally unethical by many humanists who reject behavioral engineering out of hand. To humanists, behavioral engineering is simply a form of dehumanized psychology that reduces human beings to the level of cogs in a machine to be manipulated at the will of political and corporate elites. Thus, reductionist (behavioral) science is thought to have been encouraged because it "ultimately serves specific social and political ends, the most important being the maintenance of existing structural arrangements"[111] (p. 114). This is both undignified and dangerous, humanists argue, because it may lend itself to authoritarian forms of government. As, Kipnis[112] points out, the armory of behavioral techniques available to modern psychology is so powerful that exponents of these methods may drift toward authoritarian attitudes, denigrating the target of their behavioral manipulations while resisting democratic controls over their own behavior. This is not to say that humanists do not have their own ethical dilemmas. They too can become involved in manipulation in, perhaps, a more subtle way. Although the claim of humanists is to encourage personal development through creating the necessary conditions for self-reflection and personal growth, their assumption is that this will lead to prosocial and proenvironmental behavior. They would not be too happy if their efforts to release a person's true self resulted in the production of an overtly psychopathic, rapacious corporate executive. The promotion of Arcadian goals through more subtle means is a form of manipulation regardless of protestations to the contrary. Thus, any attempt to encourage change involves the adoption of a value position about both the means to effect change and the goals that one is pursuing. To assess these contentions further, we need to look at the two orientations in more detail.

Behavioral Engineering

Orthodox behavioral engineering uses rewards and punishments to shape human behavior by manipulating the consequences that follow specific behaviors. Behavior that is desired by the engineer is rewarded, whereas undesirable behavior is punished. The assumption is that some judicious combination of rewards and punishments, the carrot and the stick, will encourage the recalcitrant to improve their behavior, that is, do what the en-

gineer wants. It is well known that rewards are more effective than punishments in promoting the desired behavior, but both are used in behavioral interventions.[113] As in all engineering methods, behavioral interventions work best when one is dealing with circumscribed, easily identifiable behaviors. Most of the research on behavioral engineering in the environmental area has, for example, been concerned with littering and pollution, energy conservation, and recycling,[114] where one can easily target the behavior in question and identify alternative strategies. When one turns to more complex behaviors, however, an engineering approach begins to falter. For instance, the complexities of policy making and environmental conflict resolution are less easily deconstructed into manageable chunks of behavior that can be shaped by rewards and punishments because there is persistent disagreement over what the most desirable behavior actually is.

Behavioral engineers have been very inventive in devising rewards and punishments. Among the most effective "positive motivations" are monetary rewards, social support, and the removal of barriers to the desired behavior.[113-115] Thus, in trying to encourage conservation behavior, one sees the following: cash refunds on the return of containers; bonuses for reducing residential energy and water use; tax credits for home insulation; free bus tickets; recycling contests and lotteries; educational credits for the first child, followed by punishment for additional children; rewards for taking marginal land out of production; and so on.[114,116] While the use of monetary and other rewards seems to encourage the desired behavior,[117] one of the technical problems with behavioral techniques is that a continuously rewarded behavior tends to taper off unless the reward is substantial and meaningful.

Nonmonetary rewards, such as social reinforcement, also play a significant role in behavioral interventions.[114] The recognition by one's friends and neighbors of meritorious behavior is frequently a source of satisfaction and a strong motivation to behave properly. Good citizen awards, commendations for bravery, and recognition of charitable work, as one sees in the Queen's Honors List in Britain, all contribute to the reinforcement of socially (and environmentally) desirable behavior, at least for those individuals who find these awards and social recognitions meaningful. At a more mundane but no less important level, the group support and encouragement that underlie the success of such organizations as Alcoholics Anonymous and Weight Watchers are examples of the impact of social reinforcers.

Turning to the use of coercion in behavioral engineering, although punishment can shape the behavior of rats in a psychological laboratory and be effective with small children or the severely mentally challenged, it is less effective in the world of human adults. Adults are inclined to resent punishment,[113] becoming angry at the punisher (the behavioral engineer or the class of society whom he or she represents), and seek ways of avoiding the punishment. Nevertheless, ruling elites have used punishment as a way of controlling the masses since the beginning of time. Today, our legal, educational, and religious systems all punish the wayward in one form or another. There is no doubt that punishment, or the threat of punishment, is effective in containing excesses among *some* adults. For instance, introverted personalities are thought to be more responsive to the threat of punishment than are extraverted individuals.[24] In contrast, the extremely extraverted individual seems to be impervious to such threats and will persist with antisocial and antienvironmental behavior even after repeated doses of punishment. Because extremes of extraversion have been linked with criminal behavior,[118] one can see why the use of punishment by itself may not be an effective basis for behavioral intervention in community-change programs. However, punishment and coercion of various kinds are a part of behavior shaping in the modern world and will continue to be a central part of behavior control. Thus, a variety of monetary punishments or disincentives are used to coerce people into desirable behaviors.[113,119] For example, one sees consumption taxes (to reduce resource consumption); the withdrawal of educational and other credits

(to control family size); mandatory fines (for transgression of pollution and emissions limits); fines (for importation and sale of endangered species); and so on. In addition, social pressure to behave in an acceptable, proper way is an ancient and effective behavioral tool, especially in small, tight-knit communities. Social disapproval and ostracism are powerful behavior shapers in villages, suburban neighborhoods, workgroups, and basically anywhere that there is some social permanence and cohesion. It is less effective in the urban environment, however, where relationships are more transient.

For a variety of reasons, interest in orthodox behavioral interventions peaked in the 1970s and decreased subsequently.[114] First, behavioral engineering takes a reductive approach to environmental problems, focusing attention on changing some specific individual behavior, thereby ignoring the broader sociopolitical context that may impose that behavior on the individual. Thus, behavioral engineering places the blame for environmental problems on individuals and seeks to shape individual behavior, rather than attending to the role of social forces, not only in causing problems but also in remedying them.[120,121] This bias is unacceptable to those who take a broader sociopolitical perspective on change.[122] Second, behavioral engineering has been criticized for its inability to affect anything other than changes in marginal, trivial behaviors. As mentioned earlier, most research on behavioral control has focused on such matters as individual littering, recycling, and domestic energy conservation.[114] Though this is useful, it is a far cry from changing the pervasive ideological beliefs and political institutions of Imperial society and the material greed that fuels our assault on the environment. Third, behavioral changes are ephemeral, they taper off unless rewards and punishments are intensified as people become satiated by them.[113,114] Unfortunately, systems of rewards can be financially very expensive and difficult to sustain over the long periods required for effective behavior control. Finally, there is the thorny issue of who controls the controller. The basic assumptions in behavioral engineering are that controllers and engineers do not need to change their own behavior, they know what is best for the masses, and it is legitimate for them to try to change the behavior of others. All of this is dubious in the extreme, of course, and has generated a lively debate over the role of professionals in this kind of behavior control. Oskamp,[123] for instance, recognizes the role that social scientists can play in resolving environmental problems but suggests caution in making "too simple or speedy prescriptions for public policies." Others worry about expert manipulation of the public and argue for safeguards against excesses.[112,119] As a consequence, there has been a shift away from orthodox behaviorism toward a greater acceptance that one cannot treat people as mindless machines to be manipulated at will but must attend to their inner thoughts, feelings, and motives. The result of these various problems with orthodox behavioral engineering has been the rise to prominence of *cognitive-behavioral* approaches that pay greater attention to the antecedent aspects of human behavior, that is, to the characteristics of the person that precede the behavior in question.

Cognitive-behavioral engineering seeks to influence behavior by using information and persuasion. Some techniques appeal primarily to the cognitive aspects of the person, others attempt to manipulate emotions and motivations. Thus, this is the world of education, persuasion, propaganda, and advertising. There is less talk of behavioral engineering and more of intervention and behavior change. Nonetheless, the essential strategy is one in which attempts are made by elites and their proxies to manipulate and control the behavior of others. Cognitive behaviorists maintain the same scientific detachment from their targets that characterizes their orthodox colleagues, but instead of being preoccupied with stimulus-response relationships, they seek to describe, and then change, the cognitive mechanisms lurking in the black box of the human mind. Through mapping the *cognitive-affective structures* of the mind (attitudes, beliefs, and values) and determining how these influence prosocial and proenvironmental behavior, it is then possible to intervene in these psychological structures by using appropriate behavioral technology.

Unfortunately, this is far from being a simple task. One of the most intractable problems with this kind of strategy is that there is often a tenuous link between what we think or feel and how we behave. A person may have very strong feelings of concern about some aspect of the environment but may remain inactive because of inhibiting personal or sociopolitical reasons. For example, a person may lack the knowledge and skills needed to engage in environmental action or, perhaps, fear that overt action may result in retribution from "them," the powerful people who run things. This is another way of saying that knowledge of the beliefs and values held by a person doesn't necessarily mean that you can predict with certainty how they are going to behave. Behavior is influenced by too many other factors to lend itself to easy prediction. Mapping the structure of a person's mind, therefore, is not only difficult but, even if it were to be accomplished, provides information on only one of the many factors that influence behavior.

Not unexpectedly, therefore, research on the personal characteristics that underlie proenvironmental behavior offers mixed messages. On the one hand, there are those who find that no single or easily identifiable set of personal dispositions (attitudes, beliefs, and values) can account for such proenvironmental behaviors[124,125] as reducing personal consumption of resources, green shopping, and encouraging others to be more environmentally sensitive.[126] The practical implications of these findings is that cognitive-behavioral engineers are unlikely to find some simple relation between personal disposition and proenvironmental behavior that they can manipulate with confidence in the outcome. Instead, they will have to undertake extensive research to identify the specific precursors of the behavior in question before they embark on any behavioral intervention.[126] The implication is that there are severe practical limits on what might be accomplished by cognitive-behavioral engineering given the constraints imposed by time and money. In contrast, there are those whose research leads them to believe that it is possible to identify a few, broad sets of personal dispositions that underlie a wide variety of proenvironmental behaviors.[127,128] For instance, Stern and Dietz[129] and Stern et al.[130] have found that environmental concern is related to egoistic, social-altruistic, and biospheric (ecocentric) values and that biospheric and altruistic values, in particular, predict a willingness to engage in proenvironmental action. However, their research does not tell us whether people will actually follow through on their good intentions.

The two opposing viewpoints present cognitive-behavioral engineers with a dilemma. The first view, that there is no single set of dispositions underlying proenvironmental behavior, implies that interventions by experts would have to be limited, targetting specific proenvironmental actions in a piecemeal manner. On the other hand, those who believe that underlying values promote environmental activism can be more ambitious in their intervention programs, reasonably confident that their attempts to influence proenvironmental beliefs and values would have a subsequent impact on a broad array of environmental intentions, if not actions. An interesting compromise between the two schools of thought has been proposed by Geller,[116] who argues that active caring is a result of the interaction of five *person states*, or feelings of self-effectiveness, personal control, optimism, self-esteem, belongingness, and empowerment. The extent to which you are willing to *actively* care about the environment (empowerment) is influenced by your feelings of self-worth (self-esteem) and your optimism about being able to have an effect on the world around you (personal control and self-effectiveness) as well as the degree to which you belong to some relevant group (belongingness). Each of these person states can be targetted by behavioral engineers with a view to creating a strong predisposition toward proenvironmental behavior. For instance[116] (p. 192):

One's self esteem . . . can be influenced dramatically by certain communication strategies, reinforcement and punishment contingencies, leadership styles and opportunities to learn. In a corporate culture, a sense of belonging can be facilitated among employees by increasing team-building discussions, group goal setting and feedback, group celebrations for achievements, and the use of self-managed (or self-directed) work teams. Community activities, ranging from picnics, sports events, church bazaars, block parties, and club meetings, increase an individual's perception of belonging to a community or neighborhood and consequently his or her propensity to actively care on behalf of the shared environment.

More generally, the techniques used by cognitive-behavioral engineers emphasize information and persuasion rather than rewards and punishment, although the latter may still be used as part of an intervention package.[113,114] Provision of information about environmental problems may remove some of the barriers to action by helping in the development of the skills needed to change one's own or other people's behavior. In their most simple form, these informational strategies can take the form of *behavioral prompts,* which seek to draw attention to the desired behavior. They are the attention grabbers that remind us to switch off lights, put out camp fires, reduce speed on highways, and so on. In addition, direct feedback about a particular behavior provides information that might give you pause for thought and stimulate a proenvironmental change. The most common of these is information provided regularly to homeowners on residential energy and water consumption that allows people to make informed changes in their behavior (if they are motivated to do so).

A somewhat more complex strategy, one that involves both cognition and motivation, uses commitment and modeling to encourage the desired behavior. Attention to motivational issues is crucial because information by itself seems to have little effect on behavior change.[131] It has been observed that both public commitment and social involvement act as strong motivations for subsequent action. Commitment, for example, can take the form of a sworn pledge to do better by committing oneself, in writing, and often as a public avowal, to follow a more environmentally or socially acceptable form of behavior. As mentioned earlier, Alcoholics Anonymous and Weight Watchers use the strategy when, for instance, a goal is set for weight reduction and one asserts a commitment to it in front of the support group. Modeling also provides information on the desirable behavior as well as tugging at one's emotions and motives.[113] A model farm or forest, or a green neighbor, can provide the ideal toward which one can work, as well as the information and skills needed to achieve that goal.

All of these simple behavioral interventions are relatively noncontroversial but tend to suffer the same fate as many other behavioral strategies, namely, that the public become habituated to these various prompts and admonitions and, after a while, ignore them. At the same time, commitments begin to waver as social contacts are lost or dissipate. A more lasting alternative is to try to change the person's underlying beliefs and values. It is here that we enter the controversial areas of public relations, persuasion, and propaganda through the use of the mass media and various forms of education. This is a vast area, to which I cannot do justice here. Instead, I shall focus on one of the basic problems that permeates environmental education and that is reflected in all behavioral engineering.

There are two approaches to environmental education: Imperial and Arcadian. As one might expect, the Imperial approach is based on the assumption that reforms should not change the sociopolitical status quo to any significant extent and has the behavioral goals of shaping human behavior. In the research that supports Imperial environmental education, "teachers and pupils are seen as manipulable by the researchers. It is considered proper to apply 'behavioural intervention strategies' and to 'manipulate situational factors in order to produce desired behavioural changes . . . Thus, positivist research in environmental education has a strongly deterministic character"[122] (p. 6). More significantly, however, Imperial environmental education emphasizes the role of the individual in both causing environmental problems as well as being responsible for their solution. This is consistent with the pride of place given to individualism in the dominant Imperial ideology.[122] One can observe this way of thinking in environmental controversies in which political and corporate establishments emphasize lifestyle (i.e., individual behavior)

causes of disease, whereas environmentalists and others place the blame on industrial pollution and other social conditions. Likewise, the recent plethora of "green consumer guides" and "personal action guides" about how to save the earth promote the idea of individual responsibility, while ignoring it seems the inexorable pressure from the marketplace for us to consume.[122] As Robottom and Hart[122] (p. 8) conclude, the behavioral research orientation that informs the dominant approach to environmental education assumes that:

The way to improve environmental education is to identify and control the variables that are thought to shape . . . the individual's environmental behavior, yet it rarely takes into account the historical, social and political context within which the environmental acts of individuals and groups have meaning . . . it tends to create a sense of individual agency and responsibility that is unrealistic in light of a range of sociopolitical constraints in the community; and . . . it misrepresents the nature of environmental issues by emphasizing individual human agency as the key factor in issue resolution."

Thus the Imperial version of environmental education, like other approaches to behavior change, overemphasizes the possibility of individual choice while downplaying the enormous pressure to conform to the demands of the consumer society in which we are all embedded.

Self-development

Unlike the Imperial approach to psychological change, which tends to assume the worst about human nature, Arcadian (humanistic) psychology follows the Romantic tradition, perceiving innate goodness and sense among the common folk. Where intractability and irrationality occurs, it is thought to reflect people's reaction to the dysfunctional society in which we are forced to live. If things are to change for the better, this widespread psychological damage will have to be undone, so that people can start to rebuild a more sane society. This kind of Arcadian thinking is prominent among deep ecologists and ecopsychologists, many of whom draw their inspiration from the work of Freud and Jung on the vagaries of human personality. The aim of both of these eminent scholars was to engage their patients in an intensive self-analysis that it was assumed, would release them from the stultifying grip of their neuroses. Ecopsychologists are attracted to this kind of therapeutic model in their pursuit of self-development and personality change. Various forms of ecotherapy and eco-learning are proposed, therefore, in an attempt to encourage greater identification with non-human nature. In so doing, it is hoped that an ecological self will develop out of the more constricted, damaged self that society has imposed upon us. This ecological self is seen as the primary requirement for adaptive change, the elemental building block upon which a sane society can be constructed. Without a shift toward this broader sense of self, the prospects for any fundamental change in the way we interact with the natural world is considered to be remote.[132,133]

The notion of *self* is used in psychology to refer to how people define what they are—how they perceive themselves. Most important in this regard is with whom, and what, a person identifies, both socially and environmentally, because the less identification there is with others, the less one is likely to be concerned about their fate. The Arcadian viewpoint is that, because of the egocentric, competitive nature of the Imperial society in which we live, there is little incentive to extend our sense of identity to others beyond our immediate family or to the nonhuman world. An *ecological self*, on the other hand, is one that broadens the sense of who we are by extending identification to include these missing elements: strangers and other creatures. The goal of ecopsychologists is to promote such a broader concern, not out of a sense of duty, but because we come to recognize that they are part of us (our self) and we are part of them. As Joanna Macy observes[134] (p. 79), this development of an ecological self:

is essential to our survival at this point in history precisely because it can serve in lieu of morality and because moralizing is ineffective. Sermons seldom hinder us from pursuing our self-interest, so we need to be a little more enlightened about what our

self-interest is. It would not occur to me, for example, to exhort you to refrain from cutting off your leg. That wouldn't occur to me or to you, because your leg is part of you. Well, so are the trees in the Amazon Basin; they are our external lungs. We are just beginning to wake up to that. We are gradually discovering that we *are* our world.

Development of an ecological self, therefore, involves a change from being independent and self-centered into someone who relates to others in terms of empathy and understanding, rather than in terms of egocentric goals.[132] Although this change in social identity is a step in the right direction, the development of an ecological self requires that these same sensitivities be extended to the nonhuman world.[132] Simple things like developing a sense of place, of belonging to a geographical region, are thought to be useful starting points for this identification process.[135]

For some ecopsychologists, such as Roszak,[136] there exists in all human beings an *ecological unconscious*, a sort of repository of innate wisdom about the reciprocity between humans and nature. It can be seen in its most pure form in the newborn's sense of an enchanted world and in the animistic quality of children's experience.[135] It is this ecological unconscious that needs to be released from the constraints imposed on it by modern society through such emancipatory activities as experiential learning and ecotherapy.[135] In developing methods to encourage this psychological transformation, ecopsychologists often draw on the work of Carl Jung and his followers who incorporate religious myths and symbols into their therapeutic work, thereby introducing a religious and spiritual element into psychology. This kind of thinking is common among those who see the need for a religious or spiritual conversion if we are to shift from the competitive materialism that afflicts us. In pursuit of this conversion, they advocate a form of self-development that stresses, among other things, exposure to the kind of spiritual experiences found in Zen Buddhism and Tao, as well as in the rituals of nature worship.[132] Thus, Thomashow[137] describes a form of meditation that he has found useful in helping students develop a heightened sense of self in natural settings. In the early fall, he hikes with his students up into the hills of New Hampshire where they are surrounded by the glories of the changing season. At the top of Gap Mountain, sitting in a circle, he proceeds to:

introduce what I call a *sense-of-place meditation,* guiding the students through a series of observations that allow them to focus on their senses in relation to the landscape. Feeling the air as it moves through our bodies, we contemplate the prevailing weather system. Listening to the sounds of the insects and birds, we become acquainted with the animal species. There are many variations on this theme. The point is simply to cultivate an awareness of ourselves in this wild place, to slow down for awhile and cherish the surroundings. Even those who are uncomfortable with the idea of meditation come to appreciate the experience (p. 15).

It is assumed that this and other similar exercises will have the cumulative effect of helping a person develop greater sensitivity to themselves in relation to the natural world. This sensitivity would then help in the development of a sense of belonging, of being part of the greater whole. In turn, such feelings might produce a sense of compassion for the people and things with whom one shares the biosphere. Unfortunately, the links in this chain of events are tenuous, to say the least, and extremely difficult to validate. To begin with, it is difficult to be precise about the goals of such spiritual education. The ambition of enhancing an ecological self is too nebulous to be formulated in some testable manner. Even if definitional precision were to be achieved, the typical manner of evaluation is to ask people if they benefitted in some way from the experience, a strategy which is open to a strong placebo effect. That is, participants may have a positive experience in such wilderness activity and genuinely feel that they have changed for the better, but then they proceed to behave much as before in their everyday lives. My own experience with Arcadian educators is that they are, by and large, resistant to the evaluation of their methods. On one occasion, I remember asking a workshop facilitator how he went about evaluating the effectiveness of his work, to which he replied that he didn't. He pointed out that his services were in great demand and, as far as he was concerned, that indicated he must be doing something right.

Ecopsychologists also use more overtly therapeutic methods in their pursuit of an expanded self. Various forms of ecotherapy and counseling are used to facilitate cognitive and emotional development, often in group settings or workshops.[133] One particular approach, that of "human relations training," became very popular in the 1960s as the counterculture emerged to challenge the status quo. Human-relations training, in its various forms (sensitivity training, encounter groups, and group dynamics), seeks to help people develop greater sensitivity to the inner life of both themselves and others. By becoming more aware of one's own feelings and the effect of one's own behavior on others, it is assumed that more rewarding human relationships are possible. With the decline of the counterculture in the 1970s, and the re-emergence of conservative values, human-relations training lost some of its charisma. However, it never went away entirely and has been resurrected by ecopsychologists as a vehicle for self-development, although the traditional preoccupation with human relationships has been replaced by a concern for human-nature relationships. For example, a key element in human-relations training is that of *role-playing,* assuming the identity of another to explore, in a very personal way, the thoughts and feeling of that person. Ecotherapists have, among other things, modified their role-playing exercises to include nonhuman creatures or objects. Thus, a participant in a workshop may be required to role-play a rock, a tree, a fox, and so on, in the hope that they will develop some sense of what it is like to be such a creature or thing.[132] Such procedures play an important part in a workshop process referred to as the Council of All Beings, a form of human-relations experience that has been developed to help people deal with their dread and despair about the changing nature of the world and the apparently inexorable decline in environmental health.[133,137] Each member of the Council (the workshop participants) role-plays a different creature, speaking on its behalf in the first person, a format that allows each participant to explore and express their own feelings about the nature of change in the world around them.[132] What is of interest is the extent to which these activities enhance the development of an ecological self in any significant or lasting way. Bragg[132] has attempted to answer this question by questionnaire and interview studies of participants in Council workshops. What she found was that participants in the workshops already had higher levels of ecological self-references, compared to nonparticipants, *before* they embarked on the workshop activities. This implies that they were attuned to the goals of the workshop and, perhaps, were more willing to listen to its message than ordinary people. In other words, they were more prone to the placebo effect. Generally, "participants did have powerful experiences of ecological self during the workshop, but . . . these experiences were relatively difficult to integrate into daily life and especially hard to explain to friends and family" (p. 104). However, where a person was able to arrange continuing moral support by joining an environmental or social group with compatible aims, it was possible to sustain whatever insights had been learned in the workshops. Bragg concluded that the workshops were effective primarily in the sense of preaching to the converted, reaching those who were already primed to its message by previous experiences. As such, she suggests, it may not be an effective vehicle for social change but can serve as a way of empowering those already committed to ecological activism.

In conclusion, ecopsychology has a number of weaknesses or difficulties that limit its acceptability and its use in promoting prosocial behavior.[132,133,135] First, it is often seen as a form of New Age pseudoscience, more like a religious stance than psychology.[135] As mentioned earlier, this is due, in part, to the difficulty in evaluating the effects of experiential workshops and similar learning methods. Often, the changes sought in such workshops are not precisely defined or operationalized. Because ideas such as *ecological self* are fuzzy concepts, vague psychological entities that are difficult to measure in some meaningful way, it is not easy to apply conventional positivist and scientific methods in evaluating the effects of experiential learning in workshops and wilderness outings. Also, much of the learning is tacit or emotional and, as a consequence, difficult to explain verbally. This is a common problem in the evaluation of therapy and human-relations training, one that has led to the dismissal of these endeavors as unscientific by the more Imperial psychologist. Even more difficult to assess is the extent to which some internal awakening translates into observable changes in environmental behavior. It may take some time, even years, for a person to work through the lessons learned in workshops and even longer to start making changes in their life. The bottom line may be that education aimed at self-development will only reach the converted, those who are already primed by virtue of their personality and life circumstances, to be receptive to such admonitions.

Second, in ecopsychology, there is the tendency to downplay the way in which we protect ourselves by self-serving mechanisms.[132,135] That is, we use psychological defenses to protect ourselves from the anxiety provoked by everyday life and are loath to give up such an effective way of reducing psychic pain. Ecopsychology, with some notable exceptions, may simply skirt around the kind of critical self-examination that allows us to confront our defenses. For instance, the experiential technique of *thinking like a rock* is used to promote oneness with nature but may, in fact, be a form of *projective identification* in which one projects one's own anthropocentric thoughts into inanimate matter.[132] In doing so, it is possible that the psychological process known as *identifying with the aggressor* is taking place. One identifies with something that is frightening so as to assuage one's fears. By identifying with inanimate and animate nature, one becomes less frightened of it. However, this is hardly a step toward self-understanding but, rather, the construction of another psychological defense.

Finally, there is a tendency in ecopsychology, just as there is in behavioral engineering, to ignore the sociopolitical context in which we struggle to survive, thereby making the implicit assumption that psychological change must precede sociopolitical change in our dysfunctional society. As Fisher[133] (p. 23) observes:

The general reluctance in ecopsychology to bring radical social analysis into its theorizing seems to have been carried over from mainstream psychology . . . (which) has traditionally profited from treating decontextualized, hollowed-out individuals, while not challenging the social arrangements that cause personal distress in the first place . . . in upholding the ideological *status quo,* psychology has in many ways actually *retarded* social change . . . (Thus) by neglecting social analysis, ecopsychology has also avoided reflecting on its own politics.

This politically conservative orientation of many psychologists, their reluctance to acknowledge the need for political change, is a serious limitation on the effectiveness of both Imperial and Arcadian approaches to reform.[120,138]

Conclusions on Single Vision Reforms

In this chapter, we have looked at six different approaches to the reform of conventional problem solving through the development of more adaptive forms of environmental behavior. For the sake of discussion, I have portrayed these various reforms as *single visions,* each limited by its ideological and disciplinary biases. In practice, reformers are inclined to combine parts of these six single visions into a more integrated package than implied in the preceding discussion. However, a major ideological chasm still exists, limiting the effectiveness of the proposed reforms. In the context of sustainable forestry, Franklin[139] (pp. 128–129) puts it this way:

Unfortunately, many participants in the resource management debates have had limited visions of what needs to be done, whether they emphasize commodities or environmental values . . . Typically, both sides want to isolate the biodiversity issue in some 'preserves' or 'set-asides,' perhaps viewing that as a much simpler solution than trying to integrate their objectives and to collaborate in achieving them . . . All participants must reexamine old assumptions. There is a lot of history and personal bias present—among academic scientists, managers, corporate executives, agency heads, environmental activists, and so forth. We need to examine these assumptions in the light of evolving societal objectives for these forest lands and of our new knowledge about the functioning of ecosystems and landscapes.

In other words, adaptive approaches to environmental problem solving will be possible only if we recognize the need to integrate the various single visions into a more comprehensive approach to our numerous environmental dilemmas. Some basic principles for moving in this direction are discussed in the next chapter.

References

1 Bell, A. 1994. Non-human nature and the ecosystem approach: The limits of anthropocentrism in Great Lakes Management. *Alternatives* 20:20–25.
2 Cawley, R., and J. Freemuth. 1993. Tree farms, mother earth, and other dilemmas: The politics of ecosystem management in Greater Yellowstone. *Society and Natural Resources* 6:41–53.
3 Haeuber, R. 1996. Setting the environmental policy agenda: the case of ecosystem management. *Natural Resources Journal* 36:1–28.
4 Stanley, T. 1995. Ecosystem management and the arrogance of humanism. *Conservation Biology* 9:255–262.
5 Born, S., and W. Sonzogni. 1995. Integrated environmental management: Strengthening the conceptualization. *Environmental Management* 19:167–181.
6 Worster, D. 1977. *Nature's economy: A history of ecological ideas.* Cambridge: Cambridge University Press.
7 Cortner, H., and M. Moote. 1994. Trends and issues in land and water resources management: Setting the agenda for change. *Environmental Management* 18:167–173.
8 Caldwell, L. 1994. Disharmony in the Great Lakes Basin: Institutional jurisdictions frustrate the ecosystem approach. *Alternatives* 20:26–31.
9 Grumbine, R. 1994. What is ecosystem management? *Conservation Biology* 8:27–38.
10 Slocombe, D. 1993. Environmental planning, ecosystem science, and ecosystem approaches for integrating environment and development. *Environmental Management* 17:289–303.
11 Palladino, P. 1989. Entomology and Ecology: The Ecology of Entomology. The 'Insecticide Crisis' and Entomological Research in the United States in the 1960s and 1970s: Political, Institutional, and Conceptual Dimensions. Ph.D. thesis, University of Minnesota.

12 Perkins, J. 1982. *Insects, experts, and the insecticide crisis: The quest for new pest management strategies.* New York: Plenum.

13 Brown, J., and N. MacLeod. 1996. Integrating ecology into natural resource management policy. *Environmental Management* 20:289–296.

14 Bocking, S. 1994. Visions of nature and society: A history of the ecosystem concept. *Alternatives* 20:12–18.

15 Fitzsimmons, A. 1996. Sound policy or smoke and mirrors: Does ecosystem management make sense? *Water Resources Bulletin* 32:217–227.

16 Botkin, D. 1990. *Discordant harmonies: A new ecology for the twenty-first century.* New York: Oxford University Press.

17 Lee, R., et al. 1992. Integrating sustainable development and environmental vitality: A landscape ecology approach. In *Watershed management: Balancing sustainability and environmental change,* ed. R. Naiman, 499–521. New York: Springer-Verlag.

18 Thompson, I., and D. Welsh. 1993. Integrated resource management in boreal forest ecosystems—impediments and solutions. *The Forestry Chronicle* 69:32–39.

19 Dryzek, J. 1987. *Rational ecology: Environment and political economy.* New York: Blackwell.

20 Gadow, S. 1992. Existential ecology: The human/natural world. *Social Science and Medicine* 35:597–602.

21 Segerstrale, U. 1992. Reductionism, "bad science," and politics: A critique of anti-reductionist reasoning. *Politics and the Life Sciences* 11:199–214.

22 Kay, J., and E. Schneider. 1994. Embracing complexity: The challenge of the ecosystem approach. *Alternatives* 20:32–39.

23 More, T. 1996. Forestry's fuzzy concepts: An examination of ecosystem management. *Journal of Forestry* 94:19–23.

24 Miller, A. 1991. *Personality types: A modern synthesis.* Calgary: University of Calgary Press.

25 Jensen, M., et al. 1996. Ecosystem management: A landscape ecology perspective. *Water Resources Bulletin* 32:203–216.

26 Frissell, C., and D. Bayles. 1996. Ecosystem management and the conservation of aquatic biodiversity and ecological integrity. *Water Resources Bulletin* 32:229–240.

27 Slobodkin, L. 1993. Scientific goals require literal empirical assumptions. *Ecological Applications* 3:571–573.

28 Hilborn, R., and D. Ludwig. 1993. The limits of applied ecological research. *Ecological Applications* 3:550–552.

29 Moir, W., and H. Mowrer. 1995. Unsustainability. *Forest Ecology and Management* 73:239–248.

30 Pitelka, L., and F. Pitelka. 1993. Environmental decision making: Multidimensional dilemmas. *Ecological Applications* 3:566–568.

31 Policansky, D. 1993. Uncertainty, knowledge, and resource management. *Ecological Applications* 3:583–584.

32 Rubenstein, D. 1993. Science and the pursuit of a sustainable world. *Ecological Applications* 3:585–587.

33 Socolow, R. 1993. Achieving sustainable development that is mindful of human imperfections. *Ecological Applications* 3:581–583.

34 Zedler, J. 1993. Lessons on preventing overexploitation? *Ecological Applications* 3:577–578.

35 Mooney, H., and O. Sala. 1993. Science and sustainable use. *Ecological Applications* 3:564–566.

36 Foran, B., and K. Wardle. 1995. Transitions in land use and the problems of planning: a case study from the mountainlands of New Zealand. *Journal of Environmental Management* 43:97–127.

37 Costanza, R. 1993. Developing ecological research that is relevant for achieving sustainability. *Ecological Applications* 3:579–581.

38 Halbert, C. 1993. How adaptive is adaptive management? Implementing adaptive management in Washington State and British Columbia. *Reviews in Fisheries Science* 1:261–283.

39 Holling, C. 1978. *Adaptive environmental assessment and management.* Chichester: Wiley.

40 Sonntag, N., et al. 1986. *Pest management in plantations: A consultative approach.* Vancouver, B.C.: ESSA Ltd.

41 Briassoulis, H. 1989. Theoretical orientations in environmental planning: An inquiry into alternative approaches. *Environmental Management* 13:381–392.

42 Holling, C. 1995. What barriers? What bridges? In *Barriers and bridges to the renewal of ecosystems and institutions,* ed. L. Gunderson, C. Holling, and S. Light, 3–34. New York: Columbia University Press.

43 McLain, R., and R. Lee. 1996. Adaptive management: Promises and pitfalls. *Environmental Management* 20:437–448.

44 ESSA, Ltd. 1982. *Review and evaluation of Adaptive Environmental Assessment and Management.* Vancouver: Environment Canada.

45 Gunderson, L., C. Holling, and S. Light, ed. 1995. *Barriers and bridges to the renewal of ecosys-*

tems and institutions. New York: Columbia University Press.

46 Lee, K. 1993. *Compass and gyroscope: Integrating science and politics for the environment.* Washington, D.C.: Island Press.

47 Dewhurst, S., W. Covington, and D. Wood. 1995. Developing a model for adaptive ecosystem management. *Journal of Forestry* 93:35–41.

48 Grayson, R., J. Doolan, and T. Blake. 1994. Application of AEAM (Adaptive Environmental Assessment and Management) to water quality in the Latrobe River catchment. *Journal of Environmental Management* 41:245–258.

49 Stanford, J., and G. Poole. 1996. A protocol for ecosystem management. *Ecological Applications* 6:741–744.

50 Cooperrider, A. 1996. Science as a model for ecosystem management—panacea or problem? *Ecological Applications* 6:736–737.

51 Miller, A. 1993. The role of citizen scientist in natural resource decision making. *The Environmentalist* 13:47–59.

52 Baskerville, G., and P. Duinker. 1986. Pest management in plantations: An institutional analysis. In *Pest management in plantations: A consultative approach,* ed. N. Sonntag et al. Vancouver, B.C.: ESSA Ltd.

53 Hilborn, R., C. Walters, and D. Ludwig. 1995. Sustainable exploitation of renewable resources. *Annual Review of Ecology and Sytematics* 26:45–67.

54 Meffe, G. 1992. Techno-arrogance and halfway technologies: Salmon hatcheries on the Pacific coast of North America. *Conservation Biology* 6:350–354.

55 Christensen, N., et al. 1996. The report of the Ecological Society of America Committee on the scientific basis for ecosystem management. *Ecological Applications* 6:665–691.

56 Christopherson, J., S. Lewis, and M. Havercamp. 1996. Lake Tahoe's forest health consensus group. *Journal of Forestry* 94:10–12.

57 Keiter, R. 1996. Toward legitimizing ecosystem management on the public domain. *Ecological Applications* 6:727–730.

58 Lee, R. 1992. Ecologically effective social organization as a requirement for sustaining watershed ecosystems. In *Watershed management: Balancing sustainability and environmental change,* ed. R. Naiman, 73–89. New York: Springer-Verlag.

59 Holling, C., and G. Meffe. 1996. Command and control and the pathology of natural resource management. *Conservation Biology* 10:328–337.

60 Heatherington, J., T. Daniel, and T. Brown. 1994. Anything goes means everything stays: The perils of uncritical pluralism in the study of ecosystem values. *Society and Natural Resources* 7:535–546.

61 Woodley, S., and B. Freedman. 1995. The Greater Fundy Ecosystem Project: Toward ecosystem management. *The George Wright Forum* 12:7–14.

62 Wilder, R. 1993. Is this holistic ecology or just muddling through? The theory and practice of marine policy. *Coastal Management* 21:209–224.

63 Ludwig, D., R. Hilborn, and C. Walters. 1993. Uncertainty, resource exploitation and conservation: Lessons from history. *Ecological Applications* 3:547–549.

64 Meffe, G. 1994. Human population control: The missing awareness. *Conservation Biology* 8:310–313.

65 McQuillan, A. 1993. Cabbages and kings: the ethics and aesthetics of new forestry. *Environmental Values* 2:191–222.

66 Drengson, A. 1994. Beyond empire resourcism to ecoforestry. *International Journal of Ecoforestry* 10:35–39.

67 Camp, O. 1984. *The forest farmer's handbook: A guide to natural selection forest management.* Ashland, Oreg.: Sky River Press.

68 Drescher, J. 1994. The battle for sustainability. *International Journal of Ecoforestry* 10:16–19.

69 Drengson, A., and D. Taylor, eds. 1997. *Ecoforestry: The art and science of sustainable forest use.* Gabriola Island, B.C.: New Society Publishers.

70 Clogg, J. 1997. Tenure Reform for Ecologically and Socially Responsible Forest Use in British Columbia. Master's thesis, York University, North York, Ontario, Canada.

71 Redclift, M. 1987. *Sustainable development: Exploring contradictions.* London: Routledge.

72 Eckersley, R. 1992. *Environmentalism and political theory: Toward an ecocentric approach.* Albany: State University of New York Press.

73 Clark, J. 1995. Economic development vs. sustainable societies: Reflections on the players in a crucial contest. *Annual Review of Ecology and Systematics* 26:225–248.

74 Walker, K. 1988. The environmental crisis: A critique of neo-Hobbesian responses. *Polity* 21:67–81.

75 Meadows, D. H., et al. 1972. *The limits to growth.* New York: Universe Books.

76 Paehlke, R. 1989. *Environmentalism and the future of progressive politics.* New Haven: Yale University Press.

77 Lewis, M. 1992. *Green delusions: An environmen-*

talist critique of radical environmentalism. Durham, N.C.: Duke University Press.

78 Gardner, G., and P. Stern. 1996. *Environmental problems and human behavior.* Boston: Allyn and Bacon.

79 Hardin, G. 1993. *Living within limits: Ecology, economics, and population taboos.* New York: Oxford University Press.

80 Myers, N. 1993. *Ultimate security: The environmental basis of political stability.* New York: Norton.

81 Reid, E. 1993. Global perspectives and impacts on Canada's environment. *The Forestry Chronicle* 69:146–150.

82 Royal Society of Canada. 1993. *Global change and Canadians.* Canadian Global Change Program Secretariat, Ottawa, Canada.

83 Zimmerman, M. 1995. The threat of ecofascism. *Social Theory and Practice* 21:207–238.

84 Fischer, F. 1990. *Technocracy and the politics of expertise.* Newbury Park, Calif.: Sage.

85 Parsons, W. 1995. *Public policy: An introduction to the theory and practice of policy analysis.* Aldershot, U.K.: Edward Elgar.

86 Ophuls, W. 1977. *Ecology and the politics of scarcity.* San Francisco: W. H. Freeman.

87 Pacey, A. 1983. *The culture of technology.* Cambridge, Mass.: MIT Press.

88 Arney, W. 1991. *Experts in the age of systems.* Albuquerque: University of New Mexico Press.

89 Smil, V. 1993. *Global Ecology: Environmental change and social flexibility.* London: Routledge.

90 Wong, Y. 1994. Impotence and intransigence: State behavior in the throes of deepening global crisis. *Politics and the Life Sciences* 13:3–14.

91 Gillis, R., and T. Roach. 1986. *Lost initiatives: Canada's forest industries, forest policy and forest conservation.* Westport, Conn.: Greenwood Press.

92 Sandberg, L., ed. 1992. *Trouble in the woods: Forest policy and social conflict in Nova Scotia and New Brunswick.* Fredericton, N.B.: Acadiensis Press.

93 Burrill, G., and I. McKay, eds. 1987. *People, resources, and power: Critical perspectives on underdevelopment and primary industries in the Atlantic region.* Fredericton, N.B.: Acadiensis Press.

94 Parenteau, W. 1989. Pulp, paper and poverty: Then and now: Past and present in the New Brunswick woods. *New Maritimes* 7:20–26.

95 Peluso, N., C. Humphrey, and L. Fortmann. 1994. The rock, the beach, and the tidal pool: People and poverty in natural resource-dependent areas. *Society and Natural Resources* 7:23–38.

96 West, P. 1994. Natural resources and the persistence of rural poverty in America: A Weberian

perspective on the role of power, domination, and natural resource bureaucracy. *Society and Natural Resources* 7:415–427.

97 Sessions, G. 1995. Political correctness, ecological realities and the future of the ecology movement. *The Trumpeter* 12:191–196.

98 Betts, M., and D. Coon. 1996. *Working with the woods: Restoring forests and community in New Brunswick.* Fredericton: Conservation Council of New Brunswick.

99 Dunster, J. 1989. *Establishing the Geraldton community forest. Phase I: Concepts and background information.* Guelph, Ontario: Dunster & Associates.

100 Harvey, J., and D. Coon. 1997. *Beyond crisis in the fisheries: A proposal for community-based ecological fisheries management.* Fredericton, N.B.: Conservation Council of New Brunswick.

101 Staff Reporter. 1995. N.B. community land trust established. *Daily Gleaner,* November 10. Fredericton, N.B., Canada.

102 Conservation Council of New Brunswick 1994. *Public Lands in public hands, managing Crown forests in the public Interest, a Conservation Council proposal.* Fredericton, N.B.: Conservation Council of New Brunswick.

103 Hammond, H. 1991. *Seeing the forest among the trees: The case for wholistic forest use.* Vancouver: Polestar Press.

104 Hausler, S. 1993. Community forestry: A critical assessment. The case of Nepal. *The Ecologist* 23:84–90.

105 Shepherd, A. 1995. Participatory environmental management: contradiction of process, project and bureaucracy in the Himalayan foothills. *Public Administration and Development* 15:465–479.

106 Pinkerton, E. 1994. Local fisheries co-management: A review of international experiences and their implications for salmon management in British Columbia. *Canadian Journal of Fisheries and Aquatic Science* 51:2363–2378.

107 Wilson, J., et al. 1994. Chaos, complexity and community management of fisheries. *Marine Policy* 18:291–305.

108 Cassells, D., and P. Valentine. 1988. From conflict to consensus—towards a framework for community control of the public forests and wildlands. *Australian Forestry* 51:47–56.

109 Summit News Service. 1997. Concerns expressed about N.B. wood supply. *Daily Gleaner,* March 14, Fredericton, N.B., Canada.

110 McLaughlin, D. 1995. *Grounds for change: Linking experience with a vision of sustainable agricul-*

ture. Fredericton, N.B.: Conservation Council of New Brunswick.

111 Lyng, S. 1988. Theoretical observations on applied behavioral science: Holism and reductionism within applied behavioral science: The problem of clinical medicine. *The Journal of Applied Behavioral Science* 24:101–117.

112 Kipnis, D. 1987. Psychology and behavioral technology. *American Psychologist* 42:30–36.

113 De Young, R. 1993. Changing behavior and making it stick: The conceptualization and management of conservation behavior. *Environment and Behavior* 25:485–505.

114 Dwyer, W., et al. 1993. Critical review of behavioral interventions to preserve the environment: Research since 1980. *Environment and Behavior* 25:275–321.

115 De Young, R., et al. 1993. Promoting source reduction behavior: The role of motivational information. *Environment and Behavior* 25:70–85.

116 Geller, S. 1995. Actively caring for the environment: An integration of behaviorism and humanism. *Environment and Behavior* 27:184–195.

117 Oskamp, S. 1995. Resource conservation and recycling: Behavior and policy. *Journal of Social Issues* 51:157–177.

118 Eysenck, H., and M. Eysenck. 1985. *Personality and individual differences.* New York: Plenum.

119 Kempton, W., J. Darley, and P. Stern. 1992. Psychological research for new energy problems: Strategies and opportunities. *American Psychologist* 47:1213–1223.

120 Kidner, D. 1994. Why psychology is mute about the environmental crisis. *Environmental Ethics* 16:359–377.

121 Jaeger, C., et al. 1993. Determinants of environmental action with regard to climatic change. *Climatic Change* 23:193–211.

122 Robottom, I., and P. Hart. 1996. Behaviorist EE research: Environmentalism and individualism. *Journal of Environmental Education* 27:5–9.

123 Oskamp, S. 1995. Applying social psychology to avoid ecological disaster. *Journal of Social Issues* 51:217–239.

124 Hallin, P. 1995. Environmental concern and environmental behavior in Foley, a small town in Minnesota. *Environment and Behavior* 27:558–578.

125 Hamid, P., and S. Cheng. 1995. Predicting antipollution behavior: The role of molar behavioral intentions, past behavior, and locus of control. *Environment and Behavior* 27:679–698.

126 McKenzie-Mohr, D., et al. 1995. Determinants of responsible environmental behavior. *Journal of Social Issues* 51:139–156.

127 Grob, A. 1995. A structural model of environmental attitudes and behavior. *Journal of Environmental Psychology* 15:209–220.

128 Karp, D. 1996. Values and their effect on proenvironmental behavior. *Environment and Behavior* 28:111–133.

129 Stern, P., and T. Dietz. 1994. The value basis of environmental concern. *Journal of Social Issues* 50:65–84.

130 Stern, P., et al. 1995. Values, beliefs, and proenvironmental action: Attitude formation toward emergent attitude objects. *Journal of Applied Social Psychology* 25:1611–1636.

131 Finger, M., and P. Verlaan. 1995. Learning our way out: A conceptual framework for social-environmental learning. *World Development* 23:503–513.

132 Bragg, E. 1996. Towards ecological self: Deep ecology meets constructionist self-theory. *Journal of Environmental Psychology* 16:93–108.

133 Fisher, A. 1996. Toward a more radical ecopsychology. *Alternatives Journal* 22:20–26.

134 Fox, W. 1990. Transpersonalizing ecology: "Psychologizing" ecophilosophy. *The Journal of Transpersonal Psychology* 22:59–96.

135 Reser, J. 1995. Whither environmental psychology?: The transpersonal ecopsychology crossroads. *Journal of Environmental Psychology* 15:235–257.

136 Roszak, T. 1992. *The voice of the earth: An exploration of ecopsychology.* New York: Simon & Schuster.

137 Thomashow, M. 1995. *Ecological identity: Becoming a reflective environmetalist.* Cambridge, Mass.: MIT Press.

138 Fox, D. 1985. Psychology, ideology, utopia and the commons. *American Psychologist* 40:48–58.

139 Franklin, J. 1993. The fundamentals of ecosystem management with applications in the Pacific northwest. In *Defining sustainable forestry,* ed. G. Aplet, et al., 127–143. Washington, D.C.: Island Press.

7

Toward Adaptive Problem Solving

It is a mistake to believe that adaptive problem solving can be achieved by the adoption of some clever new method. The search for technical fixes or adroit planning procedures is not going to resolve the environmental problems that confront us. Yet method-centered thinking is seductive in the face of the apparently intractable psychosocial barriers to change, which I have described at some length. When confronted by such political inequalities and entrenched power structures, professionals are inclined to attend to technical matters in lieu of seeking political change.[1] It can be argued, however, that a preoccupation with the methods and techniques of problem solving, to the exclusion of the surrounding sociopolitical context, is a form of defensive avoidance. It is a way of doing something to justify one's professional existence, while dodging the more contentious issues. As mentioned earlier, however, if adaptive solutions to our predicament are to be achieved, they will likely come through the more effective application of relatively simple, well-known technologies rather than through some technological breakthrough,[2] a strategy that can only be implemented when there is sufficient political will to engage in fundamental sociopolitical change. What I propose, therefore, is that planning for adaptive problem solving needs to reverse conventional approaches, turning away from political rhetoric and technical legerdemain toward more systemic psychosocial

change. We need to think in terms of developing new problem-solving systems rather than being content with injecting technical innovations into conventional muddling along. Some examples might help clarify what I mean.

Traditionally, the elites responsible for managing environmental problems have retained control of the problem-solving process, occasionally seeking technical and other information from their expert advisors, while excluding the broader public from meaningful involvement. The net result is that, in response to some pressing environmental issue, the problem-solving process frequently starts with the wrong question. Even recent attempts to innovate are prone to this difficulty. For instance, the Fundy Model Forest project in Canada, which is discussed more fully later, was heavily influenced at the outset by the scientific and industrial communities.[3] The questions addressed and the structure of the subsequent planning process reflected the interests of these two groups, namely, scientific understanding of the region's ecology and operational innovations in the industrial use of the land. Both sets of interests required a primary focus on data-collection and technical studies. As a consequence, the project has taken on the air of a research study, heavily weighted toward scientific and operational research with citizens playing a limited consultative role. Although this may satisfy the professionals involved in the project, it does little to empower local citizens or to develop the permanent community networks that might survive the termination of government funding. In other words, it remains a typical Imperial project, the ecosocial consequences of which may well be ephemeral.

A more adaptive response, I would suggest, is to ask a fundamentally different question at the outset, namely: How can we help local communities develop the skills and motivation to take responsibility for, and learn to manage, the problem at hand? In other words, the model on which to base adaptive problem solving is that of *community development*[4,5] rather than decision making by political and scientific elites. Community development requires political devolution and the empowerment of grassroots activism through strengthening of institutions at the local level. Emphasis is placed on the development of networks of people willing to put in the time and develop the skills needed to take responsibility for the issue at hand. As a process of human resource development, this is person-oriented rather than the more task-oriented approach commonly adopted by natural resource professionals. The establishment of these networks, and the team building that ensues would take priority over data collection. Indeed, the collection of data would be postponed until the community networks were well-enough established to begin asking questions that required research input. In practice, such work can be thankless. Gowdy,[6] for instance, notes how difficult empowering the powerless can be. After decades or even centuries of being told what to do by elites, the great mass of people may feel oppressed, deeply suspicious of efforts to engage them in any decision-making process controlled by officials, hence the frequent observation of public apathy and the persistent reluctance to engage in community endeavors under the aegis of experts.[1,7-9] Nor do they have what Gowdy refers to as a "participatory consciousness," the sense of efficacy built into their identities that would encourage them to become involved in environmental decision making.

A community development approach to adaptive environmental problem solving is time-consuming, prone to logistical difficulties, and extremely frustrating. However, the alternatives are, in all probability, much less effective. In this chapter, I intend to outline some of the ways in which such an approach to problem solving might be facilitated. The basic theme is that small groups are at the core of adaptive problem solving, and whatever promotes their effectiveness is to the benefit of the problem-solving process. The chapter begins with the kinds of social experiments needed to enhance the functioning of small groups, before looking at the personal qualities needed by key players in group problem solving, and closes by looking at the kinds of problem-solving methods of most use in these contexts.

Social Experimentation

Social change is not easy. Social experimentation is even harder. Despite the posturing of everyday life, most of us are fearful creatures, unable and unwilling to tolerate all but the most modest changes in the world around us. We seem to require a great deal of predictable routine, even while we grumble about the boredom of that routine. Any fundamental change in the institutions and social arrangements in which we are comfortably ensconced is resisted with grim determination. It is unlikely, therefore, that there will be any overwhelming enthusiasm for adaptive social experimentation. Yet we seem to have little choice in light of the inadequacies of conventional practices, the question being which direction these social experiments should take. Previously, I described how Imperial reformers seek to gather unto themselves the power to impose problem solutions on what they are inclined to see as a feckless public. In contrast, Arcadian reformers promote the devolution of sociopolitical institutions and foster local participation in the search for problem solutions. Adaptive institutions, based on new forms of collective problem solving, would have to combine these two ideological stances in innovative ways. This would not appear to be a very profound observation, as efforts to involve the public in decision making are now ubiquitous. Various forms of public hearings, round tables, planning workshops, and so on, seem to indicate the willingness of central authorities to include a wider variety of stakeholders in decision making. However, most of these efforts appear to be mere public relations exercises aimed at defusing public resentment rather than genuine efforts to develop adaptive problem-solving processes. What would more adaptive social change look like?

Becker and Ostrom[10] observe that, in the past, there has been a tendency to believe that the overexploitation of natural resources could only be stopped by either imposing government control (Imperial) or seeking to constrain individual behavior through recognizing private property rights (a step toward Arcadian devolution). However, the rigid imposition of either extreme may not necessarily work in practice. For example, planning by local communities might satisfy the needs of those involved but jeopardize the livelihood of others who were not included in the exercise. Thus, a third-world community might develop a timber-harvesting program that promoted their logging interests but that, when implemented, was found to destroy other forest products, such as medicinal herbs, which provided a livelihood for others.[10] On the other hand, government efforts to manage resources based on estimates of maximum sustainable yield have led to the overexploitation and collapse of a variety of natural resources. Becker and Ostrom[10] conclude that "imposing rigid private property or central authority on a multispecies resource use system that has evolved over centuries may not only adversely affect the very human groups who have been responsible for successfully husbanding resources but may also adversely affect the ecological function of the resource systems" (p. 125). In their view, what we need is greater flexibility in combining these various approaches to management in light of local ecological conditions. That is, "to achieve long-term economic sustainability, we need more than ever before a combination of institutions that restrain shortsighted and selfish behavior and that make rules based on flexible and cau-

tious models of the ecology of complex biological systems"[10] (p. 129). Thus, flexibility in combining public and private, centralized and decentralized social arrangements to suit local conditions seems to be the way ahead.

In support of this contention, Becker and Ostrom[10] offer three examples of radically different but equally successful arrangements for managing water supplies, some of which have been in operation for hundreds of years. Thus, "the *zanjeras* of Northern Philippines are self-organized systems in which farmers obtain use-rights to previously unirrigated land from a large landowner by building a canal that irrigates the landowner's land and that of a *zanjera*" (p. 120). The *Thulo Kulo* system in Nepal, however, is more akin to a cooperative arrangement in which households invest in the development of an irrigation system and receive an allocation of water in direct proportion to their investment. In Los Angeles County, concerns about the sustainability of underground reservoirs led to yet another set of institutional arrangements. In the face of a pumping race that threatened to deplete the resource, the major water producers bargained to obtain an equitable and sustainable share of the resource while water users at the local level assumed responsibility for managing the water basins by assessing pump taxes and undertaking replenishment programs.

In all of the cases reviewed by Becker and Ostrom,[10] success was achieved when participants were willing to adopt a long-term, less exploitative relationship to the resource and when institutional arrangements incorporated the six design principles outlined in Table 7.1. Sensible management decisions are likely to be taken when those most directly involved in resource use have clear title to some portion of the resource that they are allowed to manage without unnecessary government interference. In these small-scale management units, all of those affected by the rules governing resource use are included in the rule-making process. In addition, they need to be directly involved in monitoring harvesting practices and be willing to implement, or accept, community sanctions for rule violation. Where disputes arise, they need to be dealt with speedily

7.1. Design principles for adaptive institutions.

1. *Political devolution*. The basic management unit is a well-defined local community with clearly established rights to devise their own management practices with minimal interference from government.
2. *Collective-choice mechanism*. Most of the individuals affected by resource management decisions and rules participate in the group that can modify these rules.
3. *Equating costs and benefits*. Individuals benefit in proportion to their commitment, in time and money, to the joint management venture. Their benefits should be consistent with their costs.
4. *Monitoring and sanctions*. Monitoring of resource use and violations of management rules should involve users themselves. Violations by users and officials should be met with graduated sanctions.
5. *Conflict resolution*. Users and officials should have access to low cost, locally available procedures for resolving disputes.
6. *Nested enterprises*. Local institutions should be nested in a broader network of medium to large institutions, thereby ensuring that large-scale problems are addressed in addition to local issues.

Modified from Becker and Ostrom.[10] With permission, from the *Annual Review of Ecology and Systematics*, Volume 26, © 1995, by Annual Reviews, Inc.

through local, affordable conflict-resolution procedures. Allocation of benefits from the resource should be in direct proportion to the costs incurred by the individual in harvesting and protection. Finally, local community groups do not exist in a vacuum but are surrounded by a variety of economic and social forces. Local management groups, therefore, need to be nested within a hierarchical arrangement of medium and large-scale organizations so that broader issues, and threats, can be dealt with. Because very small community-based arrangements do not succeed in all situations, experimentation is needed to determine how these principles might form the basis of new arrangements in light of local social and ecological conditions. To illustrate some of the ways that this might be achieved, the first two principles from Table 7.1 are discussed in more detail.

Political Devolution

The use and abuse of power by established elites is an intractable problem for those seeking to reform environmental and natural resource management. A common response is to avoid the issue by attending to the technical aspects of a problem or by focusing on less politically challenging reforms such as public participation. Yet public involvement in decision making by the relatively powerless, those who have little economic or political clout, is an empty exercise. Often, the retention of power by elites is justified on the basis of a misanthropic (Imperial) belief about the uselessness of ordinary people. Hausler[1] suggests that, in many developing countries, postcolonial managers adopted the older colonial view that mismanagement of resources was due to the laziness, ignorance, and apathy of local farmers, as well as their unwillingness to control their reproduction. This justified the imposition of centrally planned, highly technical management policies that ignored local knowledge and circumstances. In doing so, local bureaucracies were aided and abetted by western technical experts intent on exporting the western technologies with which they were familiar. Thus, in Hausler's experience, attempts to foster community forestry in third-world countries began with the assumption that local farmers had to be "taught" and "motivated" to plant trees. However[1] (p. 84):

Western foresters gradually recognized that local farmers have been using and growing trees for centuries in locally-adapted ways. Perceiving the need to bring into the community forestry discourse excluded voices, particularly those of women, yet unable to speak out on politically sensitive subjects such as nepotism and corruption, which reduced people's interest in participating in community forestry development projects, foresters helped to promote "user-centred" and "participatory" approaches to forestry. These approaches created new openings for change in the discourse as well as some political space for local people to regain control over their forests. But the overall technical and institutional framework in which the foresters worked tended to reinforce the power structures which fostered environmental degradation.

In other words, while participatory methods provided the powerless with a greater voice in *talking* about forestry, the absence of any significant change in the prevailing power structures ensured that practices remained much as before. Clogg[11] makes the same point in her discussion of forestry in western Canada. In her view, "the rights granted through the forest tenure system in British Columbia have resulted in relations of power and exclusion." Government efforts toward developing collaborative decision making are unlikely to lead to sustainable forestry, she believes, "without altering the structural conditions which shape the power of different actors involved." By "structural conditions" she means the social and economic arrangements that determine the harvesting and marketing of forest products.

Needless to say, technocorporate elites, confident in their ability to manage, are not enthusiastic about sharing power with those whom they consider to be feckless. They are likely to, and do, resist any such fundamental change in power relations. As Durning[12] (p. 301) observes, "citizen empowerment would decrease the power of the people and groups that now exert the most influence on policymaking, and they possess vast resources with which to resist these efforts." In a similar vein, Betts and Coon[13] (p. 25) conclude: "Given the tremendous political and economic power of those who favor the status quo in forest policy in New Brunswick, a move toward community forestry will not be easy." Despite this resistance to power sharing, it is difficult to see how more adaptive behavior can be developed without a radical shift in current political systems. A more productive balance between Imperial control and Arcadian devolution appears to be necessary, but the form this would take would vary from place to place. We need to experiment, therefore, with different mixes of central control and devolved, small-scale practices.[10,14-16]

Proposals along these lines are now commonplace in Canadian forestry, where schemes for combining private ownership, community forests, industrial tenure, and state management into a regional management system have started to appear.[11,13,17,18] One set of proposals, mentioned earlier, suggests that the Crown (public) lands of New Brunswick, which are currently licensed to large corporations, could be subdivided into community forests, industrial forest, and protected areas.[17] The community forests would be adjacent to the local communities that managed them, while the industrial zones would be managed by state agencies in cooperation with public groups. It follows that the current system of licensing of Crown lands to the major pulp and paper mills would be abrogated in favor of management boards under an umbrella of provincial control, thereby reducing the political role of the major forestry corporations.

The implementation of such schemes is beset with a variety of problems, not the least of which is the assumption that a change in tenure arrangements will promote sustainable practices.[11] That is, it is assumed that by transferring the responsibility for managing forests from large corporations to individuals and communities, there will be a concomitant increase in environmental stewardship. Personal involvement with one's own land is believed to be a salutary experience, one that leads to caring for the land. However, as Clogg[11] found in her study of forest practices in British Columbia, there is a common perception among forestry professionals that some small-woodlot owners can act like vandals without any concern for the land that they own. Thus, advocates of private ownership as a route to sustainability are only too aware of the prodigal behavior of some private owners. A variety of constraints have been suggested for controlling such behavior, the most idealistic of which is a kind of latter-day homesteading in which owners live on or near their land, their continuing rights to the land being contingent on their performance in managing it. Thus, private ownership would require active and informed management of the land for which one is responsible.[11] The problem of restraining recklessness is, perhaps, more easily achieved within a community setting where peer pressure can be effective in reducing the incidence of unwise exploitation. Clogg[11] recounts a case in point, in which a woodlot owner clear-cut a steep hillside over the protests of local people only to find that no one was willing to transport his wood nor would any local mill accept it. Although this did not stop the original clearcutting, it probably acted as a salutary lesson to others. In general, however, it remains to be seen whether private and community ownership will encourage stewardship of the land.

Turning to the role of the corporate sector, the Imperial tradition of resource exploitation, especially within forestry, has favored volume production over value-added production. In New Brunswick, for instance, the heavy emphasis on pulp and paper manufacturing has guided both investment in and the development of industrial infrastructure. The Province is now committed to and dependent on this massive but low value-added enterprise. Fortunately, the major forestry corporations are very effective at this large-scale enterprise. So much so, in fact, that one commentator concludes: " the larger companies provide us with incredible economic power and drive, and they are best suited to doing that. Their expertise is in systems management, in getting things done, in innovation, in marketing and responding to market pressures."[11] Thus, there is a role for the big companies in any new mix of tenures that evolves through political devolution.

Yet, in the global marketplace, volume production is shifting increasingly to the South, to the southern United States, and to third-world countries where timber grows more quickly than in northern climates. Canadian forestry, therefore, is faced with the need to shift to value-added production rather than fill its traditional role as the source of raw materials for export.[11,19] Whether the impetus comes from individual initiative or government incentive, the switch to value-added production presents formidable obstacles. Unfortunately, anyone interested in developing new, value-added wood products would have to create their own production facilities and markets, all the while struggling against the prevailing tide of pulp and lumber production to which the forest industry is geared. At the same time, any significant inroads into new uses of the forest would require extensive re-education of the workforce together with considerable social dislocation.[11] It is difficult to see how this can be achieved in resource-dependent communities that are, typically, very socially conservative.

Political devolution in resource-dependent societies, therefore, is hampered by prevailing industrial infrastructures and the absence of a reservoir of skills needed to implement new ventures. It follows that social experiments in political devolution will be resisted, will stumble when tried, and will be constantly at jeopardy in face of economic globalization. Yet it seems that we have no choice but to try.

Collective Choice Mechanisms

Experiments in political devolution would increase the fragmentation of jurisdictional responsibility that already exists, thereby exacerbating the problem of coordination. New forms of collective choice and decision making would have to be developed if there was to be any hope of implementing adaptive problem solving. The implication of previous chapters is that whatever forms such mechanisms take, they would have to be more genuinely participatory than current efforts. In other words, there needs to be experimentation with a variety of public participation exercises to allow those affected by decisions the opportunity to have some genuine influence on the decision-making process.[10] Although there is considerable support, in principle, for greater public participation in decision making, actual practice is bedeviled by ideological differences over what is entailed.[20] The Imperial preference for representative democracy, with its delegation of responsibility to officials, often results in a kind of public participation that is criticized for being a public relations and education exercise, an occasion on which the public can be informed about decisions made on their behalf by elites. For instance, Bush[21] (p. 3) observed this way of thinking in a statement by the Canadian Nuclear Association to the effect that the primary purpose of public participation "should be to demonstrate to the public that right decisions are being made for, on balance, the right reasons." Such Imperial views have led to cries of tokenism and widespread disillusionment with the available forms of public involvement on the part of citizens' groups. Nevertheless, the Imperial style persists as Macadam et al.[7] (p. 317) found in their recent attempts to introduce innovative participatory methods into agricultural decision making in Nepal:

The workshop participants were nominees of ministries and other government agencies selected by the consultants . . . on the grounds that they had an interest in the development of the livestock sector. The few private sector representatives were either large-scale commercial farmers or agro-processors. These people are influential in policy-making but in doing so are likely to be acting on their own largely common interests or, at best, on 'behalf' of farm families whose situation and interests are vastly different.

More commonly, however, proponents of representative democracy are willing to tolerate a modest role for citizens in the decision process. This kind of public participation allows citizens to express their views on the situation, thereby providing decision makers with valuable information about public opinion and needs.[12] Such "consultative" forums,[22] however, usually involve little real dialogue between the rulers and the ruled, nor is their any significant shift in decision-making power. At worst, consultative procedures may be a way of diffusing potential conflict, preventing public backlash and costly delays in construction projects.[21,23]

The Arcadian ideal of a participatory democracy, in which citizens govern themselves by decisions made in "open forums," remains an unrealized dream. Such utopian visions of an informed citizenry, aided by expert advisors, openly debating issues in a meaningful way with victory going to the most persuasive, are part of an ancient political tradition. Recent versions promote the virtue of "discursive participatory democracy" and "discourse communities" as a means of involving the public in environmental and social decision making.[12,24,25] How this kind of decision making might be encouraged within a technocorporate state remains unresolved. As mentioned earlier, ruling elites in any society are unlikely

to forego their positions of status and privilege by empowering the common folk beyond manageable, and largely token, levels. Any expectation of a more extensive power sharing is politically naive. At the same time, discursive participatory democracy requires an informed, active citizenry. However, as Durning[12] (p. 301) observes, the "declining newspaper readership, participation in elections, and attentiveness to public affairs makes it difficult to be optimistic about how well such a participatory democracy would function." Along similar lines, Smil[2] (p. 208) argues that "the challenges of racial discord, deep income disparities, economic stagnation, budget deficits, crime and drug addiction, mass functional illiteracy, poor education, family breakdown and public apathy combine to create conditions hardly conducive to . . . elevating environmental management to a national priority." Yet Smil (p. 213) goes on to argue that some form of participatory process is crucial:

Nothing is . . . more important than opting for flexible strategies growing from vigorous free dialogue. Glimpses of free societies may be disquieting . . . but this uncertain, unruly, seemingly directionless assemblage of contending interests wary of the future is also still the best known self-correcting adaptive arrangement of human society.

Therefore, the way ahead seems to require (as I've mentioned frequently) some judicious mix of Imperial representative and Arcadian participatory methods. Achieving this ambition, however, faces us with a number of additional problems.[26]

First, not every problem requires extensive public involvement. Indeed, insisting on such a requirement is likely to spread limited resources too thinly over the seemingly endless array of environmental problems that parade themselves before us. The question, then, is how to determine when fully fledged participatory involvement is necessary. Sample[22] suggests a "contingent model" for achieving this juggling act. That is, if one recognizes a continuum of decision-making styles ranging from autocratic, through consultative, to the more genuinely participatory, then the particular form of decision making used should depend on the situation and circumstance. The solution lies in finding some means of distinguishing those issues that require heavy public involvement from those that do not. Both Matland[27] and Thomas[28] suggest that one way of doing so is to use more participatory methods for issues over which there are doubts about public acceptability. Where the latter is less of an issue, and policy implementation becomes more of a technical matter, then more autocratic methods of decision making would save time and potential aggravation. While this sounds reasonable in principle, the question is, of course, who decides how each problem is to be approached?

Second, opening up decision making (problem solving) to an unlimited number of interested parties "might explode the analysis well beyond manageable bounds."[26] As we have seen, this is especially significant in the problem recognition and construction stages, when the conflict between the political establishment and dissenting minorities is most acute. By opening up these deliberations to a variety of groups, one may become engaged in verbal dueling that hinders the decision process. Yet deLeon[26] and Kimmins[29] argue that it is precisely these formative stages when public involvement is crucial. Subsequent implementation could be a matter left to technical managers. How the conflict that arises in such participatory exercises might be bounded remains uncertain, although there are some promising approaches that will be discussed shortly.

Third, in multiparty problem solving, it isn't clear to whom one should listen. The dilemma is acute for professionals who seek to maintain some semblance of neutrality. As deLeon[26] points out, should the professional listen to well-organized, vocal interest groups or to those groups with whom he or she has some affinity? What, then, of the unorganized, marginal individuals who have little voice in the proceedings but who are directly affected by the decisions being made? Should the professional adopt an advocacy role on their behalf or seek to maintain some modicum of impartiality? Thus, in the face of the information overload that accompanies participatory activ-

ities, the selection of information from the ensuing mayhem remains problematic. Because some degree of selective attention is a practical necessity, professionals are always open to the charge of bias, of not listening to some crucial point being made by one group or another.[26]

In conclusion, because social experimentation is always resisted by the majority, social change can only be achieved by the efforts of so-called change agents working with organized groups. Because of their access to education and information, I believe that professionals of all stripes are well placed to play such advocacy roles. Although, as I explain more fully later, I do not wish to promote the idea of a new professional priesthood to act as Guardians leading us to salvation, I would suggest the need for a radical shift in professional roles toward greater ecosocial militancy. Whether professionals are capable of this is something to which we now turn.

Psychological Development

Adaptive problem solving requires adaptive people. That is, we have to be willing to change to some degree. However, this does not mean that everyone has to undergo a radical transformation in order to contribute to adaptive problem solving. A more practical view is implied in Belbin's[30] research on effective management teams. It seems that effective teams are those in which team members are able to fill a set of essential roles (see chapter 3). In effect, these roles complement one another by achieving a balance between orthodoxy and dissent. The more orthodox roles provide stability and practicality, whereas dissenting roles introduce the elements of criticism and novelty that push deliberations beyond the status quo. Gunderson et al.[31] have come to similar conclusions based on their analysis of adaptive change in resource management. In their view, adaptive systems experience continuous, evolutionary change in a cycle of "birth, growth and maturation, death and renewal" (p. 510). Different players assume significant roles at each stage. For instance, bureaucrats and managers are particularly important during the growth and maturation stages of policy in the sense that they must implement current plans. The birth of new policy, however, lies in the hands of decision makers who construct new systems of resource exploitation. When policies are clearly near death, it is activists, loyal heretics, and rebel bureaucrats who draw attention to the potential dangers of maintaining the status quo. Finally, the renewal of policy requires the new visions that often emanate from epistemic communities and shadow networks of scientists and others who provide the intellectual basis for innovation.

While all of these subgroups are important elements in adaptive problem solving, effective collaboration requires that their various contributions be integrated into a coherent whole. The key players in this process have been referred to as " wise integrators,"[31] "transformers,"[32] and "boundary spanners,"[33] wise individuals who work behind the scenes stitching together compromises and collective agreements. They are successful by virtue of their professional competence and strength of character. For example, the "wise integrator is respected by players both inside and outside the system and is able to utilize traits of honesty while connecting knowledge to power in spite of countervailing political winds"[31] (p. 505). If the wise integrator also possesses visionary qualities ("capable of transforming myths among a wide group of people, spanning a variety of communities—technical, institutional, and political") and is a loyal heretic ("critically important in preparing bureaucracies and agencies for change by maintaining strong personal contacts both inside and outside the organization"), then the individual is a major force for constructive change.

My argument, then, is that we can encourage adaptive problem solving by attending to the structure of problem-solving groups[34] and ensuring, among other things, that wise integrators are included. Unfortunately, this is easier said than done for there is a dearth of wise integrators, or sages, as I prefer to call them. While the traditional educational system continues to produce orthodox thinkers in profusion, the same cannot be said for sages. Yet, without these wise individuals, who seem to

be able to reconcile competing visions while maneuvering through vested interests and political minefields, then adaptive problem solving will remain a fond hope. In what follows, I outline what is known about the personality of sages, as well as the steps that might be taken to encourage their development. However, I would like to add two caveats before proceeding. First, I reiterate that, in advocating a central role for sages in adaptive problem solving, I'm not offering yet another version of the elitist vision of an intellectual priesthood that will solve all of our problems. To the contrary, sages can arise in all walks of life, out of all socioeconomic classes and educational backgrounds. Their defining characteristic is wisdom, not membership of a privileged group. Second, the focus on sages does not absolve the rest of us from the need to become more flexible and self-reflective, something that I shall discuss later in the chapter. For the moment, however, let us turn to what is known about these interesting people.

In chapter 2, I described four personality prototypes, each of which exhibits specialized ways of thinking and striving developed after years of struggling to cope with the demands of our internal (psychological) and external (social and environmental) worlds. Our *personality*. therefore, is the more or less stable structure that we have taken a lifetime to construct. For some of us, this structure is relatively inflexible. We may be unable to adopt new ways of thinking or believing, even when the demand to do so is paramount. The changes involved are too disruptive to the whole system, causing levels of anxiety with which we cannot cope. This is another way of saying that many individuals are incapable of significant change, and attempts to impose such change is both a waste of time and potentially damaging to the person in question[35] It is for this reason that I suggest we pay most attention to the *personality generalist*,[36] the flexible individual who has the capacity to shift from one prototypic style to another depending on the demands of the situation. It is this generalist who has the potential to develop into a sage. What then are the characteristics of such a person?

Cognition

Cognitive Content (knowledge)

Sages exhibit wisdom, the capacity to make sensible judgments in the face of great uncertainty while being acutely aware of their fallibility in making such decisions.[37] Although there are many viewpoints about the kinds of knowledge that provide the cognitive basis for wisdom, one of the most interesting is Habermas's notion of three cognitive interests: technical, practical, and emancipatory.[38] Technical knowledge provides us with the scientific and technical information which allows us to cope with the biophysical environment. Practical knowledge is the social intelligence that helps us communicate effectively with others and get things done in the sociopolitical world. Emancipatory knowledge is the self and social awareness that allows us to understand how we are controlled by both our inner psychological world, as well as the social and natural forces that surround us. In this way of thinking, a judicious blend of all three kinds of knowledge lays the foundations for wisdom.[38] According to Chandler and Holliday,[38] however, this integration appears to be in short supply, because recent intellectual history has been "made up of a string of intellectual disasters . . . brought to a fine point by a variety of 'positivistic' philosophers of science whose legislative brand of 'scientism' . . . succeeded in largely ruling out of court all possible forms of knowledge save a neutered brand of technical expertise that renders the concept of wisdom essentially meaningless" (p. 123). Thus, an excessive concern for technical knowledge, at the expense of practical and emancipatory knowledge, has limited the development of wisdom, particularly among professionals.

Technical Knowledge

Wise choices cannot be made in the absence of substantive technical knowledge about the problem at hand. Because environmental problems are not amenable to solutions based on single disciplines, it follows that sages would need to be knowledgeable in, and comfortable with, an interdisciplinary context.

What this means in practice is not entirely clear, however, because there is considerable difference of opinion, even confusion, over the meaning of interdisciplinarity. Interdisciplinary cooperation can involve something as simple as a joint effort between a few, closely related disciplines or, in contrast, a more extensive collaboration across diverse disciplines and paradigms. In addition, collaboration can range from a multidisciplinary level, in which interactions are additive rather than integrative, to the more transdisciplinary efforts in which there is a mutual interpenetration of ideas.[39] Because these differences are important in understanding the intellectual capacities of the sage, they warrant closer scrutiny.

Multidisciplinarity is additive in the sense that the various bodies of knowledge remain separate. "Their relationship may be mutual and cumulative but not interactive, for there is no apparent connection, no real cooperation . . . The participating disciplines are neither changed nor enriched, and the lack of a well-defined matrix of interactions means disciplinary relationships are likely to be limited and transitory"[39] (p. 56). In other words, there is little mutual learning. A typical multidisciplinary situation is where a group of natural scientists and engineers enlist the support of a social scientist to provide additional information on the human dimension of the problem being addressed. There is no real provision for determining how this social science information might be used. Nor is there any mutual learning in the sense that the interaction between disciplines might encourage participants to change their view of the problem or their preferred, disciplinary ways of approaching it. Interaction is limited and safe, for there is no requirement to change how one thinks.

When multidisciplinarity exists within someone's head through education or experience across a number of disciplines, the same problem with compartmentalization can occur. "Exposure" to a variety of disciplines is no guarantee of integrative thinking. For example, human dimensions research that provides technical professionals with information on the attitudes of the public to a project is often safe because it does not require the professionals involved to question their own function within the prevailing socioeconomic system. Nor does the simple juxtaposing of natural and social science necessarily encourage professionals to begin reflecting on their own biases and prejudices. In other words, multidisciplinary thinking is a form of single-loop thinking, the pursuit of instrumental ends by adding together bits and pieces of information.

In contrast, transdisciplinary thinking is much more akin to double-loop learning.[31,40] That is, not only is information combined in an additive form but a second loop of activity occurs in which one monitors and reflects on the way one is approaching the problem, thereby exposing oneself to self-criticism and evaluation. To clarify, some years ago I entered into a joint appointment with the Forestry faculty of my university. I took on the rather intimidating task of bringing some social science expertise to the forest resource management stream of the program. My initial impression was that I was required to function in an additive role, simply providing information to students and faculty that they believed might be of use in coping with the growing number of people problems in forestry. In effect, I was to be added to the curriculum. As time went by, I began to realize that a more integrative role for psychosocial ideas would be as a critical mirror on the assumptions underlying traditional Forestry. Accordingly, I tried to interest students in a critical evaluation of the assumptions of political neutrality and scientific objectivity that permeate the discipline. However, this did not prove to be a successful endeavor, to say the least, in large part because of the clumsy way in which I tried to implement it. On the other hand, I did learn from this exposure to forestry that many of the traditional psychological skills and methods I had taken great pains to learn were essentially useless in the practical arena of forestry. Indeed, I came to the conclusion that a major portion of academic (positivist) psychology is engaged in fiddling while Rome burns; this was not a comforting thought. Thus, I was reduced to a more humble station in life through self-reflection on my failure to engage the students in meaningful dialogue. Sages would need to

go through this and similar learning experiences to develop their ability at transdisciplinary thinking. Hopefully, this would help them become more effective in interdisciplinary settings than I proved to be.

Practical Knowledge

Practical (social) knowledge and skills are essential in establishing and maintaining meaningful communication with others, as well as in getting things done in the social world. Knowing who to contact, how to approach a sensitive issue, what not to do, how to avoid political pitfalls, and so on, are all part of the social competence needed to be effective in problem solving. Social intelligence of this kind can be seen just as much in the streetwise youths of the inner city as in the battle-scarred bureaucrat, a survivor of the internecine warfare of government agencies. As one might expect, practical knowledge and social intelligence are not necessarily related to the kind of academic intelligence that is much prized in our educational systems. Yet this practical wisdom is a crucial element in problem solving. A case in point, albeit fictional, is the contrasting images portrayed in the movie *Mississippi Burning*, in which two FBI agents are dispatched to the southern states to unravel the mystery of missing civil-rights activists. The leader, a northerner with little experience in the South, approaches the problem by letting loose a tide of FBI agents to interview recalcitrant locals and to search likely spots where bodies might have been buried—a fruitless endeavor that provided the local villains with a great deal of amusement. All the while, our other hero, a grizzled veteran of the Bureau with local roots, went about his own investigation using what might be described as unorthodox methods. These included manipulating a disgruntled housewife and, ultimately, using intimidation of the town's mayor (one of the villains) to obtain the precise location of the burial site. This is not to say that the use of practical knowledge must necessarily result in unsavory or dubious practices, but it is a very political kind of competence that is used in both ethical and unethical ways. One would hope that sages would use their social skills in ethical ways—indeed, their very credibility depends on their reputation for honesty and trustworthiness.

The world of environmental politics provides us with many examples of the way in which practical knowledge is used. For example, many environmental activists have developed a worldly wise understanding of the way in which decisions are made within the technocorporate state, as well as how to intervene effectively in this process. This practical know-how is often published in manuals that provide practical guides for effective social involvement in environmental controversies.[41,42] One manual, mentioned previously, sketches the "Battle Plan" of industrial polluters and offers suggestions on how to cope effectively with the political tactics that the author of the manual believes are sometimes used by industry to diminish the effectiveness of grassroots groups.[41] In Collette's experience, these tactics include seeking to divide local groups by enlisting some of them on to citizen-advisory committees, and infiltrating grassroots organizations with spies and provocateurs whose aim it is to disrupt and derail the group's activities. "The 'spy' reports back to the opponent on internal conditions and strategies within the group. The 'provocateur' attempts to steer the citizens' group into taking unwise action . . . Another form the 'provocateur' could take is a person who disrupts the group by sowing dissension and getting members, especially top leadership, fighting with each other" (p. 27).

Although Collette believes these more nefarious tactics are relatively infrequent in the United States, local groups do worry about their being used. The manual offers sage advice on how to avoid becoming too paranoid about such issues, while providing practical steps to monitor and control any such situations, if they were to arise. This advice is the product of many years of experience in environmental politics and, though some may react to this and similar manuals with distaste, as an example of confrontational politics, it is an attempt to condense the wisdom gained by marginalized players in trying to influence the powerful. However, this kind of knowledge is

difficult to convey even in well-written manuals. Almost every page contains examples of situations where judgments have to be made, judgments which may be easy enough for the experienced but problematic for the novitiate[41] (p. 29):

I sometimes get calls from leaders who wonder if that 'long-haired, hippie-looking guy who always recommends radical stuff' is an agent provocateur from the opponent. I always doubt it and advise leaders that it would be a big mistake to stifle youthful enthusiasm. Instead, make sure that you have a clear and democratic process for making decisions so that the group can work through whether or not these ' radical' ideas appropriately fit into the group's plan. There's simply no substitution for paying attention to structural detail in a group. Its your best defense.

The same practical sense is needed when working inside a system, such as a bureaucracy, to reform it from within. A case in point is that of Gordon Baskerville's role in reforming forest policy in New Brunswick.[15,31] Up until the 1970s, policymakers had always assumed that there was enough wood supply in New Brunswick to justify both the expansion of industrial capacity and an increase in levels of harvesting. In the mid-1970s, however, Baskerville was one of a group of researchers who began to develop sophisticated simulation models of forest dynamics that appeared to show that the province would soon experience serious shortfalls in wood supply. At first, this ominous conclusion was resisted on a number of grounds, and the province continued with business as usual[43] (pp. 68–69):

The notion that the limit of timber supply had been reached was not readily accepted in the forest community, and that community worked at not believing . . . No one wanted to believe the bad news. This was partly unwillingness to admit that the problem had occurred under what were supposed to have been controlled forest management conditions.

This resistance to unpalatable news continued for a number of years. However, the information from the simulation models, together with the findings of two influential task forces on forest policy, seems to have swayed opinion toward a more realistic policy based on the recognition of wood-supply shortages. Although Baskerville[43] does not say so in his narration of the story, the changes that did take place were due, in no small part, to his efforts as a "wise integrator."[31] As a native of the province, he had spent most of his life in the forestry community in a variety of roles, ranging from that of a research manager with the Canadian Forestry Service, to a professor of forest ecology, and to an assistant deputy minister with the provincial government. With a foot in so many camps, and extensive experience in both technical and political forestry, he was well placed to play the role of change agent. His efforts would have been unsuccessful, however, in the absence of the kind of practical skills with which we are concerned here.

Yet many people have similar occupational experiences to that of Baskerville but remain abrasive, socially inept, and politically clumsy. One learns while the other doesn't. Reasons for this difference become evident when we examine the nature of practical knowledge more closely. Ford[44] recognizes two sets of skills that together contribute to practical competence: *prosocial* (social sensitivity, empathy, and social concern) and *social-instrumental* (ability to communicate, social know-how, ability to handle conflict, and leadership ability). All of these skills involve tacit knowledge, in the sense that they are difficult to teach formally and develop as a result of many years of appropriate experience. The predisposition to empathy and social sensitivity, for example, is learned at one's parents' knee over many years of exposure to parents who teach by example.[45] Without this formative experience, subsequent learning may remain stunted and one's capacity for social relations impoverished. Thus, streetwise children in urban ghetto's may develop street smarts by virtue of the demands of daily survival, but whether they become empathetic human beings would depend very much on their family circumstances.

The moral of the story is that although everyone has a modicum of practical wisdom, only a few have the capacity to exhibit these qualities to any significant extent. Attempts to enhance these qualities through education, for instance, would have to take into account the tacit nature of practical knowledge as well as its origins in personality development over the years preceding higher education. As I mentioned in chapter 2, this presents some problems in the education of scientific and technical professionals because many of them exhibit the characteristics of the objective-analytic and objective-holistic personality prototypes. Both types experience varying degrees of social and emotional detachment that does not augur well for the development of practical (social) knowledge, at least to the level required by adaptive problem solving. It follows, therefore, that proposals for a new kind of planning and management, which makes more use of professional judgment and practical, intuitive wisdom,[46-48] may founder in the absence of people with appropriate skill levels.

Emancipatory Knowledge

Emancipatory knowledge offers the means to free ourselves "from both the arbitrary forces of nature and the social structures that limit self-understanding"[38] (p. 122). That is, it provides the kind of critical insight into how our behavior is influenced both by the world around us and by our inner world of psychological biases and defenses. Recognition of the importance of this kind of knowledge has a long history,[49] and it is well known that its attainment requires us to engage in critical self-reflection.[50] However, self-analysis is not to everyone's taste because habitual, unreflective behavior is comforting, a sanctuary in troubled times. For others, this unreflective life is tantamount to living in a self-imposed prison constructed of our beliefs and prejudices. Doris Lessing[51] believes that the tendency to think in black and white terms, to split the world into "us" and "them," to submerge ourselves unthinkingly into groups, and the unwillingness to learn from the past are all ubiquitous among human beings. Most conceptions of wisdom, however, imply that this blinkered existence is counterproductive and, most certainly, a barrier to the development of adaptive forms of thinking. Professional myopia is a case in point.[52]

The status of professionals in society is based, at least in part, on the claim that they possess expert knowledge beyond the reach of ordinary mortals.[53] Many professionals actually believe this to be the case and, as a result of professional socialization, sally forth to take their rightful place in society. However, in chapters 2 and 3, and elsewhere, I've pointed out the way in which professional expertise is limited by psychological factors and controlled by political pressures. Professional expertise has many failings, therefore, especially when it is based primarily on technical rationality. Professionals who prefer to avoid this insight are, in the words of a visiting speaker addressing a graduating class of forestry students, walking "time bombs" waiting to wreak havoc on the unwary. Avoiding critical self-awareness of one's personal and professional limitations, and the fallibility of expertise, does not lend itself to the development of wisdom.[50]

It is also common for many professionals to regard themselves as being ideologically neutral, beyond the influence of ideology. As a result, one of the most common charges leveled at environmental activists is that they are *ideologues,* driven not by reason but by ideological fervor. The implication is that there is little point in trying to open up dialogue with them because it is impossible to break through the wall of ideology with which they are surrounded. Yet, as I've explained in earlier chapters, professionals are, like everyone else, profoundly ideological, riddled with value judgments and other unexamined assumptions. Refusal to acknowledge this simply means that these various biases will continue to influence the person's behavior despite the posture of objectivity. Thus, professionals are apt to believe that they are politically neutral, acting as impartial technical advisors and managers. This mantle of political detachment is commonly justified on the grounds that they are part of the overall process of representative democracy, and those who don't like the policies they implement can express their displeasure at the ballot box. Such a view, however, ignores the nature of power in society (chapter 4), especially the way in which elite groups have assumed and held on to power.[54] It also ignores the fact that bureaucracies are not elected and resist change even by the most powerful of leaders, elected or otherwise. The net effect is that professionals, especially those working in natural resource agencies, may be inclined to overlook the way in which their work is embedded within, and acts to maintain, the sociopolitical status quo.

Finally, many professionals also decry the "emotionality" of environmentalists, a defect they claim to have overcome in themselves by virtue of professional training. The ideal professional, therefore, is impartial and emotionally balanced, capable of withstanding the emotionalism of environmentalists and enraged citizens. We saw examples of this viewpoint in the professional response to the residents of Love Canal described at length by Levine.[55] However, the passions that swirl around scientific research,[56] the interpersonal conflicts and lasting animosities that arise during research projects,[57] the more mundane frictions of everyday organizational politics,[58] and so on, attest to the contrary. My own experience is that most technical professionals are convergers (chapter 2) and, as such, try to keep their wayward emotions tightly under control but succeed only in pushing them to some inner place where they proceed to fester. The subsequent impact of unacknowledged emotional conflict on both personal mental health and professional practice is well documented.[58-60]

The purpose of these examples is not to single out professionals for unwarranted criticism, for similar shortcomings can be found in all of the groups involved in environmental disputes, but, rather, to make the point that the unexamined life, one that is devoid of critical self-reflection, condemns one to live within the mental prison referred to by Lessing. On the other hand, self-analysis is fraught with danger.[35] It is not an easy or comfortable experience. Perhaps this is why it is avoided and even resented by so many. For instance, early in my career, I assumed, rather naively, that other people would be as interested as I was in learning about their problem-solving styles and the various psychological biases that impeded their problem-solving effectiveness. I also believed that they would be keen to divest themselves of these impediments. The first stage in my disillusionment came when I found myself teaching in a small agricultural college where I was expected to teach budding agriculturalists some ideas and skills that might be of use in their jobs as advisors to local farmers. I anticipated that my charges would benefit from training in interpersonal skills with some emphasis on introspective analysis of their problem-solving styles. This splendid exercise was to be conducted in the human relations manner that was so popular in the 1960s. The reaction to this was predictable, although, at the time, I was dumbfounded. About one-third of the students showed some interest, a further third were apathetic, and the remainder quickly developed such a sullen hostility toward me and my activities that the classes degenerated into a farce.

Matters came to a head when my colleagues and I decided that we might be more effective if we knew something about the learning styles of the students. Like the well-meaning scientific psychologists we were, we attempted to collect appropriate data using questionnaires. At this point, the antagonistic third started to flex their muscles. The more malevolent among them began a campaign of harassment made more effective by the fact that the college was residential with many faculty living on campus. We were spat upon, we had various objects thrown at us, our car tires were deflated, our sleep was interrupted, and we were subjected to verbal abuse. The pointed nature of this derision finally caused the penny to drop as far as I was concerned. I had just been reading Liam Hudson's[61,62] work on cognitive styles in which he described the links between cognitive convergence, authoritarianism, and masculinity. Because one of the more notable characteristics of the malevolent third was their strutting machismo, it occurred to me that at the root of their violent reaction was an offended masculinity. "Real" men do not spend their time discussing their feelings and values. I came away from this experience with some salutary lessons about psychology in the real world. Obviously, taking psychology to the people was more complex than I had imagined. Also, it could be potentially dangerous. Thus, what seemed at the time to be relatively innocuous subject matter was, evidently, profoundly threatening to many of these young men. In addition, attempting to conduct psychological research under those conditions was obviously farcical. Although nowhere as vicious as my first experiences, over the past 25 or so years I have had the same range of reactions to my efforts to encourage self-analysis and the examination of basic problem-solving styles. As a psychologist, this should not have surprised me. I should have been well aware that few, if any, people actually enjoy a fundamental reappraisal of their habitual modes of thinking.

In conclusion, very few people can manage to develop the sophisticated mix of technical, practical, and self-knowledge that one finds in sages. More commonly, we seem to emphasize one set of knowledge and skills over the others. It would help, of course, if education were to attend to these three components of wisdom, but, rather than doing so, conventional education concentrates on development in the technical area at the expense of the more psychosocial skills. The latter are assumed to develop by themselves in the absence of educational attention. This, of course, is a fond hope.

Cognitive Styles

In chapter 2, I outlined the strengths and weaknesses of two ubiquitous cognitive styles (analytic and holistic), concluding that adaptive problem solving required their integration into a more comprehensive way of thinking. Analytic thinking is effective in dealing with well-structured, tame problems but cannot cope with ill-structured messes. On the other hand, holistic thinking can provide useful intuitive insights and creative guesses about messy problem situations that, however, remain only promises rather than solutions without further analytic development. Massey and Wallace[63] are correct, I believe, when they point out that complex, ill-structured problems "do not lend themselves to structuring and formulation by quantitative models, nor by simple intuitive problem solving. Rather, making sense of these situations necessitates considering and oftentimes negotiating alternative models of the ill-structured situation" (p. 253). The question is, what sort of cognitive style would be effective in this rather fluid, uncertain atmosphere? What sort of cognitive style would our sage need to develop? Some guidelines are provided by research on post-formal reasoning.

Cognitive psychologists, who study the development of thinking skills in young people, make a distinction between what they call "concrete" and "formal" operations. Concrete thinking, which is typical of preadolescent children, allows the child to recognize simple cause-effect relationships in a very restricted time frame. Concrete children live very much in an egocentric present, observing the world through relatively superficial eyes, unable to delve below the surface or to go beyond the immediately obvious. With the onset of adolescence, however, there is a gradual shift toward more complex thinking in which one "sees the beginning of the hypothetico-deductive method of the scientist, the use of hypotheses formulated and tested as part of a gradual elimination of alternative explanations. During this final period of intellectual development the adolescent gradually acquires the ability to consider inter-relationships between several concrete properties and also to think logically about abstract principles"[64] (p. 169). This kind of thinking is referred to as *formal operations*.

For many years, as implied in the above quotation, formal operational thinking was considered to be the final stage in cognitive development. In other words, the abstract, analytic, scientific mode of thinking involved in formal operations was seen as the epitome of intellectual development. Nothing of any significance was thought to occur after this achievement, other than the further refinement of formal skills. Given this belief, one can see why analytic-positivist modes of thought dominate and are promoted by so many academic disciplines. In more recent years, however, as the shortcomings of formal operational thinking have become more evident, there has been growing awareness that important kinds of intellectual development take place beyond that of formal thinking. This *postformal thinking* is an expression of intellectual maturity in which there is a sophisticated integration of the analytic skills of formal operations with a highly developed kind of holistic thought. What results is a capacity to cope with conflict, uncertainty, and doubt.

The sequence of development in which a formal operational capability is transformed into postformal thought is one in which the belief in absolute truth, based on factual information, is replaced by a more relativistic mode in which absolutes are rejected in favor of the arbitrariness of "truth."[37,65] One can see this transition in some undergraduate students who arrive at a university seeking the absolute truth and are confused when professors offer them contradictory interpretations. At the formal operational stage, students may see alternative viewpoints as a situation where authorities "can't get their stories straight," a temporary confusion that will be remedied by further study.[66] This same way of thinking is commonly found among many scientifically and technically trained professionals who believe that there is a single correct way of conceptualizing a problem and that further research will not only determine what this is but also provide the means for its solution.

Although many students (and other adults) remain at this absolutist stage of thinking, others begin to realize that truth is relative, a reflection of the standpoint of the observer. This *relativistic* thinking brings with it a great deal of confusion as students thrash around trying to cope with the numerous contradictions with which they are faced. Uncertainty and a considerable degree of cynicism are a consequence of the release from absolutist thought.[65] What happens next is of interest to us. It seems that by virtue of a willingness to remain open to contradiction, together with liberal doses of critical self-reflection, a few fortunate souls develop what Kramer[65] (p. 291) refers to as *dialectical* thinking, an advanced capacity to resolve contradiction:

Relativistic and dialectical thinking are . . . well suited to the resolution of ill-structured problems, by taking into account indeterminism, contradiction, and change. However . . . these forms of thinking are not purely cognitive in nature but involve a great deal of affective involvement, as they are invoked in affectively laden situations and require conscious-unconscious integration for their development. In order to engage them . . . one must have an incisive understanding of oneself and one's interactions with others, the functioning of social systems, and the nature of change. Without an awareness of one's inner emotional life, such understanding is unlikely to ensue.

In other words, dialectical thinking is an extremely sophisticated achievement, something that requires a great deal of self-knowledge, as well as understanding of the social and interpersonal worlds with which we seek to solve problems.

In conclusion, sages have managed to acquire a formidable array of cognitive knowledge and skills that allow them to cope effectively with the complexity of wicked problems. Again, this is an achievement that most of us will never manage. How their skills can be nurtured and put to good use is discussed further in the next chapter. However, my sense of how these wise individuals develop is that they do so despite their incarceration in higher education rather than because of it. Indeed, the way to promote the development of sages is to protect them, as best one can, from the damage that can occur during conventional education.

Motivation

The wise integrator, the sage whose qualities I'm trying to elucidate, is not a remote Zen master, confounding his disciples with conundrums, but an environmental activist (in the broadest sense of the term) deeply embroiled in policy making. As such, the sage is faced with the problem of tempering his own sense of agency (Imperial power and assertiveness) with communion (Arcadian caring and empathy). Unmitigated agency leads, at minimum, to a gross insensitivity to others that, in extreme cases, becomes simple brutality. An excess of communion leads to misplaced oversensitivity and a dithering inability to act. The trick, if one can put it that way, is to find ways of problem solving that combine decisive action with sensitivity.[67] Wise actions, therefore, "blend intelligent autonomy and adaptability with kindness, human caring, empathic com-

munion, and compassion"[68] (p. 266). When couched in such general terms, this prescription for wise action sounds sensible enough. Who can disagree with the admonition that one should attempt to temper egocentric desires with concern for others? However, in trying to abide by these principles, one is confronted with a number of practical problems, not the least of which are the moral dilemmas involved.

Professionals, for example, differ in their perceptions of morality. They offer a variety of justifications for distinguishing between acceptable and unacceptable professional behavior. Endy and Vesilind[69] have summarized some of these differences among engineers within a moral development framework (Table 7.2). The first two stages of moral development are primarily egocentric. Professionals operating at this level put their own careers before anything else, grabbing whatever opportunities that arise to advance their prospects, regardless of the consequences for their firm, their profession, or society. In other words, their agentic striving is not mitigated by any concerns for others. At stages 3 and 4, one sees a beginning of a wider concern beyond oneself to an identification with that of the firm and one's profession. What is morally acceptable is behavior that is good for the firm (stage 3) and one's profession (stage 4). We see here the beginnings of moral conflicts between, for instance, one's professional standards and what one is required to do by one's bosses. These conflicts are further exacerbated at stage 5, where concerns are broadened to include the welfare of society as a whole. Morally acceptable behavior is defined not so much in terms of one's obligations to the firm or one's profession but to the well-being of society more generally. As long as professional behavior is kept within the bounds of legal codes and environmental regulations, and is socially acceptable, then it is considered good. Moral conflicts arise when professional standards and the requirements of one's job appear to contravene socially acceptable practices. Morally acceptable behavior in stages 3 to 5, therefore, is justified in terms of acceptance of, and adherence to, standards and norms established

7.2. Professional moral development.

Level 1. Preprofessional

Stage 1. At this level, the engineer is not concerned with social or professional responsibilities. Professional conduct is dictated by the gain to the individual, with no thought to how such conduct would affect the firm, profession, or society.

Stage 2. Recognition that there is some personal advantage to be gained from appropriate "professional" behavior. While there is some understanding of the notions of loyalty to the firm and client conficence, ethical professinal behavior is based on the motive of self-advancement

Level 2. Professional

Stage 3. At this stage, the engineer puts loyalty to the firm above any other consideration. The firm can dictate the proper action, and the engineer is freed from further ethical considerations. The engineer buries him or herself in technical matters and becomes a teamplayer who ignores the ramifications of the job on society and on the environment.

Stage 4. The engineer recognizes that the firm is part of the larger profession, so that loyalty to the profession enhances the reputation of the firm and brings rewards to the engineer. Engineering practice is viewed from a purely professional perspective, with no thought toward the larger issues of professional responsibility and social welfare.

Level 3. Principled Professional

Stage 5. Service to human welfare is considered paramount, and it is recognized that such service will also bring credit to the firm and profession. It is the rules of society that determine professional conduct, as long as these rules have been arrived at by democratic process.

Stage 6. Professional conduct is dictated by universal rules of justice, fairness and caring for fellow human beings and the whole of nature. This sometimes brings the engineer into conflict with the prevailing social order as deeply felt principles override social and professional expediency.

legally by one's social group, be it their firm, their profession, or society. In the final stage 6, however, moral judgments extend beyond these conventions to more universal principles of justice and fairness, as well as reaching back into one's own personal feelings of concern and compassion. Thus, stage 6 moral behavior reflects concerns with both justice (agency) and caring (communion). A case analysis provided by Endy and Vesilind[69] (pp. 117–118) might help clarify these differences:

Large consulting firms commonly have many offices, and often communication among the offices is less than efficient. Engineer Stan, in the Atlanta office, is retained by a neighborhood association to write an environmental impact study which concludes that the plans by a private oil company to build a petrochemical complex would harm the habitat of several endangered species. The client has already reviewed draft copies of the report and is planning to hold a press conference when the final report is delivered. A few days before the final report is presented to the client, the New York office becomes aware of the study and tells Stan to postpone its delivery. Engineer Bruce from New York flies down to explain to Stan that the oil company is one of the firm's most valued clients, and that the president of the oil company has found out about Stan's report and threatens to pull all of their business should the report be delivered to the neighborhood association. Bruce tells Stan to rewrite the report in such a way as to show that there would be no significant damage to the environment.

In this example, Bruce, the New York engineer, is recommending a course of action that puts the best interests of the firm before social concerns (stage 2 or 3). Stan could follow in his footsteps or take a higher road by withdrawing from the project on the grounds of conflict of interest, thereby asserting the priority of his professional obligations (stage 4) and taking his chances for continued employment. An even more courageous moral stance would be to refuse to change the report on the grounds that it would obscure potential harm to the community (stage 5) or continue working with the neighborhood group because of his concern for their welfare (stage 6).[69]

Many professionals, possibly the majority, are entangled in a web of professional and career obligations that defines acceptable behavior (stages 3 and 4). This professional stance limits the incidence of moral conflict in one's professional life, as illustrated by the engineers at two corporations who "claimed never to have encountered a conflict between their professional obligations and their roles as 'company men' "[70] (p. 81). Commenting on the same issue, however, Jack and Jack[71] argue that, at times, professionals may adopt a posture of neutrality, hiding behind rules and regulations that serve as a convenient way of avoiding the more unpleasant social consequences of their actions. Thus, conventional morality may obscure environmental injustice. It follows that stages 5 and 6, especially the latter, might be seen as the preferred moral positions, combining as they do a sense of justice with the more personal ethic of caring.[71] The question is how might these higher stages of morality be promoted?

A sense of fair play, decency, and caring all develop at an early age as a result of parental example.[72] In principle, the process is very simple. If a child is exposed to parents who show all of these desirable qualities in dealing with others and who treat their children with patient affection within a set of understandable and fair rules, then they have a chance of internalizing their parents' values. The children will want to be like their parents, at first out of a desire to please their parents and, as time goes by, because they love their parents. As a result, children will have a powerful reason to try to control their more agentic impulses (anger, aggression, and self-centeredness) in favor of social and emotional bonding with their parents and others. In the process, they may learn to have genuine feelings of empathy for those less fortunate than themselves. In contrast, many children are raised by relatively cold, authoritarian parents who may verge on being abusive. Under such circumstances, the emotional bonding that is so important in the development of a caring disposition simply does not take place, leaving troubled adolescents, unable to control their impulses and who neither understand notions of fair play nor care what others think.[73] Similar, though less extreme, experiences with parents may underlie the emotional and social detachment of the engineers studied by Hyde and Mc-Lean.[74] In their view, many engineers are "alienated," being only loosely attached to family and society, and motivated more by material considerations and the pleasures of technical virtuosity than anything else.

The point being made here is that the twin bases of higher levels of moral sensibility, a sense of fair play and emotional caring, start to develop early in life. Indeed, there seems to be a window of opportunity in the early years during which emotional bonding with others occurs, after which it is difficult to create genuine feelings of concern for others. As pointed out in chapter 2, it is possible that relatively few children are raised in the ideal family setting noted above, most parents being less than perfect. The result is that the basic requirements for stage 6 morality, deep-seated concerns for both fair play and caring, may not be too widespread.[75] This is another reason for believing that sages are likely to be few in number and that they will have to learn to cope with the rest of us who are inclined to be morally delinquent, at least from time to time.

In conclusion, wisdom involves a rare mix of cognitive and motivational attributes. While few of us seem to achieve this heady mixture, some do. If, as I suggest, sages play a crucial role in developing adaptive problem solving, then we need to pay more attention to the kinds of social upbringing and educational experiences that will encourage more sages to develop. This topic is discussed further in the next chapter.

Adaptive Ecosocial Management

In this final section, group problem-solving methods that hold promise for adaptive management are considered. There is a great deal of interest within the management and policy-making communities in this topic, especially finding ways to incorporate the public more meaningfully into the decision process. One sees a variety of proposals for new modes of public participation in "learning communities" and planning groups.[22,76-79] However, most of these recommendations do not deal explicitly with the thorny question of political devolution. In what follows, I assume a best-case scenario in which some degree of political devolution has occurred and community groups are engaged in participatory forms of decision making. This assumption allows us to explore the kinds of group problem-solving methods that would help such groups move toward adaptive problem solving. Community groups, being diverse in composition, will experience all of the psychosocial problems discussed in previous chapters. These range from the intellectual problems in combining lay and expert viewpoints to the more emotional conflicts involved in personality and ideological conflicts. It might be useful to start with ways of dealing with the intellectual problems faced by community groups.

As explained previously, the Imperial tradition in environmental management sought to deal with the problem of knowledge integration by adopting a systems orientation based on a positivist (analytic-reductive) epistemology. The result has been an emphasis on quantitative systems modeling and simulation. Of the numerous problems inherent in this hard-systems approach,[80] the most notable is the difficulty in incorporating soft data from the social sciences and lay experience. Because the latter represents a repository of wisdom deemed crucial in the Arcadian tradition, more effective ways of utilizing such knowledge need to be developed.

Again, as noted in chapter 6, Imperial reformers have turned to such strategies as Adaptive Environmental Management (AEM) in an effort to integrate some aspects of this qualitative (Arcadian) knowledge into more traditional quantitative (Imperial) methods. The problem formulation and modeling exercises at the core of AEM are but one example of scenario building, a ubiquitous strategy found in the field of planning. Scenario building is an attempt to link cause and effect in complex problem situations by generating predictions about the likely outcome of specific actions. It can be based primarily on quantitative modeling,[81] qualitative expert opinion,[82-84] or a combination of the two.[85] In practice, however, all of these methods succumb to the influence of positivism and the assumptions underlying hard-systems thinking, so typical of the objective-analytic and objective-holistic personality types who dominate these proceedings.

For example, Canada's Model Forest projects are part of the Canadian Government's "Green Plan" for promoting sustainable use of natural resources. One of these, the Fundy Model Forest (mentioned earlier), covers an area of 420,000 ha in southeastern New Brunswick, which has been logged intensively for about 200 years. Currently, 35,000 people live in the region that includes a large national park together with a mosaic of public and private landholdings. A variety of forest-related activities occur, including tourism, recreation, and the production of Christmas trees, timber, pulpwood, fuelwood, and maple syrup.[86] The stated aims of the Model Forest project are to make use of the forest's full social and economic potential while ensuring environmental sustainability. To do so, efforts have been made to develop an integrated management plan based on deliberations by a wide variety of "partners" drawn from government, industry, academia, environmental interest groups, and associations. On the face of it, the 5-year project appeared to be a reasonable attempt to develop sustainable practices by engaging the broader public in innovative planning methods. However, in practice, the design of the project was influenced strongly by traditional Imperial assumptions from the outset. This bias originated, in part, from scientific concerns about the ecological viability of Fundy National Park, a popular tourist destination located within the Model Forest's domain. Environmental managers and scientists in Parks Canada (a federal agency) have been concerned about the effect on the park of the habitat fragmentation that was occurring in the surrounding forest hinterland, due in part to industrial forestry. The response of Parks Canada was to mount an ecological research effort (the Greater Fundy Ecosystem project) that would provide the scientific information thought to be essential in determining how the hinterland might be managed to preserve the ecological integrity of the Park. Subsequently, members of the project played a significant role in obtaining the funding necessary to establish the larger Fundy Model Forest exercise, so that when the agency (through its research project) became a partner in the Model Forest, it brought along its substantive concerns and research orientation.[3] This introduced into the Model Forest deliberations the rational-comprehensive assumption that problems are to be solved by the accumulation of large quantities of new scientific data. At the same time, another major player, a multinational forestry company with significant landholdings in the area, was concerned with the viability of industrial forestry in the face of growing demands from other users. This succeeded in establishing a central role for mercantile interests in the project, thereby creating a basic tension between ecological research aimed at preservation of the integrity of the Park and the more operational concerns of the forest industry. Both, however, were couched within the Imperial tradition in the sense that it was assumed that scientific research would provide answers to the Park's problems and that industrial forestry should continue to be the mainstay of economic operations in the region. One can see the influence of these fundamental concerns reflected in the project's budget. For instance, during the period 1994–1995, approximately 25 percent of the budget was allocated to primarily ecological research while 35 percent was spent on industrial operations such as silviculture, new harvesting methods, and wood-supply modeling. The psychosocial aspects of the project received short shrift, with only about 11 percent of the total budget going to communications and education, and a tiny fraction of research funds going to socioeconomic research.[87] Thus, although broad public involvement in an innovative planning process was a central aim of the project, the allocation of funds was decidedly Imperial in orientation. Indeed, a study of the project's decision-making process concluded that the level of community involvement achieved was very modest.[88] It is no wonder, then, that a survey of local awareness of the project indicated that 72 percent of respondents were not at all familiar with the project and only 4 percent saw themselves as being very familiar with it.[89] Midway through the project, therefore, those involved in the more psychosocial aspects seemed to be struggling to find a meaningful role for themselves. For instance, the October 14, 1994 minutes of the Information and Education Committee recorded that it was still debating whether or not to change its name yet again (there had already been four name changes).

The prospects for a full-blown exploration of more Arcadian approaches to the problems of the Fundy area, such as community forestry and industrial devolution, would seem to have withered within this Imperial context. Instead, the project managers are attempting to cope with the information overload characteristic of the rational-comprehensive approach:[87]

> One of the big questions at this stage of the model forest, is how the knowledge gained, from the various research and operational projects now under way, will be brought together and used in the development of an integrated management plan? . . . A GIS based forest simulation model will be used to develop a strategic plan on a small portion of the model forest. By involving partners that represent the range of resource interests, it is hoped that we can focus in on how the model can be used to aid in making an integrated resource plan.

To some of the scientists involved, however, this is not a major problem. Rather, they are gratified that it has been possible to bring a (scientific) research focus to the Model Forest project that has resulted, they believe, in a greater level of dialogue among the partners than has been possible hitherto, thereby contributing to an "enormous increase in the common level of understanding and appreciation of each others' problems"[3] (p. 13).

On the other hand, it could be argued that the use of Geographic Information Systems (GIS) and other computerized decision aids makes the inclusion of Arcadian views even more remote, because the organization of planning around quantitative modeling disadvantages and disenfranchises those with nontraditional views. This is not unique to the Fundy Model Forest project but is a ubiquitous problem. For instance, industrial development in the Columbia River basin over the last century led to a growing demand for power and the subsequent construction of numerous major dams and dozens of smaller hydroprojects.[90] As a result, the once abundant salmon runs were reduced to a fraction of their former glory. In the early 1980s, the four state governments involved were faced with the prospect of having to increase their power-generating capacity in the face of opposition from environmentalists, who favored conservation, and Indian tribes who were distressed about the threat to their culture posed by the declining fish stocks. In response, an interstate planning Council was formed to develop ways and means of coping with these pivotal problems.[90] As an appendage of state government, however, the Council was established within the Imperial mode. In its initial planning exercises, for instance, the Council used systems models that had been developed by a regional power utility and the Army Corp of Engineers.[91] However, both fisheries agencies and tribal authorities objected to this on the grounds that the models' assumptions favored the power utilities' interests. In a spirit of collaborative problem solving, the Power Council responded by building new models in cooperation with interested parties.[91] These were used to explore alternative scenarios and were available to interest groups along with expert help from the Council. As McLain and Lee[91] (p. 443) conclude, however:

Because the council is required by law to make decisions on the basis of the 'best scientific knowledge' available, scientific information has become a key weapon in the political struggles over how to manage the Columbia River Basin. Stakeholders who have the capacity to produce and analyze scientific data are in a much more powerful position than those who do not. A 'model war' has broken out as the various stakeholders compete to produce the 'best' science.

Thus, in both the Fundy and Columbia River cases, problem-solving efforts reflected Imperial assumptions about the central role of quantitative, scientific data, as well as the continuing importance of industrial production. The more radical kinds of Arcadian viewpoint, which call for a fundamental reappraisal of both assumptions, appear to have had little impact on the unfolding events. While this may be in large part due to unresolved political issues, at least one major hurdle to effective integration of ecological knowledge lies with the methods available for this purpose. What I conclude, along with others, is that hard-systems thinking needs to be replaced, or at least complemented, by a soft-systems alternative.

Soft-Systems Thinking

Soft-systems thinking is a qualitative approach to the integration of ideas that differs in some fundamental ways from the more traditional, quantitative, hard-systems methods. The latter approaches to problems use a convergent strategy in which a single correct problem definition is adopted, with subsequent analysis fleshing out the details of this single definition. In contrast, soft-systems approaches attempt a much more comprehensive, even divergent, problem definition, one that takes the form of a rich picture, not of the problem alone but also of the human context in which the problem is embedded. In addition, rather than seeking to elaborate any single correct way of looking at the problem, several contrasting or conflicting problem definitions are pursued to ensure a broad coverage of issues.[92] The distinction, then, is between a hard-systems approach, which was developed by engineers and scientists for complex but tame problems, and a more flexible soft-systems methodology developed by management professionals for coping with complex wicked problems.[80] As Brown and MacLeod[92] point out, soft-systems methods have been used primarily in dealing with management and social problems and have only recently begun to be used in natural resource and environmental management.[7,93,94] To get a better grasp of what is involved in soft-systems analysis, it might be useful to look at how it could be used in developing new approaches to the spruce budworm problem (chapter 4). In the following discussion, I draw heavily on Checkland's conception of soft-systems methodology.[80] I also assume some degree of political devolution has taken place so that the SBW problem is now the responsibility of community groups. Because SBW infestations occur on such a large scale, it follows that no single community group could deal with it in isolation. We need to think, therefore, in terms of community networks of cooperating groups and provincial organizations.

Step 1

The first task faced by the community network is to develop a rich picture of the problem and its surrounding context. Rather than seeking to delimit the problem, efforts are made to maintain as broad a perspective as time and resources permit. A rich picture of the budworm problem would include not only its ecological and forestry components but also the history of the pest, its role in forestry and the forest industry, its impact on the political economy of the province, the social groups and communities directly involved in forestry, the potential medical impacts of the use of pesticides, and so on. By developing a rich picture at this early stage in the proceedings, a holistic perspective is maintained that helps to avoid the premature, myopic narrowing of attention that often occurs in problem solving. Much of this picture may turn out to be irrelevant, but the separation of relevant from irrelevant information should come later, when a better understanding of the problem has been achieved. An additional benefit of extending deliberations in this way is that the concerns of all parties, especially lay participants, can be incorporated into the developing problem definition. In contrast, consider the traditional, hard-systems approach to modeling that played such a prominent role in spruce budworm politics a decade ago. This model was constructed by technical professionals and limited to predicting the interactions between the SBW and a few commercially important timber species under different harvesting regimes. It did not include, nor could it illuminate, the human systems involved in the problem nor, indeed, little of relevance to the noncommercial use of the forest. Such a model served its purpose within the Imperial tradition, but from an Arcadian perspective it was mute on the most important aspects of the problem. At the end of the first step, therefore, the community networks should have a comprehensive, qualitative picture of the problem in all its glory.

Step 2

The second step involves trying to identify the structures and processes at work within the problem situation, especially those that operate on different scales and at different speeds. Because there is no urgency to begin intensive analyses or rush to action, this identification process remains broad in its reach. The purpose of this stage is to distinguish those elements of the problem that lend themselves to immediate action from those that operate on a scale that is outside one's scope. Thus, the cyclical nature of SBW epidemics operates on a 30- to 40-year timescale and is a function of the fundamental structure of the New Brunswick forest. Similarly, the political economy of New Brunswick is dependent on the pulp and paper industry to such an extent that fundamental change is impossible without wrenching social dislocation. In both instances, therefore, there is little that can be done to alter these systemic givens in dealing with any new epidemic. In the short term, therefore, the province would have to fall back on the old remedy of aerial spraying or, possibly, consider the more radical possibility of using SBW as a silvicultural agent by allowing it to devastate selected areas of the forest. Longer-term possibilities to do with altering the structure of the forest and forest industry could be considered, in which case use of pesticides would be a temporary expedient buying the necessary time. However, rather than getting too involved in policy options, this early stage in a soft-systems analysis would seek only to identify the scales at which ecological and economic processes are operating and the major players or stakeholders in the drama.

Step 3

Having developed a rich picture of the problem, together with its major stakeholders, the third step involves imposing some structure on further deliberations. Checkland refers to this as selecting "root definitions" or "hypotheses concerning the eventual improvement of the problem situation by means of implemented changes which seem to both systems analyst and problem owners to be both 'feasible' and 'desirable' "[80] (p. 167). Root definitions play the important role of making explicit the basic assumptions of those studying the problem, as well as providing different starting points for alternative approaches to the same problem. For example, the four main personality types discussed in chapter 2 could well produce quite different root definitions of the SBW problem. Objective-analytic persons might start with the assumption that the problem is an insect plague that threatens the commercial wood supply and must be brought under control by biocidal methods as quickly as possible. Alternatively, the objective-holistic perspective would probably suggest that because SBW is an endemic insect that is part of the natural ecology of the forest, it would be better to use it as a natural silvicultural tool rather than attempting to exterminate it. Subjective-holistic personalities might prefer to see the problem as a reflection of an unbalanced political economy that needs to be diversified, making it less dependent on tree species that are vulnerable to periodic attack from the pest. Finally, subjective-analytic individuals would be more inclined to develop a philosophical and psychological root definition and stress the fact that there is no SBW problem as such, only an overweening desire for material consumption by the citizens of New Brunswick that needs to be restrained.

Step 4

To this point in the proceedings, the community network has operated as a single unit, developing a heterogeneous problem definition and rooting around for hypotheses. However, with the advent of different root definitions, the subsequent modeling process begins to diverge. Rather than taking steps to constrain this centrifugal tendency, as one would in a hard-systems analysis, diversity of models is encouraged in soft-systems thinking. Such diversity provides a conceptual basis for the subsequent "dialectic" debate over assumptions

and beliefs. The fourth stage of the soft-systems method, therefore, involves the elaboration of some or all of these root definitions into conceptual models of the problem and its context. Especially important is the identification of those human systems that are amenable to change and the incorporation, into the models, of ideas about how these changes might actually be implemented. In other words, models in the soft-systems approach are not attempts to depict reality as such but, rather, schemes that provide a way of thinking about how to mobilize human resources in pursuit of one's goals. On the other hand, there is no reason why quantitative methods[95] should not play an important role in refining the models being developed. For instance, elaboration of a root hypothesis that explored the idea of using SBW as a silvicultural agent would have to include a technical GIS component that helped organize harvesting and protection of the forest, as well as a more socioeconomic component dealing with the human logistics involved. On the other hand, a model dealing with changes in the political economy of New Brunswick would be an entirely different matter. Attention would have to focus on the long-term political and economic changes that would be needed to achieve a major restructuring of the forest industry. Quantitative modeling would help in this process but would be a secondary consideration to that of understanding how to develop political coalitions and institutions. Pursuit of these different options is best conducted by subgroups formed within the community network itself. Most likely, these will be self-selected as smaller groups of stakeholders with similar vested interests join together in common cause.

Step 5

As Checkland notes, it is a matter of judgment when to stop this model-building process and proceed to the next stage of evaluation and comparison. This crucial fifth stage is essentially an attempt to use the models that have been developed to generate debate over the merits of differing assumptions. This technique is not unique to soft-systems thinking but has also been advocated as a form of "decision-making experiment" by Mitroff and others.[96-98] By engaging the proponents of the different models in a debate over the assumptions underlying each approach, it is believed that each subgroup not only will learn to understand, their own and others' suppositions and concerns but will also develop a basis on which to negotiate compromise. The "dialectical debate" can be as formal or informal as the community group wishes. Each subgroup can circulate details of its model prior to a plenary session in which it presents its model, together with what its members believe to be the main assumptions underlying it. Other subgroups can then debate the merits of each model, elaborating further what they believe to be the subgroup's misconceptions. Under the best conditions, one achieves a thorough evaluation of the various stakeholders' assumptions, concerns, and vested interests. For instance, the assumed political neutrality of the scientific establishment would be difficult to maintain if a management subgroup were to present a model that implied continuation of prevailing arrangements for control of New Brunswick's Crown lands. Similarly, if an environmental subgroup argued for the use of a model that assumed the tripartite subdivision of Crown lands, it would take little discussion to reveal their deep-seated concerns about industrial forestry. Keeping such a discussion within bounds is, of course, a difficult task, one that occupies the final stage.

Step 6

Because no single approach to ill-structured problems holds the promise of adaptive solutions, the most sensible route is to look for compromise between conflicting viewpoints. The final stage of a soft-systems approach, therefore, is to settle on a plan of action by seeking an integration of the various models that have been developed. Achieving such a consensus is not easy because underlying conflicts, often of long standing, are likely to have been triggered and treasured assumptions questioned. At this stage, the proceedings can become rather brittle, to say the least. Nevertheless, there has been considerable success in resolving such conflict by introducing a formal negotiation component into the exercise.[79,93,99-102] Such mediation and negotiation can take place within the community group itself or, in some instances, an external mediator can be brought in for the purpose. In either case, compromise over tendentious environmental issues is likely to be a difficult and lengthy prospect. For instance, in the case of spruce budworm in New Brunswick, both sides in the dispute seem to recognize that aerial spraying of pesticides is undesirable and merely buys time in which to implement changes in forest management. It is the specific nature of these changes that leads to so much dispute. The political-corporate establishment seek to sustain the prevailing industrial system, while environmentalists argue for a more diversified industrial base. These aims are not necessarily incompatible. The problem faced by the players is to devise a management plan for the transition to some new state of cooperative bliss that would not result in economic disaster. A soft-systems approach could help in this regard.

In conclusion, there are many interesting new approaches to group problem solving and decision making that, in light of the available evidence, should help in the development of adaptive responses to problems. However, in my experience, a major obstacle is the resistance of groups to the adoption of these techniques. Often people feel that they are being manipulated, put through a series of hoops, the purpose of which they cannot understand and with which they quickly become frustrated. Among professionals, for instance, there is the feeling that they do not need these esoteric planning methods. Indeed, they may feel insulted at the very suggestion, because it implies that they are incapable of effective problem solving without some external assistance. Resistance to the adoption of such methods, however, is more likely to be a form of psychological avoidance. Which professional wants to have his or her assumptions and beliefs questioned by lay people in a public forum? More fundamentally still, who wants to accept that his or her beliefs are flawed or, at the very least, biased?

It is for these reasons that I have not addressed the various ways in which interpersonal problems can be dealt with in group problem solving. In previous chapters, I have pointed out the debilitating effects of personality and ideological conflicts on decision making. Community groups will be no less immune to these problems than any other. However, while psychology has a number of methods for addressing these issues, such as group therapy and human relations training, the prospect of their being adopted by problem-solving groups in the arena of environmental management is remote. Given the stubborn resistance to new *intellectual* methods, one can see that trying to use methods that paid explicit attention to the more *personal* aspects of group behavior would be extremely aversive to many, if not most, participants. However, there is some limited prospect that group training would be tolerated in the educational system by at least some students. What, then, are the prospects for the adoption of new approaches needed to encourage adaptive problem solving?

References

1 Hausler, S. 1993. Community forestry: A critical assessment—The case of Nepal. *The Ecologist* 23:84–90.

2 Smil, V. 1993. *Global Ecology: Environmental change and social flexibility.* London: Routledge.

3 Woodley, S., and B. Freedman. 1995. The Greater Fundy Ecosystem Project: Toward ecosystem management. *The George Wright Forum* 12:7–14.

4 Maser, C. 1996. *Resolving environmental conflict: Towards sustainable community development.* Delray Beach, Fla.: St. Lucie Press.

5 Maser, C. 1997. *Sustainable community development: Principles and concepts.* Delray Beach, Fla.: St. Lucie Press.

6 Gowdy, E. 1994. From technical rationality to participating consciousness. *Social Work* 39:362–370.

7 Macadam, R., et al. 1995. A case study in development planning using a systems learning approach: Generating a master plan for the livestock sector in Nepal. *Agricultural Systems* 49:299–323.

8 Shepherd, A. 1995. Participatory environmental management: contradiction of process, project and bureaucracy in the Himalayan foothills. *Public Administration and Development* 15:465–479.

9 Shindler, B., B. Steel, and P. List. 1996. Public judgments of adaptive management: A response from forest communities. *Journal of Forestry* 94:4–12.

10 Becker, C., and E. Ostrom. 1995. Human ecology and resource sustainability: The importance of institutional diversity. *Annual Review of Ecology and Systematics* 26:113–133.

11 Clogg, J. 1997. Tenure Reform for Ecologically and Socially Responsible Forest use in British Columbia. M.EnvS. thesis, York University, North York, Ontario, Canada..

12 Durning, D. 1993. Participatory policy analysis in a social service agency: A case study. *Journal of Policy Analysis and Management* 12:297–322.

13 Betts, M., and D. Coon. 1996. *Working with the woods: Restoring forests and community in New Brunswick.* Fredericton: Conservation Council of New Brunswick.

14 Hilborn, R., C. Walters, and D. Ludwig. 1995. Sustainable exploitation of renewable resources. *Annual Review of Ecology and Sytematics* 26:45–67.

15 Holling, C. 1995. What barriers? What bridges? In *Barriers and bridges to the renewal of ecosystems and institutions,* ed. L. Gunderson, C. Holling, and S. Light, 3–34. New York: Columbia University Press.

16 Lee, R. 1992. Ecologically effective social organization as a requirement for sustaining watershed ecosystems. In *Watershed management: Balancing sustainability and environmental change,* ed. R. Naiman, 73–89. New York: Springer-Verlag.

17 Conservation Council of New Brunswick. 1994. *Public lands in public hands: Managing Crown forests in the public interest: A Conservation Council proposal.* Fredericton: Conservation Council of New Brunswick.

18 Hammond, H. 1991. *Seeing the forest among the trees: The case for wholistic forest use.* Vancouver: Polestar Press.

19 Binkley, C. 1993. Creating a knowledge-based forest sector. *The Forestry Chronicle* 69:294–299.

20 Wiedemann, P., and S. Femers. 1993. Public participation in waste management decision making: Analysis and management of conflicts. *Journal of Hazardous Materials* 33:355–368.

21 Bush, M. 1990. *Public participation in resource development after project approval.* A Background Paper Prepared for the Canadian Environmen-

tal Assessment Research Council. University of Calgary, Calgary, Alberta, Canada.

22 Sample, V. 1993. A framework for public participation in natural resource decisionmaking. *Journal of Forestry* 91:22–27.

23 Higgelke, P., and P. Duinker. 1993. Open doors: Public participation in forest management in Canada. Report to the Canadian Pulp and Paper Association and Forestry Canada. Lakehead University, Thunder Bay, Ontario, Canada.

24 Dryzek, J. 1990. *Discursive democracy: Politics, policy, and political science.* Cambridge: Cambridge University Press.

25 Fischer, F. 1990. *Technocracy and the politics of expertise.* Newbury Park, Calif.: Sage.

26 deLeon, P. 1990. Participatory policy analysis: Prescriptions and precautions. *Asian Journal of Public Administration* 12:29–54.

27 Matland, R. 1995. Synthesizing the implementation literature: The ambiguity-conflict model of policy implementation. *Journal of Public Adminitration Research and Theory* 5:145–174.

28 Thomas, J. 1993. Public involvement and governmental effectiveness: a decision-making model for public managers. *Administration and Society* 24:444–469.

29 Kimmins, J. 1993. Ecology, environmentalism and green religion. *Forestry Chronicle* 69:285–289.

30 Belbin, R. 1981. *Management teams: Why they succeed or fail.* London: Heinemann.

31 Gunderson, L., C. Holling, and S. Light. 1995. Barriers broken and bridges built: A synthesis. In *Barriers and bridges to the renewal of ecosystems and institutions,* ed. L. Gunderson, C. Holling, and S. Light, 489–532. New York: Columbia University Press.

32 Perelman, L. 1976. *The global mind: Beyond the limits to growth.* New York: Mason/Charter.

33 Tushman, M., and T. Scanlan. 1981. Boundary spanning individuals: Their role in information transfer and their antecendents. *Academy of Management Journal* 24:289–305.

34 Innes, J. 1992. Group processes and the social construction of growth management. *Journal of the American Planning Association* 58:440–453.

35 Winter, D. 1996. *Ecological Psychology: Healing the split between planet and self.* New York: Harper Collins.

36 Barmark, J., and G. Wallen. 1981. The development of an interdisciplinary project. In *The social process of scientific investigation, Sociology of the sciences yearbook, volume IV (1980),* ed. K. Knorr, R. Krohn, and R. Whitley, 221–235. Dordrecht, Holland: D. Reidel Publishing Company.

37 Kitchener, K., and H. Brenner. 1990. Wisdom and reflective judgment: knowing in the face of uncertainty. In *Wisdom: Its nature, origins, and development,* ed. R. Sternberg, 212–229. Cambridge: Cambridge University Press.

38 Chandler, M., and S. Holliday. 1990. Wisdom in a postapocalyptic age. In *Wisdom: Its nature, origins, and development,* ed. R. Sternberg, 121–141. Cambridge: Cambridge University Press.

39 Klein, J. 1990. *Interdisciplinarity: History, theory, and practice.* Detroit: Wayne State University Press.

40 Knight, B. 1996. Reflecting on 'Reflective Practice'. *Studies in the Education of Adults* 28:162–184.

41 Collette, W. 1989. *The polluters' "secret" plan: And how you can mess it up."* Arlington, Va.: Citizens' Clearing House for Hazardous Wastes, Inc.

42 Merrifield, J. 1989. *Putting the scientists in their place: Participatory research in environmental and occupational health.* New Market, Tenn.: Highlander Research and Education Center.

43 Baskerville, G. 1995. The forestry problem: Adaptive lurches of renewal. In *Barriers and bridges to renewal of ecosystems and institutions,* ed. L. Gunderson, C. Holling, and S. Light, 37–102. New York: Columbia University Press.

44 Ford, M. 1986. For all practical purposes: criteria for defining and evaluating practical intelligence. In *Practical intelligence: Nature and origins of competence in the everyday world,* ed. R. Sternberg and R. Wagner, 183–200. Cambridge: Cambridge University Press.

45 McMartin, J. 1995. *Personality psychology: A student centered approach.* Thousand Oaks, Calif.: Sage.

46 Brunner, R., and T. Clark. 1997. A practice-based approach to ecosystem management. *Conservation Biology* 11:48–58.

47 Friedman, J. 1994. The utility of non-Euclidean planning. *Journal of the American Planning Association* 60:377–379.

48 Maljers, F. 1990. Strategic planning and intuition in Unilever. *Long Range Planning* 23:63–68.

49 Sternberg, R. 1990. *Wisdom: Its nature, origins, and development.* Cambridge: Cambridge University Press.

50 Fien, J. and R. Rawling. 1996. Reflective practice: A case study of professional development for environmental education. *Journal of Environmental Education* 27:11–20.

51 Lessing, D. 1986. *Prisons we choose to live inside.* Montreal: CBC Enterprises.

52 Petak, W. 1980. Environmental planning and management: The need for an integrative perspective. *Environmental Management* 4:287–295.

53 Kennedy, M. 1987. Inexact sciences: Professional education and the development of expertise. In *Review of Research in Education*, ed. E. Rothkopf , 133–167. Washington, D.C.: American Educational Research Association.

54 Chomsky, N. 1989. *Necessary illusions: Thought control in democratic societies*. Montreal: Canadian Broadcasting Corporation.

55 Levine, A. 1982. *Love Canal: Science, politics, and people*. Lexington, Mass.: Lexington Books.

56 Mahoney, M. 1976. *Scientist as subject: the psychological imperative*. Cambridge, Mass.: Ballinger.

57 Arney, W. 1991. *Experts in the age of systems*. Albuquerque: University of New Mexico Press.

58 Hirschhorn, L. 1988. *The workplace within: The psychodynamics of organizational life*. Cambridge, Mass.: MIT Press.

59 Baum, H. 1987. *The invisible bureaucracy: The unconscious in organizational problem solving*. New York: Oxford University Press.

60 Ewens, W. 1984. *Becoming free: The struggle for human development*. Wilmington, Del.: Scholarly Resources Inc.

61 Hudson, L. 1968. *Contrary imaginations*. Harmondsworth: Penguin.

62 Hudson, L. 1970. *Frames of mind*. Harmondsworth: Penguin.

63 Massey, A., and W. Wallace. 1996. Understanding and facilitating group problem structuring and formulation: Mental representations, interaction, and representation aids. *Decision Support Systems* 17:253–274.

64 Entwistle, N. 1981. *Styles of learning and teaching*. Chichester, UK: Wiley.

65 Kramer, D. 1990. Conceptualizing wisdom: the primacy of affect-cognition relations. In *Wisdom: Its nature, origins, and development*, ed. R. Sternberg, 279–313. Cambridge: Cambridge University Press.

66 Rybash, J., and P. Roodin. 1989. Making decisions about health-care problems: A comparison of formal and postformal modes of competence. In *Adult development: Comparisons and applications of developmental models*, ed. M. Commons et al., 217–235. New York: Praeger.

67 Birren, J., and L. Fisher. 1990. The elements of wisdom: overview and integration. In *Wisdom: Its nature, origins, and development*, ed. R. Sternberg, 317–332. Cambridge: Cambridge University Press.

68 Pascual-Leone, J. 1990. An essay on wisdom: toward organismic processes that make it possible. In *Wisdom: Its nature, origins, and development* ed. R. Sternberg, 244–278. Cambridge: Cambridge University Press.

69 Endy, E., and P. Vesilind. 1986. Moral development and professional engineering. In *Environmental ethics for engineers*, ed. A. Gunn and P. Vesilind, 111–121. Chelsea, Mich.: Lewis Publishers.

70 Florman, S. 1987. *The civilized engineer*. New York: St. Martin's Press.

71 Jack, R., and D. Jack. 1989. *Moral visions and professional decisions: The changing values of women and men lawyers*. Cambridge: Cambridge University Press.

72 Damon, W. 1988. *The moral child: Nuturing children's natural moral growth*. New York: The Free Press.

73 Vasey, P. 1995. *Kids in the jail: Why our young offenders do the things they do*. Windsor, Ontario: Black Moss Press.

74 Hyde, R., and G. McLean. 1992. *Alienated engineers: Part of the problem*. Paper presented at the Canadian Sociology and Anthropology Association meeting, May 30. University of Prince Edward Island, Charlottetown, P.E.I., Canada.

75 Bebeau, M., and M. Brabeck. 1987. Integrating care and justice issues in professional moral education: A gender perspective. *Journal of Moral Education* 16:189–203.

76 Diemer, J., and R. Alvarez. 1995. Sustainable community, sustainable forestry: A participatory model. *Journal of Forestry* 93:10–14.

77 Hancock, T. 1996. Health sustainable communities: Concept, fledgling practice and implications for governance. *Alternatives Journal* 22:18–23.

78 Irwin, A., S. Georg, and P. Vergragt. 1994. The social management of environmental change. *Futures* 26:323–334.

79 Selin, S., and D. Chavez. 1995. Developing a collaborative model for environmental planning and management. *Environmental Management* 19:189–195.

80 Checkland, P. 1981. *Systems thinking, systems practice*. Chichester, U.K.: Wiley.

81 Harwell, M., et al. 1996. Ecosystem management to achieve ecological sustainability: The case of south Florida. *Environmental Management* 20:497–521.

82 Branson, C., et al. 1989. Great Lakes policy exercise: Lake St. Clair feasibility study: Project completion report. University of Michigan, Ann Arbor, Michigan.

83 Toth, F. 1988. Policy exercises: Objectives and design elements. *Simulation and Games* 19:235–255.

84 Underwood, S. 1988. The policy exercise: Cooperative learning for long-run policy assess-

ment. In *Simulation/gaming in education, training, and development,* ed. A. Cecchini et al. Oxford: Pergamon Press.

85 Robinson, J. 1992. Risks, predictions and other optical illusions: Rethinking the use of science in social decision-making. *Policy Sciences* 25:237–254.

86 Anonymous. 1993. *Fundy Model Forest: An invitation to get involved.* Sussex, N.B.: Fundy Model Forest Inc..

87 Anonymous. 1994. *Fundy Model Forest: Work Plan 1994–5.* Sussex, N.B.: Fundy Model Forest Inc.

88 Arsenault, D. 1996. Community participation in forest management decision-making in New Brunswick. Research report for Forest Protection Ltd./Fundy Model Forest/Jacques Whitford, Sussex, N.B., Canada.

89 Anoynmous 1995. Public awareness of the Fundy Model Forest. Research Report, May 29, Baseline Market Research Ltd., Fredericton, N.B., Canada.

90 Lee, K. 1993. *Compass and gyroscope: Integrating science and politics for the environment.* Washington, D.C.: Island Press.

91 McLain, R., and R. Lee. 1996. Adaptive management: Promises and pitfalls. *Environmental Management* 20:437–448.

92 Brown, J., and MacLeod. 1996. Integrating ecology into natural resource management policy. *Environmental Management* 20:289–296.

93 Daniels, S., et al. 1996. Using collaborative learning in fire recovery planning. *Journal of Forestry* 94:4–9.

94 Gough, J., and J. Ward. 1996. Environmental decision-making and lake management. *Journal of Environmental Management* 48:1–15.

95 Ellis, S., and U. Seal. 1995. Tools of the trade to aid decision-making for species survival. *Biodiversity and Conservation* 4:553–572.

96 Mitroff, I., and F. Sagasti. 1973. Epistemology as general systems theory: An approach to the design of complex decision-making experiments. *Phil. Soc. Sci.* 3:117–134.

97 Mitroff, I., J. Emshoff, and R. Kilmann. 1979. Assumptional analysis: A methodology for strategic problem solving. *Management Science* 25:583–593.

98 Mitroff, I., and J. Emshoff. 1979. On strategic assumption-making: A dialectical approach to policy and planning. *Academy of Management Review* 4:1–12.

99 Christopherson, J., S. Lewis, and M. Havercamp. 1996. Lake Tahoe's forest health consensus group. *Journal of Forestry* 94:10–12.

100 Wondolleck, J., N. Manring, and J. Crowfoot. 1996. Teetering at the top of the ladder: The experience of citizen group participants in alternative dispute resolution processes. *Sociological Perspectives* 39:249–262.

101 Manring, N. 1993. Reconciling science and politics in Forest Service decision making: New tools for public administrators. *American Review of Public Administration* 23:343–359.

102 Torell, D. 1993. Viewpoint: Alternative dispute resolution in public land management. *Journal of Range Management* 46:70–73.

8

Prognosis

In the preceding chapters, I have tried to present a scholarly discussion of the issues bedeviling environmental problem solving. In this final chapter, however, I would like to offer a more personal view of the prospects for adaptive change.

Thus far we conclude that the capacity of most industrial societies to deal effectively with the environmental problems that confront us is inadequate. Not only are conventional problem-solving systems based on flawed epistemological assumptions and inappropriate cognitive styles but they are driven by a materialistic ideology that has resulted in the overexploitation of natural resources. To make matters worse, the organizational structures entrusted with natural resource management appear to be more concerned with self-protection than adaptive problem solving, in a political system controlled by elites in pursuit of their own self-interest. The reforms that have been proposed are fragmentary and uncoordinated, tending to address only isolated aspects of the complex psychosocial and ecological mess we have created for ourselves. This is not an encouraging state of affairs as we survey the rapidly deteriorating state of the natural environment that is under increasing pressure from a burgeoning human population. An adaptive response to this collective dilemma requires us to engage in radical sociopolitical change toward political devolution and cooperative problem solving. Although, in democratic societies, these changes are everyone's responsibility, it is likely that professional cadres, especially sages, will play a crucial role in any transition to new political arrangements. In what follows, I examine the prospects for such changes, starting with the problems inherent in educating sages.

Educating Sages

Many professionals who play a central role in environmental management do not have the personal qualities and skills needed for the kind of leadership role envisaged earlier.[1-3] For instance, a recent study by the U.S. Forest Service concluded that[2] (p. 596):

Forest Service biologists, while very strong in the scientific and technical areas, could not compete for leadership positions because other qualities were poorly developed. They lacked sufficient skills and knowledge in: communications and leadership; natural resources policy, social values and economics; team-building and cross-disciplinary problem-solving; and program development and administration.

As Kessler[2] notes, the traditional response to these deficiencies is to squeeze additional courses into the curriculum in the hope that the professional "tool box" might help students acquire the required competences. Unfortunately, the crowded, technically biased nature of traditional curricula virtually guarantees that this will not happen. Instead, I believe that there needs to be a shift away from the traditional preoccupation with technical-skills training toward a greater role for personality or character development,[2,4] a suggestion that always raises hackles whenever and wherever it is proposed. The reasons are straightforward enough. Current professional education in both the natural and social sciences is dominated by positivism and its obsession with objectivity, resulting in the kind of impersonal education favored by objective-analytic and objective-holistic personality types. By arguing for a greater subjective-introspective component in education, I am promoting the kind of educational experiences that would be more agreeable to subjective-analytic and subjective-holistic personalities, few of whom gravitate to science and engineering. It follows that there would be resistance to such an educational shift from orthodox faculty who, by virtue of their own personalities, training, and professional experience, are not suited to pursuing such goals.[5] Thus, in contemplating the education of sages, one is faced with the thorny problem of finding the sages needed to teach them. The kinds of difficulty that have to be overcome are outlined in the following.

Cognition

Technical Knowledge

Some of the worst excesses of traditional professional education, with its disciplinary fragmentation and rigid positivism, have been elaborated at length elsewhere, as has the negative effect of such education on the development of problem-solving capabilities.[6] One reaction has been the promotion of a more integrative kind of education, usually under the rubric of interdisciplinary, holistic, and systems approaches to learning.[7,8] Unfortunately, even where such innovations are introduced, they often remain firmly within the positivist

and Imperial traditions.[9] In other words, the need for radical change in education is subverted, often inadvertently, by the unexamined biases of the very "innovators" themselves who believe they are implementing significant change but who, in practice, are not. Educational innovation, especially in higher education, is glacial in its slowness, resisted at every stage by both faculty and students alike.[8,10]

Aiding and abetting this tardiness is a resistance to, or unawareness of, the kinds of innovation that would help in the development of adaptive problem solvers. For instance, discussions of soft-systems training, postformal and dialectical thinking, analytic and holistic cognitive styles, and so on, are confined to specialist literatures. They seldom surface in the environmental management literature and are even less evident in practice. My own experience with technical professionals is that the majority are blissfully unaware of these concepts, fixated as they are on their technical specialities. It is difficult to imagine, therefore, how they can function as effective teachers of the next generation of sages.

None of this augurs well for the adoption of problem-based learning,[11] which I believe to be the most effective way in which to facilitate the cognitive development of students in the technical area. Unlike traditional education, which emphasizes the didactic presentation of information from the established disciplines, problem-based learning is organized around broad interdisciplinary problems. Typically, small groups of students, with the aid of a tutor, tackle the problem in question over a period of 6 weeks or more, during which time they research the problem collectively using the human and other resources available to them. Few if any lectures are provided in this kind of learning environment that attempts to encourage initiative, cooperation, and problem-solving skills over and above the mere memorization of technical information. Thus, learning does not focus on disciplinary content but on the skills needed to resolve the kind of practical problems found in the real world.

Problem-based learning has been used successfully in the education of medical doctors[12,13] and is being introduced into natural-resource education.[2,11,14] As one might expect, however, there is considerable resistance to such a sweeping innovation.[13] In part, this is due to the threat it poses to disciplinary hegemony in universities. Academic departments constantly monitor their relative standing within the university and resist anything that might diminish their staffing levels or budget. Such disciplinary chauvinism has curtailed many attempts to develop interdisciplinary programs in university settings.[15] A more fundamental threat posed by problem-based learning, however, is its emphasis on an increased interaction between faculty and students. Not only is this time-consuming for faculty but it also requires a shift toward a more egalitarian teaching style, one that weakens the traditional distinction between the teacher and learner.[13] The intellectual limitations of highly specialized faculty become patently obvious in such group problem-solving contexts. Not all faculty are happy with this shift in roles, nor are all students comfortable with the changes in learning requirements. In my experience, many students prefer the more passive, rather detached didactic experience where they can sit in class without being required to take an active part in the proceedings. Problem-based learning, however, is a much more demanding exercise.

There is no recipe available that will guarantee the establishment and acceptance of problem-based learning. Where it has been developed, one sees a core group of innovative faculty backed by a university administration with the financial and political power to push through the necessary measures. Even under these conditions, resistance from conservative faculty, administrators, and students will persist, threatening to undermine the changes that have been implemented and shift the program back toward a more traditional offering. A common argument used in this regard is that problem-based learning "may result in unsystematic coverage of important scientific disciplines"[13] (p. 202). One student I heard voicing his dissatisfaction with the process reflected this same concern in suggesting that he felt he wasn't being properly "prepared" for professional life, a curiously gastronomic metaphor.

Presumably, conservative faculty and students believe that traditional didactic teaching does prepare students for the task of mess management and complex problem solving, an even more curious viewpoint in light of the available evidence to the contrary.[6,16,17] Problem-based learning is not, of course, a panacea that will resolve all the deficiencies of higher education but merely a useful component of the kind of educational experience likely to produce sages. Whether it will be implemented on a sufficiently large enough scale to have any effect depends on the extent to which it is possible to overcome widespread institutional denial about the serious problems with current didactic methods.

Practical (sociopolitical) Knowledge

Technical knowledge is of little use without the capacity to use it effectively in the practical world of people and politics. This practical knowledge about how to maneuver through the minefields and roadblocks that hinder all practical action is a crucial element in the professional tool box. Unfortunately, traditional education in environmental and natural resource management provides little opportunity for the development of the social intelligence that underlies such practical skills. For instance, the absence of the intensive group experiences that might help in such development has been recognized for decades but has received only token attention.[18,19] As a result, many natural resource professionals are particularly deficient in this area.[2,3]

In part, the relative absence of practical (sociopolitical) knowledge among natural resource professionals stems from their underlying personality and interests. As mentioned in chapter 2 and elsewhere, the majority of such professionals exhibit the characteristics of objective-analytic and objective-holistic personality types, both of which are drawn to the biophysical aspects of the natural world and tend to shy away from the more psychosocial or introspective worlds. Typically, the field-independent nature of environmental management professionals inclines them toward a task-oriented approach to problems, especially in groups.[20] The resulting preference for highly structured interactions with others does not help in the development of the social skills needed to manage the personality conflicts and intellectual confusion that invariably arise in trying to cope with wicked problems.

Practical knowledge or social intelligence, then, requires an understanding of interpersonal and group dynamics as well as a broader awareness of how organizations and social choice mechanisms work in society. Clark[21] talks about this in terms of the different kinds of knowledge needed to be effective in contributing to the policy-making process. The problem, however, is how to encourage students and professionals to acquire these intellectual skills when the personality and interests of the majority tend to avoid engagement with such areas of expertise. In my own experience with forestry students, it is clear that many are attracted to a life in the outdoors away from the complications of politics and other urban diseases. I have been informed outright, during my attempts to encourage an interest in the psychosocial world, that this latter is not what they had signed up for. If they had wanted psychosocial education, they say, they would have pursued a liberal arts degree. Students are not alone in this dislike of the psychosocial world. It is common enough in both academic faculty and professionals in the natural sciences among whom the term *politics* is used as an epithet, and politically involved scientists, especially those who adopt an advocacy role, are viewed with some disdain.[22] Thus, faculty pass on to students their own personal antipathies toward, or disinterest in, nontechnical matters, thereby reinforcing the students' predilections toward disengagement from the psychosocial world. As students graduate, they come under the additional influence of professional associations and organizational cultures that encourage this role of the detached, technical professional.[23] The question is how to break out of this vicious circle.

There will be little progress toward the widespread development of sociopolitical knowl-

edge among natural resource students and professionals as long as professional culture is dominated by Imperial attitudes. Even if there was an increase in the adoption of problem-based learning with its emphasis on group learning and practical implementation of problem solutions, the potential benefits of such innovations are likely to be thwarted by the prevailing attitudes and values of traditional faculty. It is interesting that, some years ago, I encouraged a graduate student in Forestry to undertake a historical review of the attention that had been paid to the social side of forestry education. It was depressing to find that, despite regular calls for a more prominent role for the nontechnical aspects of forestry dating back to the beginning of the century, little if anything had been done to meet this perceived deficiency in any forestry program in Canada, or elsewhere. However, in environmental studies, a more eclectic discipline, there are numerous efforts to incorporate sociopolitical experiences into the curriculum.[24]

I believe that the only way out of this dilemma is to encourage a change in the kind of people engaged in environmental and natural resource management. Something of the sort has happened already in environmental management, which has attracted a diverse professional body from a variety of intellectual backgrounds.[25] Unfortunately, the traditional natural resource fields have yet to undergo this transformation, although there are signs that it is happening. Meanwhile, attempts to encourage an interest in, and development of, practical social intelligence in natural resource students by exposure to such activities as group learning and electives in the social sciences may be a waste of time. Only the minority who are already interested may benefit. The futility of trying to impose learning experiences on students who, at a fundamental personality level, reject the subject matter is reflected in a study of the electives selected by engineering students. "When . . . an independent investigator evaluated engineering students' humanities selections, it was apparent that the major factor governing the students' choice in these areas was the proximity of classrooms. In other words, the student as-

signed such negligible value to the humanities
. . . that he responded by minimizing his walking distance"[26] (pp. II-11).

Emancipatory Knowledge

When we turn to emancipatory knowledge,
the kind of self-insight that frees us from the
shackles of unacknowledged bias, we are faced
with apparently insurmountable educational
obstacles. Delving into the introspective world
of beliefs, values, and motives is not to the
taste of the objective personality types who
make up the majority of environmental professionals. The reaction of such individuals to
introspection can be extreme, as I have found
to my chagrin. Any attempt to force the more
objective person to engage in introspection can
result in the kind of viciousness exhibited by a
minority of students described earlier. In essence, such students are protecting their self-
identity from the threat of introspection the
only way they know how.

Self-insight is not achieved in one fell swoop
but develops over time at different levels of
sophistication.[27,28] Perhaps the most elemen-
tary form of self-insight is the ability, and willingness, to acknowledge technical error. If one
can take this one step further and identify the
cause of error in, for example, the lack of
knowledge or inappropriate methodology,
then a further step in emancipatory thinking
is achieved. However, even this modest form
of self-insight is difficult for some professionals, as we saw in the previous discussion of
adaptive management that has, as one of its
cardinal rules, the need to acknowledge error
in learning about complex ecosystems. More
sophisticated forms of self-insight that delve
more deeply into the psychological roots of
bias and irrationality, by trying to understand
the role of one's beliefs, values, and emotions
on one's behavior, are more difficult to encourage. A necessary prerequisite for such an
introspective journey is facility with the kind
of psychological language that allows reflection on emotions and values. Unfortunately,
many adolescents are incapable of meaningful
introspection because they cannot describe to
themselves, or others, what they are feeling or

how their emotions influence what they do.
Their accounts are incoherent and uninformed.[28] It is only through exposure to parents or counselors who can help them develop
such a language that they may begin to develop some self-insight. At the same time, the
looming prospect of uncovering aspects of
oneself or one's job that are unappealing or
downright disturbing is likely to prevent most
professionals from engaging in too much introspection.[29,30]

The problem posed by all of this for the education of sages is that many natural resource
professionals exhibit the varying degrees of resistance to or loathing of introspection characteristic of the objective personality types. As
explained in chapter 2, this is a deeply embedded characteristic that cannot be easily overcome. It is an attitude of mind that develops
over a person's formative years and is not amenable to change by exposure to a few well-
meaning educational exercises. Any ambition
to turn most professionals into effective self-
reflective practitioners is likely to be, therefore, a fond hope. As Knight[31] concludes, "reflective practice seems to have been very
quickly adopted on a widespread scale by
many professional groups, and the speed and
scale of this adoption appears to have rendered
it as part of the traditional culture of higher
and professional education . . . (but) it may be
the case that widespread professional acceptance of it has either been relatively superficial, and/or is in need of reformulation and review." Thus, many professional groups may
not really understand what is involved in "reflective practice." For a minority of students,
none of this is a problem. Their personality
structure is such (a strong subjective component) that they already engage in effective introspection and need only encouragement to
extend this to their professional behavior. It is
the reluctant majority that pose a more difficult challenge, those who see introspection as
either useless rumination on past mistakes[31] or
have a more deep-seated animosity to any
kind of self-exploration.

I'm inclined to view the latter as beyond
reach. They will not engage willingly in any
such educational exercises, and, indeed, it is

clearly unethical to try to force them to do so, as well as being psychologically damaging to them. For the less intractable student, however, the methods used in values clarification are a reasonably innocuous introduction to an exploration of value conflicts in professional affairs.[32] Beyond that point one enters uncharted territory, at least in the field of environmental management, for the most effective methods to achieve introspective insight are those of group dynamics and human relations training. Although the latter have been used to good effect in in-service training,[27,33] their use as an integral part of undergraduate education is more tendentious. In part, this is because human relations training involves intensive workshop experiences that focus on interpersonal interactions and feelings. The atmosphere in some groups can become very impassioned as beliefs and interpersonal styles are explored. This can be an upsetting experience for anyone who is unused to interpersonal honesty and the forthright expression of emotion. My own experience suggests that such intensive methods would be rejected out of hand by the majority of science and engineering students, just as they were by a group of graduate students in psychology whom I tried to engage in a human relations group some years ago. They endured the ordeal because it was a required part of their program, but they hated every minute of it and couldn't see how all this interpersonal twaddle helped them become the scientific psychologists they aspired to be.

Motivation

One of the most daunting challenges facing any society is that of controlling its young men. The natural ebullience of youth, coupled with the culture of machismo that permeates many western societies, makes this task difficult to say the least. Young people who exhibit the more subjective personality types are unlikely to be part of this problem and, given the right family circumstances, will develop the sensitivity and moral outlook that makes their socialization easier to accomplish. In contrast,

it is the extremely objective personality types, especially the psychopathic element among them, who are potentially the most problematic.[34] Extremely objective persons, those who prize the more brutal kinds of masculine power and its accompanying callow insensitivity, can wreak havoc on those around them.

As discussed in chapter 2, objective personality types are attracted disproportionately to science and engineering. Not all of them will be extremely objective of course. For instance, in a study of moral reasoning by forestry students,[35] it was found that half of the sample of students exhibited egocentric ways of thinking while the remainder were more conventional in their moral concerns. In other words, a substantial proportion of these students appeared to be primarily concerned with themselves and their own welfare. This is consistent with the conclusions of Hyde and McLean,[5] mentioned earlier, to the effect that many engineers are to a large extent detached from the influence of friends and family. Their lack of interest in people and the concomitant social isolation is compounded by the nature of the engineering curriculum and the way it is taught. Hyde and McLean characterize engineers as "marginalized and inarticulate in a world where discourse and personal connections are the predominant strategies for coping with the sterility of life in the technological society" (p. 10) and engineering schools as "centers which recruit alienated people, train them in alienating ways and send them out to perpetuate alienation through their promotion of unreflecting development of technology" (p. 12). *Alienation*, the process of becoming mentally and emotionally detached from others, is not unique to engineering but is a common feature of all disciplines that foster positivist, Imperial modes of thinking. The fact that objective personalities are attracted to these disciplines—becoming, in the fullness of time, the next generation of professionals and teachers—makes it difficult to implement change. What hope is there, under these circumstances, for a better balance between Imperial and Arcadian ideologies and for the development of sages?

The quick answer, I'm afraid, is not much.

Attempts to affect changes on the objective-subjective dimension of human personality is very difficult. There is no easy or effective way to persuade the more objective person (such as the alienated engineer) to adopt a more subjective mode of behavior. Individuals who enjoy the technical over the personal, who are used to an impersonal way of dealing with others, who carefully hide or suppress their true feelings, whose empathy and caring is limited, who relish power, and so on, are not going to shift into a more subjective mode simply because someone asks them to do so. Objectivity and subjectivity are such an integral part of personality, one that has developed over a lifetime, that whatever change is possible must necessarily be slow and, sometimes, painful. This is not to say that shifts along the objective-subjective dimension, from detachment to a more caring disposition, do not occur during the college years. Rest[36] concludes, for instance, that moral development occurs in concert with general social development:

The growing awareness of the social world and one's place in it seems to be more important than (exposure to) specific moral crises . . . the people who develop in moral judgment are those who love to learn, who seek new challenges, who enjoy intellectually stimulating environments, who are (self) reflective . . . who see themselves in the larger social contexts of history and institutions and broad cultural trends, who take responsibility for themselves and their environs. (p. 192)

The dilemma facing educators in environmental and natural resource management is how to attract students of this caliber to their programs and, once enrolled, how to provide opportunities for this broad involvement in and exposure to social and cultural issues. If more subjective persons were to enter environmental and resource management, especially women, a different voice would be added to the prevailing masculine (objective) discourse in professional schools. Bebeau and Brabeck[37] have found, for instance, that "women dental students showed significantly greater sensitivity to the ethical issues contained in professional dilemmas than did their male colleagues." In addition, women are more inclined than men to focus on the caring aspects of morality, in contrast to men who tend to focus on more rational, abstract aspects of moral dilemmas. It can be argued that caring and compassion provide a more fundamental basis for moral action than a facility in abstract moral reasoning.[38] For this reason, the typical moral education course, which emphasizes rational exercises in moral reasoning and judgment, is unlikely to promote any significant shift in moral behavior.

The impact of moral education is also diminished when teaching faculty continued to behave in an objective, Imperial manner. It is well known in psychology that the young learn most effectively through identification and imitation. If faculty were to begin behaving in a more subjective, socially involved and morally concerned way, this would be a powerful incentive for at least some budding professionals to act similarly. In my experience, many faculty (and professionals) believe they are acting in a socially responsible manner already and are offended by the implication that they are not. Perhaps they miss the point. I am suggesting that being socially responsible within the Imperial mode is a form of ecological irresponsibility, one that Arcadians never tire in pointing out. As Yeazell[39] points out, we need to attend to the moral sensibility of teachers, an often neglected competency. The need, therefore, is to find a more effective way of combining the best of objective (Imperial) and subjective (Arcadian) ways of thinking and behaving—and imbuing this in the young.

Sages have this capacity to integrate conflicting ideologies within themselves in a constructive way. It enables them to work effectively with those whom they may disagree and, by doing so, make connections between antagonistic camps. Hudson[40,41] suggests that students with the potential to become sages have already developed the kind of balanced integrative personality that lends itself to such wise behavior, at an early stage in their lives, long before they arrive at college. Presumably, the best way to encourage their further development is to engage in damage control, to limit as far as possible the damage done to these students by higher education. What I mean is that socialization into any discipline is often a ritual in which a set of rigid intellectual prescriptions are forced on to the student in the name of professional development. Those who can tolerate these constraints prosper; those who cannot may drop out.[42] Even among faculty, orthodoxy rewards itself, whereas innovators have difficulty in prospering.[43,44]

It must be obvious that I have a profoundly jaundiced view of orthodoxy in higher education. Along with many others,[45-47] I see it as a major impediment to adaptive change. Unfortunately, this observation has been made for decades, to little effect.

Anxiety and Defense

According to many theorists who study human personality, myself included, we are all riddled with psychic conflicts as well as being afflicted by perpetual anxiety and fear.[48] Our response to this is to erect psychological defenses that allow us to carry on with the daily round without being immobilized by our terrors.[30] Taken to an extreme, the net result is an inability to face the reality of our own lives or the problems that arise in the world around us. Needless to say this is not a promising basis for adaptive problem solving, which requires, as a first step, that we acknowledge that problems actually exist. It seems that objective and subjective personalities fear different things. The more objective types fear that their inner world of emotion and impulse may begin to disrupt their orderly lives. On the other hand, subjective types fear that a breakdown in human relationships may leave them vulnerable to the dangers posed by the physical world. Sages, however, appear to be able to tolerate these fears without undue defensiveness. They are comfortable operating in the impersonal mode of the objective personality without losing sight of the need for compassion, and they are not perplexed by the emotional expressiveness associated with subjectivity. This capacity, as do so many desirable personal qualities, develops during the formative years of childhood in family settings that encourage emotional honesty and intellectual curiosity. As I've mentioned repeatedly, by the time students reach higher education, their characteristic ways of coping with anxiety and fear are well established and difficult to change.

Any attempt to explore these matters in educational settings is very controversial. A person's anxieties, and how they cope with them through psychological defense, are such a fundamental aspect of personality, so fraught with emotional overtones, that it is doubtful whether educational intervention should be contemplated. In the past, I have used human relations training and autobiographical approaches in my teaching that touch on these aspects of the person's life, but I have been very careful to keep the proceedings at an innocuous level. Anything more intensive really requires a psychotherapeutic approach, which is inappropriate in traditional educational settings. In any case, most teachers of environmental and natural resource management are not equipped to deal with such matters, nor are their students who would, as I've explained earlier, react very badly to any attempt to have them delve into their psyches. I've experienced this kind of repugnance in discussions with professionals whenever I offer the suggestion that professional development should include a modicum of self-analysis. Even psychologists (of the positivist/Imperial tradition) resist the notion that they would benefit from such activity.

According to Orr,[46] the academic establishment has been slow to respond to the crises facing us because of a number of problems inherent in institutions of higher education. The professionalization of learning, for instance, with its rules and rituals has narrowed intellectual discourse and has resulted in "a crippling alienation from each other and what is healthy for ourselves." Thus, professionals see themselves as part of the established order rather than critics of it. Orr also sees a failure in leadership within the universities, an absence of vision that has turned universities into knowledge factories rather than institutions that help students cope with the fundamental ecosocial problems confronting us. I agree with his view that higher education must be revamped to recognize the growing environmental crisis by, among other things, fostering the "ecological wisdom and practical wherewithal that will enable the young" to rebuild society along more ecologically sustainable lines. Another way of putting this is that enhancing the sagaciousness of students requires a fundamental reform of higher education rather than the tinkering that currently occurs.

Prospects for Sociopolitical Change

Previously, I've suggested that the kind of collaborative behavior that leads to adaptive problem solving is more likely to occur under conditions of increased political devolution. Sages would play a crucial role in facilitating a transition to these new political arrangements as well as helping to sustain them once in place. Unfortunately, these ideas may turn out to be idealistic dreams in face of the harsh reality of life in the technocorporate state. The growing influence of transnational corporations, the dislocation of work forces through downsizing and the search for cheap labor, together with the de-skilling of the work force as it is polarized into experts and drones, is part of a trend that has been unfolding for much of the century. Ewens[49] points out that industrial development led to an inexorable loss of the control of work by working people and a concomitant de-skilling process that has left them dependent on their industrial masters. With the industrial revolution, independent craftspeople saw their craft skills appropriated and deconstructed by scientific managers into simple tasks that could be fitted into industrial production processes, each of which required little skill or training. Because of this, workers became interchangeable units in the industrial treadmill rather than skilled craftspeople in charge of their own work. Along with this change in work came the centralization of economic power in the hands of industrial conglomerates that were able to control access to markets. Both de-skilling of the work force and the concentration of economic power continue unabated as we enter the era of global free trade under the aegis of the transnationals.

Any attempt to break out of this global system toward the Arcadian ideal of self-sufficient communities living in ecological harmony within their bioregions would likely founder in the absence of appropriate levels of skill and difficulty in competing in markets—unless, of course, citizens were willing to live at a subsistence level. Clearly, this is not a prospect that would galvanize the citizens of affluent societies into action. It follows that political devolution and collaborative problem solving within the prevailing Imperial system is a charade that serves only to placate the commonsense view that we are being manipulated by political and economic forces beyond our control. Whether some meaningful sociopolitical change is possible might best be explored in the context of the forest management case dealt with in chapter 4.

We left the story with the recognition that the economy of New Brunswick is heavily dependent on revenues from the transnational pulp and paper corporations that dominate the provincial forest industry. This concentration of economic power began to build in the 1920s and has continued apace ever since. One consequence of the increased specialization of forest production has been the relative de-skilling of the work force. In earlier days, rural inhabitants were relatively self-sufficient, combining farming with woodcutting and the production of a variety of wood products. Many of the skills needed for self-sufficiency have disappeared. Even jobs in an increasingly mechanized forest industry have declined as woodcutters are replaced by massive, and expensive, machines run by a single operator. In the last 30 years, the number of loggers employed in the pulp and paper industry in New Brunswick has declined by 55 percent, with further losses evident on the horizon. Apparently, this reduction in the workforce has not resulted in any decline in timber harvesting volumes nor in any significant rise in the wages of those woodcutters who remain.[50]

Meanwhile, looming in the background is the threat of another spruce budworm epidemic, made more devastating than anything seen before by the state of the forest and the absence of any adequate control measures.

The ban imposed on the use of Fenitrothion in the aerial spraying of forests leaves the province of New Brunswick in dire straits. Unless a new aerial program is mounted with a new pesticide, the resulting devastation of pulpwood stands will reduce wood supplies to the mills, thereby exacerbating the processing overcapacity that exists in the province. All of this will underscore the uncertain future of some of the transnational companies operating in New Brunswick as they experience the vagaries of international markets for pulp and lumber products.[51] As a result, New Brunswick's rural communities continue to experience the uncertainty commonly associated with resource-based economies.[52] Betts and Coon[50] suggest that the "spiralling unemployment, stagnant rates of pay, and declining rural economies are all increasingly forcing the people of New Brunswick to question the current industrial approach to managing forests." Whether this will result in grassroots action is doubtful in light of the sense of despair about their future prospects among rural youth.[53] Despair and the loss of hope are not a satisfactory basis for adaptive action.

The provincial government's response to the continuing poverty and unemployment in rural New Brunswick, and the provincial dependence on the primary production of natural resources, has been to seek to diversify the economy. Efforts have been made to attract service industries that make use of the province's electronic infrastructure, as well as attempts to encourage value-added secondary processing of wood products.[54] Still, only " a scant five per cent of Crown wood goes into high-end, value-added products"[55] (p. 20). Both initiatives are commendable efforts but are firmly ensconced within the Imperial tradition. Neither involves any devolution of political power to communities, nor has the control of Crown lands shifted away from the pulp and paper transnationals.

Thus, the Arcadian vision of a devolved society, coping adaptively with environmental problems, is faced with some harsh realities: a difficult economic situation, an impending onslaught on the main wood supplies by an insect that is difficult to control, a rural population that may not have the skills to develop new products and markets, and a government that shows little interest in radical devolution. If, by some chance, the provincial government were to adopt an ecosocialist strategy and reorganize the Crown lands into community, industrial, and protected forests, it would be necessary to mount a massive community development exercise to help communities cope with their new responsibilities. This, of course, would not be done in one fell swoop but could build on the few community forestry projects already in existence.[50] Sages would play a crucial role in this transition as advocates working to link government agencies and local communities. All of their social skills, patience, and conflict resolution capacities would be tested in such a cauldron of social change.

Yet, I believe that none of this is about to happen. Nor do Betts and Coon,[50] who conclude that: "Given the significant barriers to community forestry in New Brunswick, it is unlikely that a comprehensive community forestry policy which devolves control of all Crown forests to municipalities and First Nations will occur in the near future" (pp. 23–24).

Conclusions

The prospects for a more widespread adoption of adaptive problem solving remain dim. Numerous educational and psychological barriers stand in the way of encouraging the development of what I've referred to as sages. Given the relative paucity of these wise individuals who can stitch together the social fabric, environmental conflicts and maladaptive problem solving will continue to worsen. At the same time, the continued dominance of the Imperial way of thinking, and its reticence in promoting political devolution, hinders the creation of the kind of sociopolitical arrangements that help foster adaptive problem solving. The net result is that nothing of significance changes except the severity of environmental problems.

It is easy to become discouraged under these circumstances, especially when one sees the apparent lack of urgency in the majority who go about their daily lives as if nothing of significance was happening. This stoicism, the human capacity to conduct business as usual in the face of overwhelming threat, may be admirable under some circumstances but could be our undoing in the years to come. Ehrenfeld[56] and Ophuls and Boyan[57] come to a similar conclusion in commenting on the lack of adequate response to the growing global scarcity that they and others alerted us to some 20 years ago. As Ophuls puts it[57] (p. 315):

I regret to say, however, that I am now less inclined than before to believe that we will respond positively to the ecological challenge. Then (in 1977), I saw us confronted with a grand opportunity to create a more humane future, and I believed that the transition to a more ecologically enlightened world view had already begun. But we have frittered away the two decades since the first Earthday without seizing this grand opportunity . . . we have spent the last 20 years doing all the easiest and least painful things. Now we must do the hard things: reshape basic attitudes and expectations, alter established lifestyles, and restructure the economy accordingly. But rather than adopt ecological principles for public policy, we seem to do everything we can to avoid facing up to the inevitability of limits and of changing our profligate way of life. In other words, time has grown shorter and the problems have become larger and more entrenched, but our resistance to dealing with them constructively has increased.

What, then, is to be done? What are the prospects for adaptive problem solving on the scale that is needed? One common view is that we must await some ecosocial cataclysm that will shock the great mass of people into an acceptance of the need for a radical change in lifestyles. Unfortunately, one of the most common responses to catastrophe is defensive avoidance and self-protection.[58] Needless to say, the social conflicts that will accompany further ecosocial decline, as many of us struggle to protect ourselves, will be extremely ugly. As Ehrenfeld[56] (p. 191) concludes:

The most terrifying thing about this disintegration for a society that believes in prediction and control will be the randomness of its violent consequences. The chaotic violence will include not only desper-

ate, ruthless struggles over the wealth that remains, but the last great rape of nature. What will make it worse is that, at least at the beginning, it will take place under a cloud of denial and cynical reassurances.

Waiting for Armageddon, in the hope that everyone will be shocked into more sensible, adaptive behavior, is not a strategy that appeals to me. Bearing in mind the current levels of ecosocial myopia, I do not believe that the majority response to terminal ecosocial deterioration will be constructive. Something needs to be done now while there is still a semblance of social order and stability, at least in some parts of the world.

I'm more inclined to agree with Ophuls and Boyan[57] that a widespread shift to ecological self-restraint needs to be a more organized response following radical political change. If we are to avoid the imposition of a monolithic totalitarian regime that will seek to resolve problems by brute force, "we need a form of government that is effective in obliging humankind to live within its ecological means but that does not require us to erect an ecological Leviathan (totalitarian state) . . . the new form of government would have to be based on an 'ecological contract' to ensure that a basic harmony between man and nature was at the core of politics" (p. 313). In his most recent book, Ophuls has become pessimistic that this political change is possible, as we have seen in the previous text. However, I believe that there is some, albeit slim, prospect for adaptive change if the professional cadre making up the bureaucratic and scientific establishments was to assume a greater leadership role. As explained earlier, I am not suggesting the rise of a new technocracy based on the old Imperial model, nor do I imply some new technical priesthood to act as guardians leading us to salvation. The professional body I have in mind would be rather different from that which currently exists, and it would adopt a radically different role. The new professionals would have many of the qualities of sages and adopt an advocacy role in promoting ecologically sane behavior in all sectors of society. They would do this by engaging others at all levels of the problem-solving system from the highest policy levels to the lowest grassroots, community project in discourse over ecological rationality. To be effective, the new professional would have to be more militant than the prevailing "neutral" or neutered individual who sees his or her role as being subservient either to political and corporate elites or to the god of scientific objectivity. Professionals have the time and opportunity, by dint of their education and status, to develop a more informed picture of the world around us than most other members of society. It behooves them to make more effective use of this privileged position in promoting what Ophuls calls "civic virtue."

Although the situation that confronts us is grim, I believe the position taken by Higgins[59] is the most sensible. He argues for "courageous pessimism," the belief that, while it may not be possible to overcome the problems that face us, the only sensible response is to try to do the best we can in the hope that, in the end, we shall prevail.

References

1 Florman, S. 1987. *The civilized engineer.* New York: St. Martin's Press.

2 Kessler, W. 1995. Wanted: A new generation of environmental problem-solvers. *Wildlife Society Bulletin* 23:594–599.

3 Magill, A. 1988. Natural resource professionals: The reluctant public servants. *The Environmental Professional* 10:295–303.

4 Rauste—von Wright, M. 1986. On personality and educational psychology. *Human Development* 29:328–340.

5 Hyde, R., and G. McLean. 1992. Alienated engineers: Part of the problem. Paper presented at meeting of the Canadian Sociology and Anthropology Association. May 30. University of Prince Edward Island, P.E.I., Canada.

6 Schon, D. 1987. *Educating the reflective practitioner.* San Francisco: Jossey-Bass.

7 Dorney, R. 1989. *The professional practice of environmental management.* New York: Springer-Verlag.

8 Malone, R. 1990. The need for global education. In *Rethinking the curriculum,* ed. M. Clark and S. Wawrytko, 167–179. New York: Greenwood Press.

9 O'Sullivan, P. 1986. Environmental science and environmental philosophy—part 1: Environmental science and environmentalism. *International Journal of Environmental Studies* 28:97–107.

10 Dalton, L. 1986. Why the rational paradigm persists—the resistance of professional education and practice to alternative forms of planning. *Journal of Planning Education and Research* 5:147–153.

11 Boud, D., ed. 1985. *Problem-based learning in education for the professions.* Kensington, Australia: Higher Education Research and Development Society of Australasia.

12 Birch, W. 1986. Towards a model for problem-based learning. *Studies in Higher Education* 11:73–82.

13 Thompson, D., and R. Williams. 1985. Barriers to the acceptance of problem-based learning in medical schools. *Studies in Higher Education* 10:199–204.

14 Zundel, P., T. Needham, and J. Kershaw. 1994. Designing and implementing a learning system in forestry to create reflective practitioners. In *Fifth National Conference on Problem Solving across the Curriculum.* June 23–25, Hobart and William Smith Colleges, Geneva, N.Y.

15 Leckie, G. 1975. Interdisciplinary research in the university setting. Center for Settlement Studies, Occasional Paper No. 9, University of Manitoba, Winnipeg, Canada.

16 Bowen, J. 1988. Science, education and the environment: Ecocentrism as the new paradigm. *Education and Society* 6:3–15.

17 Reckmeyer, W. 1990. Paradigms and progress: integrating knowledge and education for the twenty-first century. In *Rethinking the curriculum,* ed. M. Clark and S. Wawrytko, 53–64. New York: Greenwood Press.

18 Shalinsky, W., and R. Norris. 1986. The place of small group process in planning education. *Journal of Planning Education and Research* 5:119–132.

19 Vallentyne, J. 1974. Limnology and education in the next decade. *Journal of the Fisheries Research Board of Canada* 31:513–519.

20 Gruenfeld, L., and T. Lin. 1984. Social behavior of field independents and dependents in an organic group. *Human Relations* 37:721–741.

21 Clark, T. 1993. Creating and using knowledge for species and ecosystem conservation: Science, organizations, and policy. *Perspectives in Biology and Medicine* 36:497–525.

22 Baskerville, G., and K. Brown. 1985. The differ-

ent worlds of scientists and reporters. *Journal of Forestry* 83:490–493.

23 Hodges, D., and R. Durant. 1989. The professional state revisited: Twixt Scylla and Charybdis? *Public Administration Review* 49:474–484.

24 Murphy, R., J. Perkins, and T. Rainey. 1992. Political ecology and the education of environmental professionals. *Environmental Professional* 14:310–318.

25 Petulla, J. 1987. *Environmental protection in the United States.* San Francisco: San Francisco Study Center.

26 Rosenstein, A. 1968. *A study of a profession and professional education: The final publication and recommendations of the UCLA educational development program.* Report EDP 7–68. Reports Group, School of Engineering and Applied Science, University of California, Los Angeles.

27 Fien, J., and R. Rawling. 1996. Reflective practice: A case study of professional development for environmental education. *Journal of Environmental Education* 27:11–20.

28 Weinstein, G., and A. Alschuler. 1985. Educating and counseling for self-knowledge development. *Journal of Counseling and Development* 64:19–25.

29 Baum, H. 1987. *The invisible bureaucracy: The unconscious in organizational problem solving.* New York: Oxford University Press.

30 Hirschhorn, L. 1988. *The workplace within: The psychodynamics of organizational life.* Cambridge, Mass.: MIT Press.

31 Knight, B. 1996. Reflecting on 'Reflective Practice'. *Studies in the Education of Adults* 28:162–184.

32 Kinnier, R. 1995. A reconceptualization of values clarification: Values conflict resolution. *Journal of Counseling and Development* 74:18–24.

33 Henning, D. 1980. In-service training for environmental personnel: A critical analysis with innovations. *Journal of Environmental Education* 11:4–10.

34 Spiecker, B. 1988. Psychopathy: The incapacity to have moral emotions. *Journal of Moral Education* 17:98–104.

35 Miller, A. 1982. Environmental problem-solving: Psychosocial factors. *Environmental Management* 6:535–541.

36 Rest, J. 1988. Why does college promote development in moral judgement? *Journal of Moral Education* 17:183–194.

37 Bebeau, M., and M. Brabeck. 1987. Integrating care and justice issues in professional moral education: A gender perspective. *Journal of Moral Education* 16:189–203.

38 Martin, J. 1987. Transforming moral education. *Journal of Moral Education* 16:204–213.

39 Yeazell, M. 1986. The neglected competency—moral sensibility. *Contemporary Education* 57:173–176.

40 Hudson, L. 1970. *Frames of mind.* Harmondsworth: Penguin.

41 ———. 1968. *Contrary imaginations.* Harmondsworth: Penguin.

42 Witkin, H., et al. 1977. Role of the field dependent and field independent cognitive styles in academic evolution: A longitudinal study. *Journal of Educational Psychology* 69:197–211.

43 Ircha, M., and J. McLaughlin. 1988. Teaching public policy in an engineering college. *Engineering Education* (April): 703–704.

44 Wenk, E. 1987. *Trade Offs: Imperatives of choice in a high-tech world.* Baltimore: Johns Hopkins University Press.

45 Lynton, E., and S. Elman. 1987. *New priorities for the university: Meeting society's needs for applied knowledge and competent individuals.* San Francisco: Jossey-Bass.

46 Orr, D. 1996. Educating for the environment: Higher education's challenge of the next century. *Journal of Environmental Education* 27:7–10.

47 Wilshire, B. 1990. *The moral collapse of the university: Professionalism, purity and alienation.* Albany: State University of New York Press.

48 Maguire, J. 1996. The tears inside the stone: Reflections on the ecology of fear. In *Risk, environment and modernity: Towards a new ecology,* ed. S. Lash, B. Szerszynski, and B. Wynne, 169–188. London: Sage Publishers.

49 Ewens, W. 1984. *Becoming free: The struggle for human development.* Wilmington, Del.: Scholarly Resources Inc.

50 Betts, M., and D. Coon. 1996. *Working with the woods: Restoring forests and community in New Brunswick.* Fredericton: Conservation Council of New Brunswick.

51 Marchak, P. 1990. For whom the tree falls: Restructuring of the global forest industry. In *Canadian Political Science, Anthropology and Sociology joint meetings.* May. Victoria, B.C., Canada.

52 Freudenburg, W., and R. Gramling. 1994. Natural resources and rural poverty: A closer look. *Society and Natural Resources* 7:5–22.

53 Finnamore, A. 1997. Isolation, despair a way of life. *Daily Gleaner,* May 23. Fredericton, N.B., Canada.

54 Llewellyn, S. 1997. Mill owners told to keep pace or lose Crown wood. *Daily Gleaner,* March 13, Fredericton, N.B., Canada.

55 MacFarlane, B. 1997. Reinventing forestry. *The Best in the Business, Telegraph-Journal Supplement* (October): 17–20. St. John, N.B., Canada.

56 Ehrenfeld, D. 1993. *Beginning again: People and nature in the new millenium.* New York: Oxford University Press.

57 Ophuls, W., and A. Boyan. 1992. *Ecology and the politics of scarcity revisited: The unraveling of the American dream.* New York: W. H. Freeman.

58 Milbrath, L. 1989. *Envisioning a sustainable society.* Albany, N.Y.: State University of New York Press.

59 Higgins, R. 1978. *The seventh enemy: The human factor in the global crisis.* London: Hodder and Stoughton.

Index